GUENTHER O. W. KRUSE

*Guenther Kruse
25. Dec. 1979
Adelaide, S.A.*

AF335535

NATURAL HISTORY OF THE ADELAIDE REGION

edited by

C. R. Twidale, M. J. Tyler & B. P. Webb

Royal Society of South Australia Inc.

Registered in Australia for transmission by post as a book

Printed by Graphic Services Pty. Ltd., Northfield, S. Aust.

Issued December 1976

© Royal Society of South Australia, 1976

National Library of Australia card number and ISBN
0 9596627 0 7

Front cover design: Kathy Bowshall

CONTENTS

PREFACE

This year, 1976, sees the centenary of publication of the *Transactions of the Royal Society of South Australia.* To mark the occasion the Council has adopted a suggestion that the Society produce a special publication devoted to papers concerned with various aspects of the natural history of the Adelaide region. Such a commemorative volume is surely appropriate, and is not without precedent, for in an Anniversary Address published in Volume I of the *Philosophical Transactions* (as the *Transactions* then were) Professor Ralph Tate reported on progress that had been made towards "a better knowledge of the natural history of South Australia" (Tate 1878). The Society has a long involvement and concern with the natural history of the State, and it seems eminently worthwhile to produce a conspectus of the advances achieved during the last hundred years.

Such an undertaking is a formidable task, for nowadays any well-rounded and integrated natural history of any area, no matter how small, must be unsatisfactory, if not impossible: "Tout ce qui est simple est faux; tout ce qui est complexe est inutilisible." But it was not always so. Just as Leonardo da Vinci, who lived from 1452-1519, could be regarded as a universal mind in that he made major and original contributions to wide areas of both the arts and the sciences, so three centuries later such men as Alexander von Humboldt (see for example Hartshorne 1939, p. 76 *et seq.*; Achernecht 1955; Sinnhuber 1959) and Gilbert White, were in their different ways and contexts just able to span sufficient of the natural sciences to produce local studies that were and are splendid examples of integrated observations of nature. Both lived and worked some two centuries ago when it might retrospectively be argued that a man of intelligence and industry could, particularly if he were not compelled to devote too much of his time to earning a living, encompass and master sufficient of the natural sciences successfully to describe and analyse the intricacies of nature. Even so a form of specialisation was necessary. White, the author of the *Natural History of Selborne* (1789), and the greatest of the field naturalists whose work has been handed down to us, was firmly of the view that intense and prolonged study of one small area was the only way to understand nature:

> "Men that undertake only one district are much more likely to advance natural knowledge than those that grasp at more than they can possibly be acquainted with: every kingdom, every province should have its own monographer."

No one can doubt the value of long-continued and detailed observations such as those so accurately, and yet elegantly, recorded by White. He confined his activities to his native parish situated in a small corner of Hampshire near the Sussex and Surrey borders, and though they were made two centuries ago the reverend gentleman's observations still spring to life from the sterile pages. We have no difficulty picturing the Chalk Downs he loved so well, in reliving the panoply of changes of season and weather, of meadow and stream, in following phenological variations, cycles of life, and the habits of vole and viper, of swallow and stoat, of bee and bat, and of tench and toad.

Even one hundred years ago Tate had such a command of the whole range of the natural sciences that he was able unaided to compile the authoritative review referred to earlier. And during the last hundred years, devotion to one area over a long period by father and son made possible the wide-ranging (though predominantly geological and archaeological) essays published posthumously in *Worth's Dartmoor* (Worth 1953). But regrettably, though inevitably, such individual efforts are now a thing of the past. What has happened to put the study of the full range of natural history beyond the reach of any one person? Why has it taken more than a score of scientists to produce what is undoubtedly a more detailed but possibly less coherent review of what was so manifestly within the grasp of admittedly outstanding minds only one or two centuries ago?

It is surely not that modern scientists are less intelligent than their forebears; merely that times, and science, have changed. Specialisation is, for better or worse, the order of the day. No one person can produce today the delightful comprehensive studies of yesteryear. Ecology in the broadest sense of the term is now viewed in depth, and the nature of interrelations and interactions is, as it were, viewed on the molecular scale. Observations like those recorded by Humboldt and White are fundamental, and those of modern workers such as Lorenz and Goodall, remain basic and invaluable beginnings; but this is, for better or worse, the age not only of the specialist but of mobility. It is increasingly unusual for scientists to remain in one place long enough for detailed continuous observations to be made, so that the intimate detailed knowledge of any area in all its garbs is largely, though not entirely, a thing of the past. A balance has to be struck between intimate and long continued local investigations, and the wider, possibly more superficial, comparative studies which can, however, have a seminal influence.

Yet the very same factors that could well have acted to the detriment of the Royal Society, and turned it into an interesting but useless anachronism, have also and increasingly become its *raison d'être*. For the Society is dedicated to the investigation and dissemination of knowledge in all the natural sciences, and particularly in the natural history of South Australia. Specialisation and mobility have undermined these purposes, but because the Society provides the basis of continuity, because it forms a focus for work concerned with the State, and because it is a forum where various specialists can come together and hear and exchange views on matters of common interest, the Society goes on from strength to strength. Most students of natural science appreciate that our small compartments of knowledge are unreal and that no natural phenomena can be examined in isolation. Moreover it is salutary to recall that when the Society was established in 1853 many of its members who, as is the case also today, came from all walks of life and from a great range of professions and occupations, were also considered specialists in their day, though we should regard some of them as positive polymaths. Ralph Tate, to whom earlier reference has been made, epitomised his age in this respect: appointed Professor of Natural History in the University of Adelaide, Tate is perhaps best remembered for his contributions to botany and palaeontology, but he professed over such a wide range of the natural sciences

that, as he himself once remarked, he occupied not a chair but a sofa.

Our members today hail from many countries and disciplines, but they have in common an interest in natural history and in South Australia. They are sufficiently tolerant of their fellows' enthusiasms to listen to their latest ideas and discoveries, and not infrequently have their reward on earth as well as in heaven: how often does one go to a Society meeting in order to hear a particular paper in one's special field of interest only to be entranced by, and to find something pertinent to one's own research problems, in one of the "other" contributions on offer?

In this specialised world it is useful at least to try and bring together a conspectus of knowledge concerning the natural history of the Adelaide region. The Society is well placed to undertake such a task by virtue of the wide-ranging interests and expertise of its members. Sensibly a team approach has been adopted. The various individual contributions have been in some measure correlated and integrated but there are, and hopefully ever will be, differences of opinion and interpretation. This Special Publication may confuse those who expect to read an unequivocal, seemingly final, account of the natural history of the Adelaide region, those who do not realise that science is inherently controversial, those who do not appreciate that good science is provocative and questioning, and those who are unaware that it is very difficult to recognise that nebulous ideal called truth. Disputes, arguments and discussions are the very lifeblood of science, and in these collected papers, controversies have not been suppressed. Various and varied interpretations of the same evidence, explanations of the same features, have been welcomed: the "open forum" concept of the Society itself finds expression throughout these contributions. It is obvious that this approach is not entirely satisfactory because the different attitudes and treatments usually preclude any possibility of an overall view. The present writer cannot agree with some of the assertions made in some chapters, and no doubt what he himself has written will also be disputed. Show a dozen scientists the same evidence and they may devise at least a dozen explanations. But as in so many other fields of scientific endeavour, it is better to travel and not arrive than not to travel at all.

The Adelaide region is taken in very general terms to include the Mount Lofty Ranges and adjacent plains and waters. The Murray Basin and River and related plains are excluded from consideration. The natural environment embraces the rocks, landforms, soils and waters, both fresh and oceanic, and associated biota that live on and in the various terrestrial and aquatic habitats. An evolutionary view has been adopted wherever possible. The basic question asked is, "How and when did the present situation come into being?"

The aboriginal peoples who lived in the area are discussed, but European man has been excluded from consideration save insofar as he has influenced, and continues to affect, the various facets of the natural setting. To do otherwise would be to extend beyond what long have been the main interests of the Society, and into such fields as sociology, economics and politics, in which we have an interest but claim no expertise.

It is hoped that this collection of essays represents an authoritative and balanced review of the natural history of the Adelaide region as we now see it, useful and interesting for the present, and a reliable historical document capable of serving as a viable basis for comparison when our successors come to review the second century of publication of the *Transactions* of the Royal Society of South Australia a hundred years from now.

Though many people have contributed to the production of this Special Publication of the Royal Society of South Australia, particular mention must needs be made of those who undertook the thankless task of encouraging, persuading, cajoling and pressuring the various authors to complete their work by due date. Ingenious and diabolical were their ploys! I refer to Bruce Webb, Mike Tyler, John Holmes, Wayne Harris, and Bill Williams. The Society is greatly in their debt. The editors also wish to thank Jennie Bourne and Margaret Davies for their invaluable assistance in the onerous work of preparing the manuscripts for the press.

C. R. TWIDALE
President, 1975-76.

References

ACHERNECHT, E. H. (1955).—George Forster, Alexander von Humboldt and Ethnology. *Isis* **49**, 83-95.

HARTSHORNE, R. (1939).—"The Nature of Geography". (Assoc. Amer. Geogr.: Lancaster, Penn.)

SINNHUBER, K. A. (1959).—Alexander von Humboldt, 1769-1859. *Scot. Geogr. Mag.* **75**, 89-101.

TATE, R. (1878).—Anniversary address. *Trans. Phil. Soc. S. Aust.* **1**, 11-47.

WHITE, G. (1789).—"The Natural History of Selborne." (MacMillan: London, 1880.)

WORTH, R. H. (1953).—"Dartmoor." (David & Charles: Newton Abbott.)

CHAPTER 1

GEOLOGY

by B. Daily,* J. B. Firman,† B. G. Forbes† & J. M. Lindsay†

The geological history of the Adelaide region is revealed now as in the past by a study of the outcropping rocks, supplemented by information from drilling carried out in the search for underground water, petroleum and minerals. The development of geophysical and geochemical techniques, radiometric dating, the availability of aerial photographs, and modern improvements in roads and in transport generally, have made more information and more precise data available. Nevertheless, it would be remiss not to acknowledge that this summary rests firmly on foundations laid down by such pioneers as Ralph Tate, H. Y. L. Brown, R. L. Jack, Walter Howchin, L. K. Ward and Sir Douglas Mawson.

A sequence of rocks dating from early geological time through to the recent past has been identified and is described in the following text.

A. Precambrian‡

Precambrian rocks form a great part of the Mt Lofty Ranges and are exposed at the coast near Hallett Cove and south of Normanville as well as in quarries, cuttings and natural cliffs within the uplands. The disposition of these rocks is shown in Figs 1–3. As a result of compression and fracturing, the oldest and lowest rock mass, the *crystalline basement,* can be seen at the surface in anticlinal or upfolded regions such as the Houghton Inlier, near the Warren Reservoir, near Carey Gully and near Yankalilla.

The layers of Precambrian sedimentary rocks above the basement are termed *Adelaidean* and in this region range in age from 800–900 m.y. (million years) to 570 m.y., which is the age of the base of the succeeding Cambrian succession (Cooper & Compston 1971; Cooper 1975). The Precambrian rocks originated as sediments, including gravel, sand, clay, silt and carbonate mud containing primitive life forms which are commonly represented in wavy or columnar structures (stromatolites) found in limestone and dolomite rocks of the Adelaidean (Preiss 1973). They were deposited in varied and varying environments, some high energy, some low; some in shallow water, some deeper; some are of clastic origin, others primarily chemical precipitates.

These Adelaidean rocks and overlying strata were folded, heated and fractured during the *Delamerian* orogeny late in the Cambrian period 500 m.y. ago (see below); in the east recrystallisation was greatest and bodies of granite were formed early in the Ordovician 480 m.y. ago. There were three phases of folding (Offler & Fleming 1968) and of these the earliest is best developed in the western Mt Lofty Ranges where folds are overturned to the west and the associated cleavage and foliation planes dip consistently to the east. Most of the mineral occurrences, including gold, copper, silver-lead and barytes were formed in the Delamerian.

Crystalline basement

The Barossa gneisses and schists that form outcrops in the Adelaide hills are part of an extensive crystalline basement which is also exposed on Eyre Peninsula and Yorke Peninsula (Glaessner & Parkin 1958; Thomson 1969b).

Gneisses seen in the Houghton Inlier represent mainly sedimentary rocks that have been altered at temperatures of the order of 700°C, giving rise to high temperature minerals such as sillimanite. Although the gneisses have suffered later low-temperature alteration during which some earlier formed minerals were replaced by micas, fibrous sillimanite survives,

* Department of Geology and Mineralogy, University of Adelaide, Adelaide, S. Aust. 5000.
† South Australian Department of Mines, P.O. Box 151, Eastwood, S. Aust. 5063.
‡ Original draft prepared by B. G. Forbes.

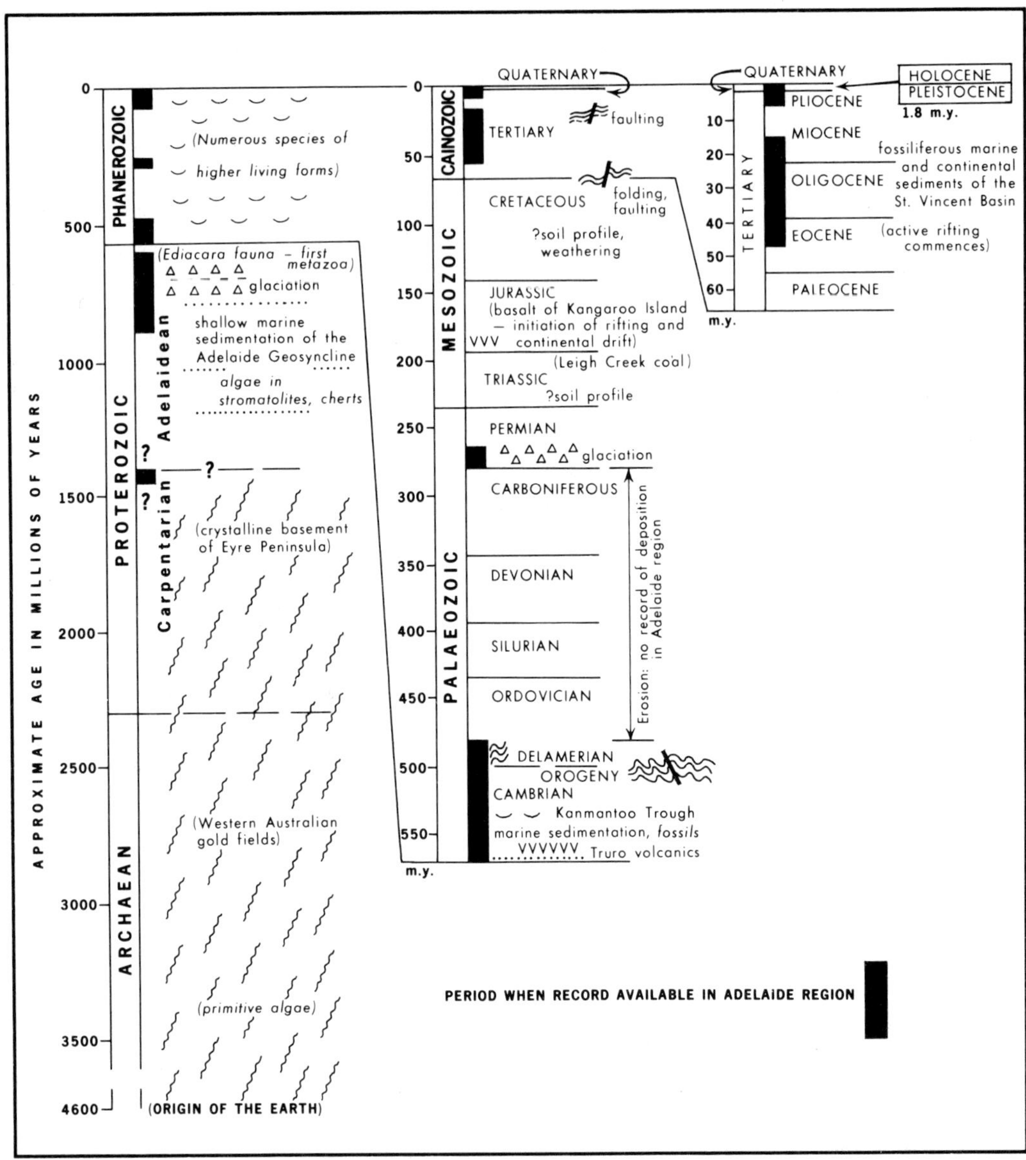

Fig. 1. Geological time scale. Left hand column shows the segments of geological time that are represented in the rocks of the Adelaide region. Some relevant events of general interest, such as commencement of crustal rifting and continental drift, are added to give some perspective. Columns to the right are enlargements of the upper part of the left hand column.

for instance in the roadside quarry 700 m west of the Torrens Gorge Kiosk (Spry 1951; Talbot 1963). Micaceous gneisses near Kangaroo Creek Reservoir (Trudinger 1973) show east-dipping planes of foliation or cleavage which are typical of folded slaty rocks in the western Mt Lofty Ranges. Greenish banded gneiss occurs at Houghton, near the Houghton Ceme-tery and near the Inglewood Hotel (Benson 1909). Low-grade uranium mineralisation is associated with feldspar-rich gneisses in this area (Dickinson et al. 1954). Similar rocks are known from the other inliers of crystalline basement to the south, and Mills (1973) has given a detailed description of the Warren Inlier.

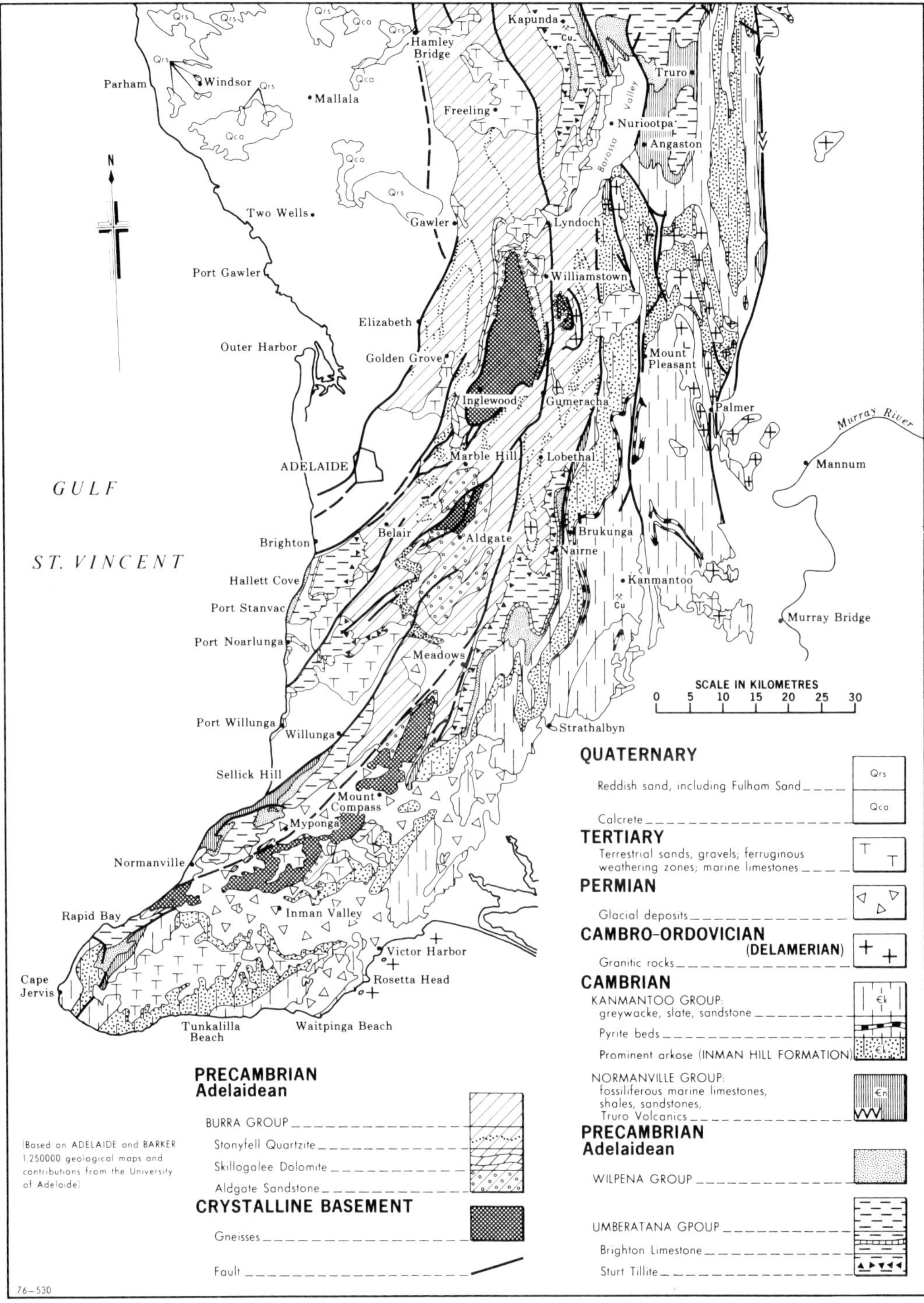

Fig. 2. Generalised geological map of the Adelaide region. (For more detailed mapping see Geological Atlas of South Australia 1:250,000 series, sheets SI 54-9, ADELAIDE and SI 54-13, BARKER.)

Adelaidean

Following the burial, consolidation, metamorphism and deformation of the basement rocks of the Mt Lofty Ranges, a new sequence of sedimentary rocks was deposited on the eroded crystalline basement. The sediments accumulated in a gradually subsiding area termed the Adelaide Geosyncline (Mawson & Sprigg 1950; Sprigg 1942, 1946; Glaessner & Parkin 1958; Thomson 1969c). This structure extends from south of Adelaide into the Flinders and Willouran Ranges and other northern areas. Fig. 4 illustrates the sequence and nature of Adelaidean rocks near Adelaide. The rock group terminology used here is based on Geological Survey mapping (see also Thomson *et al.* 1964; Taylor *et al.* 1974); a different scheme has been proposed by Daily (1963) and used by Talbot & Nesbitt (1968) and Alderman (1973).

Burra Group

The earliest product of erosion of the crystalline basement was an arenaceous sequence, the Aldgate Sandstone, which accumulated in a shallow-water environment (Howchin 1906). Evidence that these sediments were deposited on the eroded and steeply dipping layers of the crystalline basement is rarely exposed and north of Inglewood the contact is faulted; but this contact represents one of the most important unconformities and time breaks in the Adelaide region (Fig. 5a). Coarse-grained hematitic sandstones are exposed at the lookout access road, Kangaroo Creek Reservoir. Near Williamstown the Aldgate Sandstone is sufficiently hematitic to have warranted investigation as a possible iron ore (Miles 1950); a ferruginous pebbly sandstone may be seen in the parking area southeast of the spillway, South Para Reservoir. Pebbly and hematitic sandstone near the base of the Aldgate Sandstone occurs in a road cutting in its type section 800 m north-northwest of the Carey Gully stone quarries, southeast of Carey Gully; less hematitic sandstones overlying these sediments occur further to the southeast. On the beach west of Little Gorge, southwest of Normanville, pebbles in this formation have been severely distorted and the beds overturned during the Delamerian orogeny. Horwitz (1962) has, however, correlated these sandstones with the Stonyfell Quartzite.

There followed shallow marine precipitation of dolomite represented by the pale-coloured Castambul Dolomite exposed in the Torrens Gorge near Castambul. This was succeeded by argillaceous and arenaceous sediments and a darker dolomite rock, the Montacute Dolomite (Howchin 1915; Nixon 1963) characterised by the presence of light-coloured fragments of magnesite rock (Forbes 1961). In the Flinders Ranges there was algal activity during sedimentation, and black cherts enclose microscopic fossil algae (Schopf & Barghoon 1969; Preiss 1973). The Woolshed Flat Shale (Wilson 1952) and its Balhannah Shale Member (Thomson 1969a; Taylor *et al.* 1974) are partly equivalent to and partly younger than the dolomitic sequences (e.g. northeast of Balhannah, near Marble Hill and also southeast of Hamley Bridge).

The Montacute Dolomite and the more prominent Stonyfell Quartzite (Heath 1963) which overlies it, are important sources of construction material near Adelaide. The attitudes of cross-beds in the Stonyfell Quartzite in the Morialta area suggest sand transport by shallow-water currents in southeasterly and southwesterly directions.

The upper Burra Group represents a continuation of shallow marine conditions, but with occasional exposure to the air, perhaps on a tidal flat or river flood-plain, as evidenced by casts of mud cracks (Barnes & Kleeman 1934; Coats 1967; Thomson 1969c). Sandstone dykes, infillings of cracks by sandy sediments, are seen near the Belair and Blackwood railway stations (Sprigg 1946).

Umberatana Group

A dramatic change in climatic conditions, and the onset of a great ice age, is evidenced in the Umberatana Group of sediments which overlie the Burra Group. The earliest and most spectacular sediment of glacial origin in the Adelaide region is the Sturt Tillite (ca 759 m.y.) which may be seen in the type section in the Sturt Gorge and in the grounds of Flinders University (Sprigg 1942; Coats 1967). The igneous and metamorphic erratics have been transported by ice moving eastward off crystalline basement rocks which at that time were exposed in highlands to the west.

Upper Precambrian tillites equivalent to the Umberatana Group are well known from other parts of Australia and overseas. Radiometric ages of about 730 and 686 m.y. for glacigene and associated beds (Perry & Roberts 1968; Compston & Taylor 1969; Cooper *et al.* 1971) in the Alice Springs and Kimberley regions,

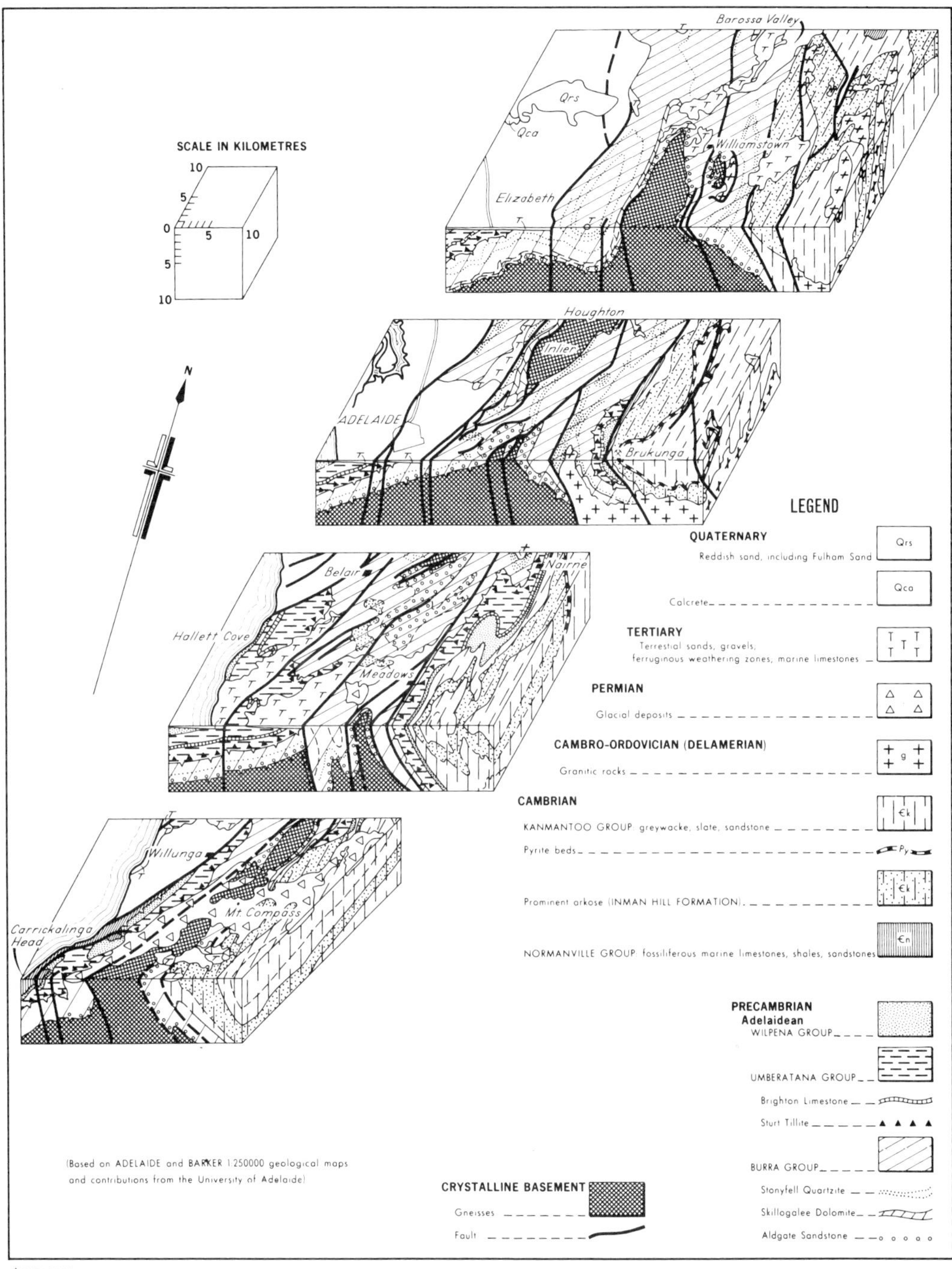

Fig. 3. Block diagrams illustrate the geological structure of the Adelaide region. Folding and fracturing of the crystalline basement and overlying Adelaidean sedimentary layers are shown. The prevalence of granitic rocks in the east is an indication of the much higher temperatures attained there during the Cambro-Ordovician Delamerian movements.

Australia, may also apply to the Umberatana Group in South Australia.

Glaciation abated and was succeeded by marine deposition of silts which gave rise to the remarkably well laminated grey siltstone. Laminated siltstones showing sedimentary ripples are exposed in highway cuttings on Tapley Hill which has given its name to the Formation. At its base, the Tindelpina Shale Member frequently contains interbedded dolomite and rests immediately above the Sturt Tillite just east of the Registry Building, Flinders University. Geochemical and petrographic studies (Sumartojo 1975) indicate a shallow marine environment with detritus again derived from the west. The copper deposits at Kapunda are disseminated in siltstones and sandstones of the Tapley Hill Formation and its Eudunda Arkose Member.

The sedimentary environment of the Brighton Limestone, which is quarried for a variety of construction purposes (Preiss 1973) was probably shallow marine to supratidal and evidently much shallower than that for the lower Tapley Hill Formation. The Rapid Bay marble, which is used as a flux in iron ore smelting, may be equivalent to the Brighton Limestone (Daily 1963). The Brighton Limestone is the formation quarried at Linwood, south of Brighton and exposed in Waterfall Creek east of Hallett Cove and at Myponga Dam. The Reynella Siltstone Member (Thomson 1966) is correlated with more obviously tillitic sequences such as the Elatina Formation and Pepuarta Tillite of the Flinders Ranges and Olary region which represent a final, weaker glaciation in Adelaidean time. It is exposed along the sea shore northwest of Hallett Cove railway station.

In parts of the Kapunda region the Brighton Limestone is absent from the rock succession, possibly as a result of generally deeper water conditions during deposition, and the Tapley Hill Formation is overlain by sandy, laminated siltstones of the Tarcowie Siltstone. This is succeeded by the Pepuarta Tillite and overlying siltstones which are the uppermost units of the Umberatana Group.

Wilpena Group

In the Hallett Cove area where Waterfall Creek enters the sea the Seacliff Sandstone Member (Thomson 1966, Thomson *et al.* 1976) locally forms the base of the Wilpena Group. The Brachina Formation is exposed on the shore platform south of Waterfall Creek, and south of Hallett Cove. The Sandison Reserve at Hallett Cove provides a unique, clear and condensed picture of local geology (Talbot & Nesbitt 1968; Cooper *et al.* 1970). It illustrates the nature of the Precambrian sedimentation, subsequent folding of these beds during the Cambro-Ordovician, and the erosional relationship with the overlying Permian glacial beds.

The Wilpena Group in the Kapunda area is composed of the basal Nuccaleena Formation, a distinctive, thin, laminated pale reddish dolomite bed, overlain by greenish laminated siltstone of the Ulupa Siltstone.

The uppermost local unit of the Wilpena Group correlated with ABC Range Quartzite of the western Flinders Ranges (Coats 1965; Thomson 1969c) may be seen just south of Port Stanvac. It is equivalent to the uppermost Precambrian quartzites and slates below the basal Cambrian Mount Terrible Formation southwest of Willunga.

Summary

The Precambrian rocks of the Adelaide region attest to continuing adjustments in the earth's crust and deep-seated chemical and physical changes. Earthquakes, volcanoes and sea floor spreading elsewhere are evidence that these processes are continuing.

The rocks of the crystalline basement probably accumulated in a shallow sea, and mineralogical changes took place in response to sinking of the sedimentary rocks into a deeper, higher-temperature zones of the crust. The resulting high-temperature minerals were later modified when the rock mass was uplifted, structurally deformed and eroded to form new deposits within the subsiding Adelaide Geosyncline. Alternations of coarse and fine detritus and chemically-precipitated carbonates in the sequence were determined by the shape of this sedimentary basin, its rate of sinking and the climate obtaining, all of which varied from time to time.

B. Early Palaeozoic*

Before the end of Late Precambrian time the 7000–8000 m thick pile of Adelaidean sediments was slightly folded by compressive earth movements. After uplift, erosion proceeded until the region was once more inundated by the sea at the beginning of the Cambrian Period. Evidence of these earth movements is demonstrated by the low-angle unconformity mapped between the Adelaidean and Cambrian

* Original draft prepared by B. Daily.

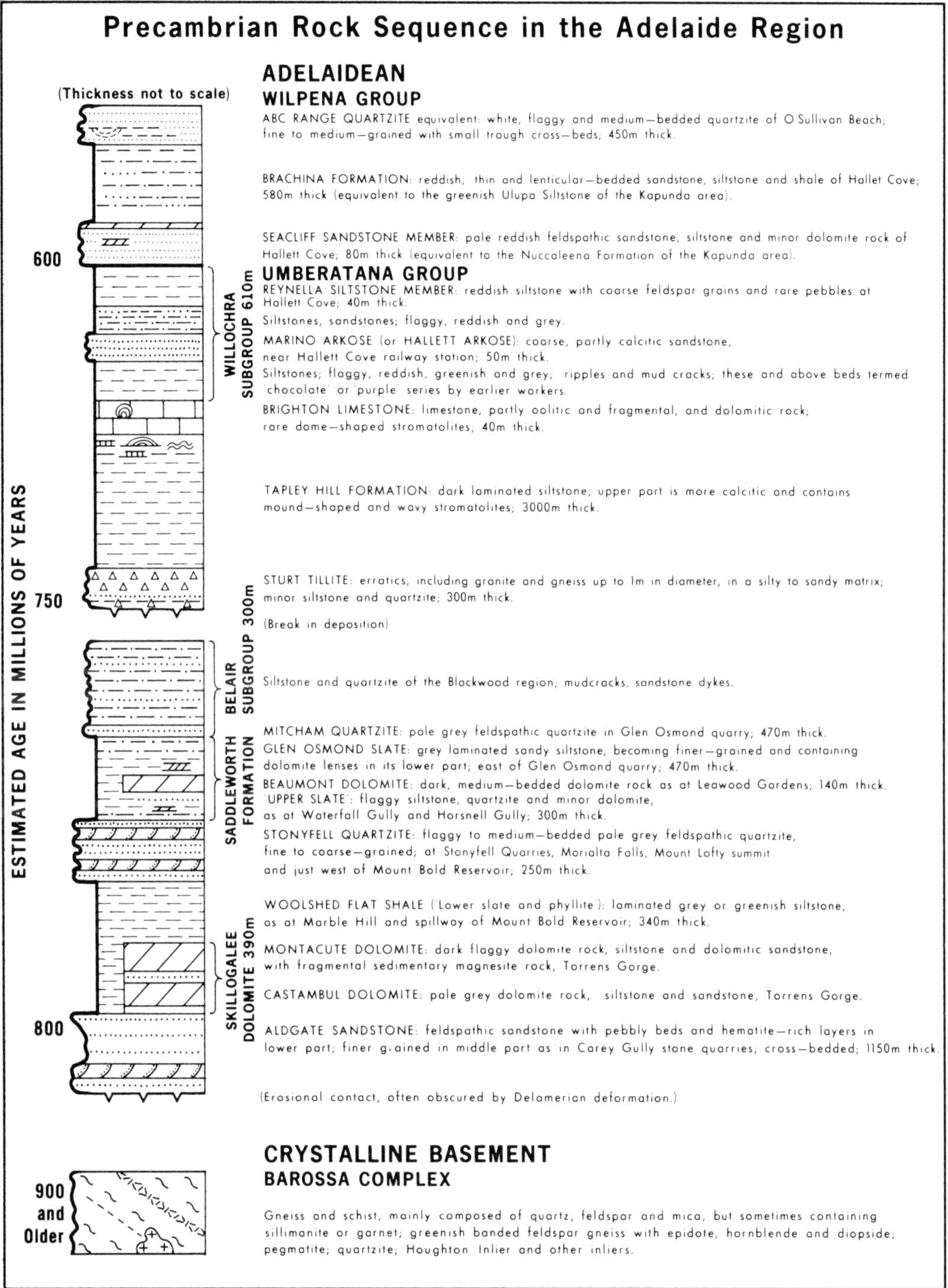

Fig. 4. Precambrian rock sequence near Adelaide before it was affected by Delamerian and later crustal movement. Thickness of the formations are not to scale. Ages in the left hand column are a tentative estimate only.

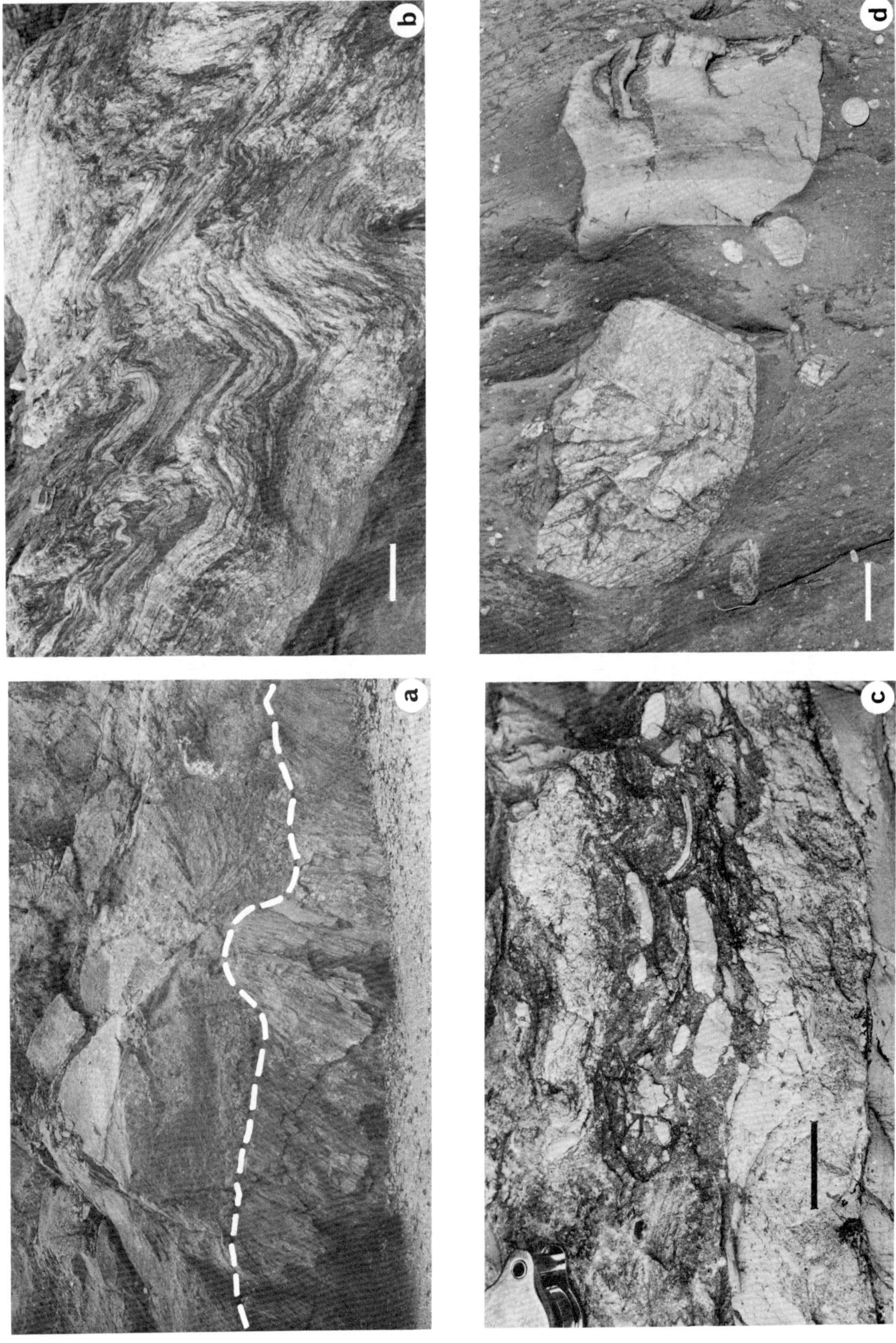

strata north of Truro in the northern Mt Lofty Ranges (Coats & Thomson 1959; Forbes *et al.* 1972) and by the unconformity well exposed in road cuttings at Sellick Hill (Fig. 8) on Fleurieu Peninsula (Thomson & Horwitz 1961).

The preserved Cambrian of the Adelaide region comprises two conformable sedimentary sequences, the older Normanville Group (Daily & Milnes 1973) and the younger Kanmantoo Group (Sprigg & Campana 1953). Their distribution and component formations are shown in Fig. 6 and Table 1.

Normanville Group

The Normanville Group is largely calcareous except in the Truro region, where interbedded lavas, mainly andesites, are locally important. The Truro Volcanics provide the only record of volcanism in the Mt Lofty Ranges and appear to have been confined in their distribution to the northeastern parts of the Adelaide region (Fig. 6).

The Normanville Group is best studied in road cuttings, in creeks and along the coastline between Sellick Hill and Carrickalinga Head for elsewhere metamorphism has obliterated many of the original sedimentary characteristics from which the environment of deposition may be deduced. In the Sellick Hill area its five exposed rock formations (Fig. 7) appear to reflect two marine cycles of shallow shelf sedimentation separated by a sharp contact marking an erosional break (a disconformity) between the Wangkonda and overlying Sellick Hill formations.

The pebbly and coarse-grained feldspathic sandstones and arkoses in the basal part of the Mt Terrible Formation (Daily 1963) fill small erosional hollows cut into the finer-grained quartzites and siltstones of Precambrian age.

Abundant trace fossils made by burrowing and sediment-ingesting organisms first appear about 5 m above the base and give the rock a characteristic spotted and churned appearance. Their presence, which contrasts markedly with their absence from the underlying Precambrian strata, is a clear reflection of the high evolutionary levels and the complex behavioural patterns attained by organisms at the turn of the Cambrian.

The oldest shelly fossils found in Australia occur in the phosphatic shales of the middle member of the Formation. According to Daily (1976), they include molluscs, conodonts, sponge spicules and shells of uncertain zoological affinities (Fig. 11a). South of Sellick Hill phosphorites (phosphate-rock) up to 4.5 m thick occur in the middle member of the Mt Terrible Formation (Blissett & Callen 1969; Callen 1970, 1971). The Koonunga Phosphorite Member of the Kapunda area (Johns 1967) and the phosphatic slates found in the Mt Magnificent area (Mawson 1939; Horwitz 1960) may be on the same stratigraphic level. The upper member, "the *Hyolithes* sandstone" of Abele & McGowran (1959), contains abundant hyolithids and small fossils such as gastropods, sponge spicules and several organisms of uncertain relationships. The same fauna occurs in a similar stratigraphic position at Delamere, 34 km southwest of Sellick Hill (Daily 1963), where it underlies Cambrian marbles. These fossils, together with remains of the long flexible tubes of the non-shelly fossil *Saarina* Sokolov found in "the *Hyolithes* sandstone", correlate the Mt Terrible Formation with the earliest Cambrian deposits of the Russian and Siberian platforms (Daily 1976).

The conformably overlying Wangkonda Formation consists of two shallowing-upward

---

Fig. 5. (a) Irregular unconformable contact-line between gneissic rocks of the Barossa Complex, extending upward about one metre from road level, and overlying pebbly sandstone of the Aldgate Sandstone, locally the base of the Adelaidean. Foliation in the Barossa Complex dips steeply to the left (easterly). Festoon-like cross-bedding can be seen in the sandstone in the upper and lower right hand side. Freeway cutting, not readily accessible to the public, about 3 km east of Stirling.
(b) Strongly folded gneisses of the Barossa Complex in the Warren Inlier (Mills 1973) near the Warren Reservoir weir. Lighter layers are composed largely of quartz and feldspar; darker layers are rich in biotite. The present crumpled appearance of these rocks results from several phases of folding. Similar rocks can be seen in Warren National Park. Scale bar = 10 cm.
(c) Sandy dolomite rock of the Montacute Dolomite (or upper Skillogalee Dolomite) from a small quarry near Torrens Gorge about 1.5 km west of Pinkerton Gully. The central darker sandy layer, and paler dolomite-rich layers above and below it, contain dolomite and magnesite fragments probably resulting from erosion of original sedimentary layers. Bar scale = 5 cm.
(d) Sturt Tillite, Sturt Gorge. Erratics composed of metamorphic rocks are scattered through a silty, sandy matrix. Traces of cleavage in the matrix extend from top to bottom. Bar scale = 5 cm.

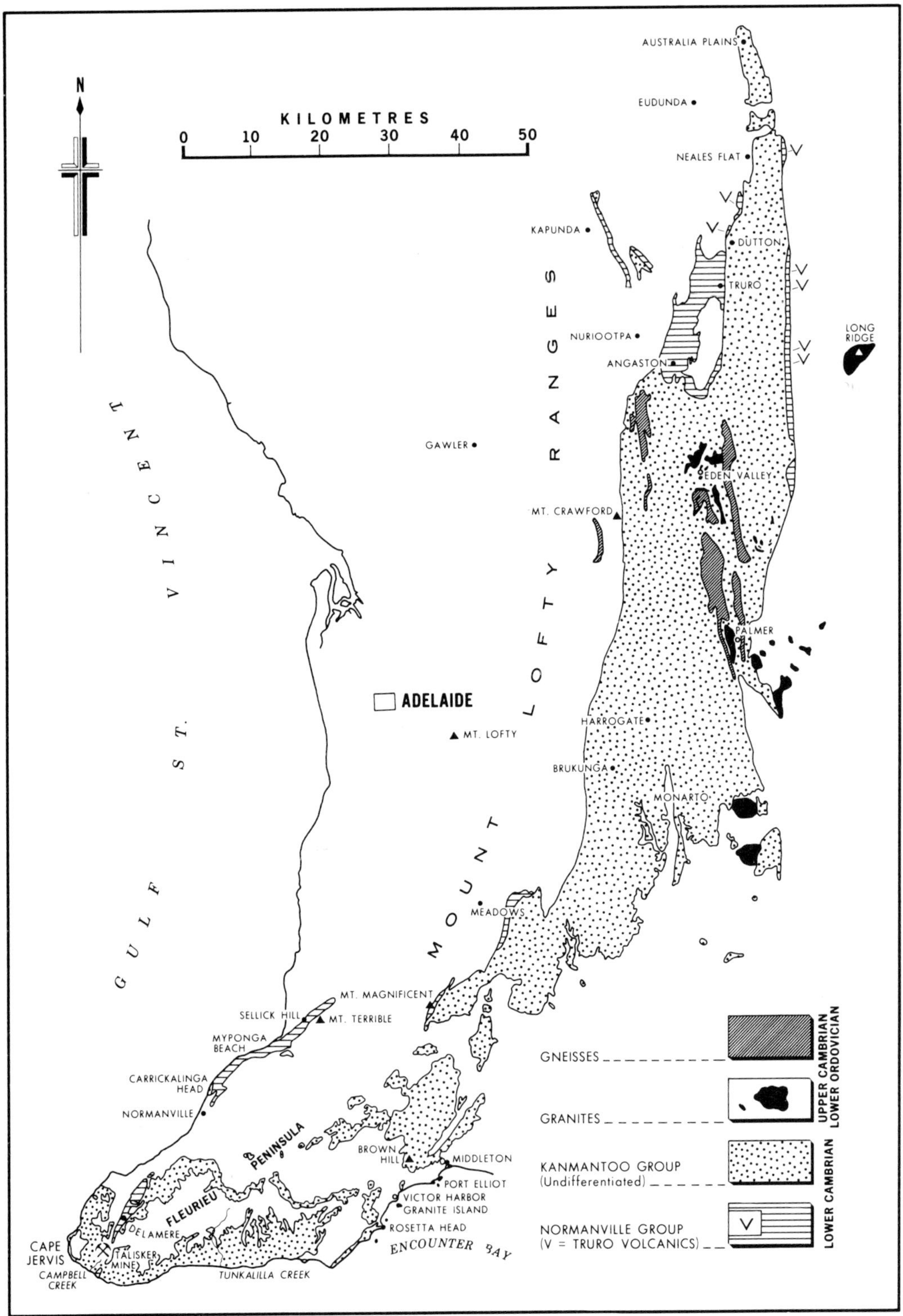
KILOMETRES
0   10   20   30   40   50
N
AUSTRALIA PLAINS
EUDUNDA
NEALES FLAT
KAPUNDA
DUTTON
TRURO
LONG RIDGE
NURIOOTPA
ANGASTON
EDEN VALLEY
GAWLER
MT. CRAWFORD
PALMER
ADELAIDE
MT. LOFTY
HARROGATE
BRUKUNGA
MONARTO
MEADOWS
MT. MAGNIFICENT
SELLICK HILL
MT. TERRIBLE
MYPONGA BEACH
CARRICKALINGA HEAD
NORMANVILLE
PENINSULA
BROWN HILL
MIDDLETON
PORT ELLIOT
VICTOR HARBOR
GRANITE ISLAND
ROSETTA HEAD
ENCOUNTER BAY
CAPE JERVIS
TALISKER MINE
DELAMERE
FLEURIEU
CAMPBELL CREEK
TUNKALILLA CREEK
GULF ST. VINCENT
MOUNT LOFTY RANGES
GNEISSES
GRANITES
KANMANTOO GROUP
(Undifferentiated)
NORMANVILLE GROUP
(V = TRURO VOLCANICS)
UPPER CAMBRIAN
LOWER ORDOVICIAN
LOWER CAMBRIAN

TABLE 1

Stratigraphic scheme for the Cambrian preserved in the Adelaide region. The regional unconformity between Precambrian and Cambrian strata represents roughly the stratigraphic interval from the ABC Range Quartzite of the Flinders Ranges to the base of the transgressive Cambrian deposits.

Cambrian System	Kanmantoo Group	Wattaberri Sub-Group	Middleton Sandstone
			Petrel Cove Formation
		Brown Hill Sub-Group	Balquhidder Formation
			Tunkalilla Formation
		Inman Hill Sub-Group	Tapanappa Formation
			Talisker Calc-siltstone
			Backstairs Passage Formation
			Carrickalinga Head Formation
	Normanville Group		Heatherdale Shale
			Fork Tree Limestone
			Sellick Hill Formation
			—— Disconformity ——
			Wangkonda Formation
			Mt. Terrible Formation

—————— Regional Unconformity ——————
Marino Group (Late Precambrian)

cycles of carbonate-rich rocks. Each commences with calcareous sandstones and siltstones which are overlain by dark grey mottled limestones and are capped by unfossiliferous pale grey limestones deposited under more restrictive marine conditions. The latter include oolitic and fragmental limestones, intraformational limestone conglomerates and breccias possibly desiccational in origin, fine-grained limestones, and "birdseye" limestone indicative of deposition in intertidal to supra-tidal environments. Shelly fossils, mainly hyolithids, are found in the lower carbonate-rich clastics, but clear evidence of the reworking of sediments by organisms (bioturbated beds) in quest of food occurs in interbedded sandstones and

siltstones. Supposed Archaeocyatha reported from the formation (Campana & Wilson 1954; Campana *et al.* 1955) are oolites.

After an erosional break the Sellick Hill Formation was laid down. The sandy and in places pebbly basal parts filled hollows eroded in already lithified "birdseye" limestones at the top of the Wangkonda Formation. Hyolithids and rare gastropods were swept into and concentrated in some of these erosional hollows by submarine currents. Above are highly bioturbated calcareous quartzose sandstones and siltstones. These are best seen on the coast immediately northwest of Myponga Beach where weathered bedding surfaces reveal a wide variety of animal burrows (Madigan 1926; Fig. 11c). Then follows scantily fossiliferous dark grey mottled and banded silty limestones and calcareous shales so distinctive of much of the Formation (Mawson 1925; Fig. 11d). Fossils, mainly the tapered and septate calcareous tubes of hyolithids (Fig. 11e), were concentrated by currents into thin bands or left as current-swept debris between nodules in some of the intraformational limestone conglomerates. Fine examples of these conglomerates are found on the northwestern point of Myponga Beach. Higher in the formation Archaeocyatha (Fig. 11b) occur in small domed mounds (Daily 1969) and in bands of paler coloured limestone near the top of the Formation as in the eastern face of the old Sellick Hill Quarry.

The major part of the Sellick Hill Formation appears to have been deposited in a marginal basin in which poor water circulation and hence stagnant bottom conditions prevailed. However, the Archaeocyatha in its upper levels argue for a general shallowing and better water circulation at that time. These conditions prevailed during the deposition of the pale coloured Archaeocyatha-rich lower member of the Fork Tree Limestone presently being exploited for aggregate in a quarry on Sellick Hill. Associated calcareous algae and stromatolites attest to the shallowness of the depositional environment. Historically, the formation is important because the Archaeocyatha discovered in it by Sir Edgeworth David in the Normanville district (Howchin 1897) provided the first proof of Cambrian rocks in the Mt Lofty Ranges. Archaeocyatha (Fig. 11b) are readily seen on the western side of the old

Fig. 6. Distribution of Cambrian rocks in the Adelaide region. Selected bodies of granite and granite gneiss of Cambrian to early Ordovician age are also indicated. Basic dykes of pre- to post-Delamerian age are not shown because of their small size.

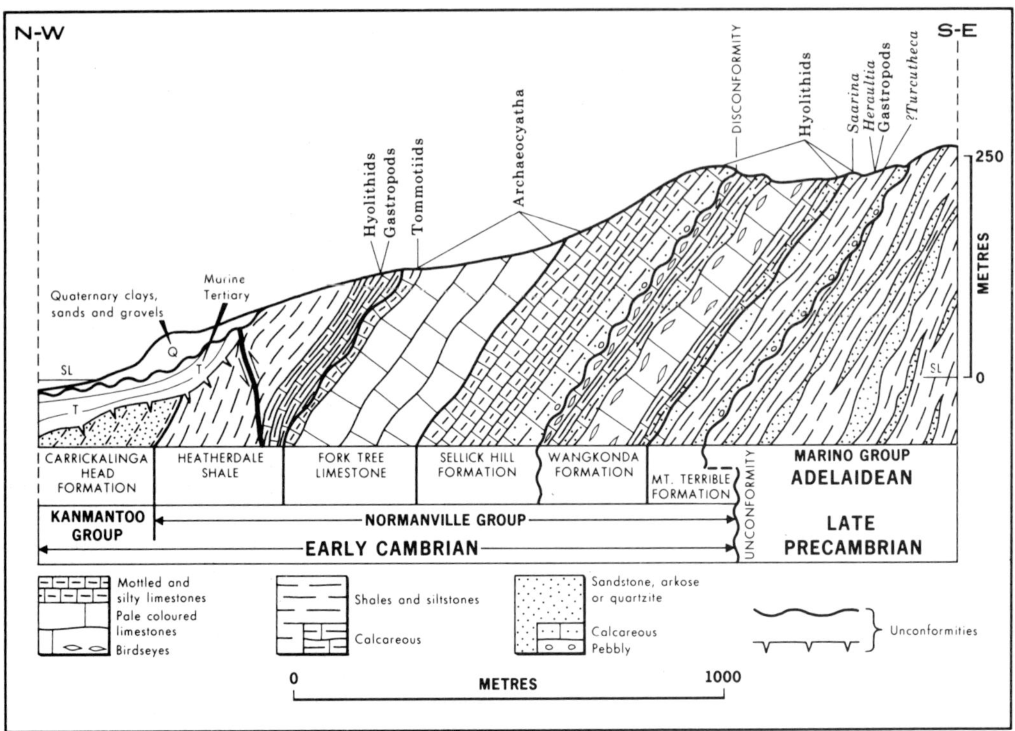

Fig. 7. Geological cross-section of the Cambrian, Sellick Hill area. The line of section lies northeast of and parallel to the new Sellick Hill road, thus avoiding the tectonic thickening of the Mt Terrible Formation seen along that road.

Sellick Hill Quarry in the lower part of the Formation. The sparsely fossiliferous, massive and strikingly mottled limestone forming the upper member of the Formation probably reflects a deeper, more reducing environment of deposition than that for the lower member.

In the Sellick Hill area the overlying Heatherdale Shale is divisible into two members (Abele & McGowran 1959), a lower calcareous member and an upper dark coloured to black shale and siltstone, generally lacking in carbonate. Nodules and irregularly shaped stringers of black phosphate, sometimes carbonaceous, occur in both members but are more conspicuous in the upper. Lateral and vertical changes in carbonate content have been reported for the Formation, the extreme case occurring just inland from Carrickalinga Head where silty limestones are developed instead of carbonate deficient shales and siltstones. It is a good example of a change in facies. Hyolithids, sponges, brachiopods and gastropods occur sparsely throughout the Formation. A curious feature, presumably related to inimical environmental conditions, is the absence of trilobites from this and all other formations in the Normanville Group, for they occur abundantly in Cambrian strata elsewhere on the mainland and on Kangaroo Island.

Kanmantoo Group

Relatively unaltered Cambrian strata younger than the Normanville Group occur inland from Myponga Beach and on the coast at Carrickalinga Head (Abele & McGowran 1959). Just north of that headland there is a sharp, conformable contact between black Heatherdale Shale and a green shale at the base of the Carrickalinga Head Formation.

This new cycle of marine deposition consists of green impure sandstones alternating with green siltstones and shales containing rare brachiopods (Daily 1963). Elsewhere the formation forms the basal unit of the approximately 9000 m thick metamorphosed Kanmantoo Group whose type section occurs along the south coast of Fleurieu Peninsula (Sprigg & Campana 1953). There metamorphism has converted these Lower Cambrian sediments to grey metasandstones, metasiltstones and phyllites respectively, the grey colour being due to

ERRATUM

Page 17 Figure 8

This figure was incorrectly positioned during printing. It should appear as shown below.

Reviews of the Kanmantoo Group have been given by Thomson (1969, 1976) and a re-appraisal of its type section and revision of its stratigraphic nomenclature have been presented by Daily & Milnes (1971, 1972, 1973). A diagrammatic structural cross-section and stratigraphic units recognised along this 70 km of coastline are presented in Fig. 9 and Table 1. The metasediments making up the Group reflect an almost continuous supply of erosive products derived from uplifted fault blocks in Investigator Strait, and land lying to the west and southwest of Kangaroo Island. Most were impure sands, occasionally pebbly, separated by thin beds of silt and mud. These ill-sorted sediments were generally deposited so rapidly or bottom conditions were such that organisms were able to rework them only rarely.

The occurrence of thick or distinctive intervals of phyllites, schists and metasiltstones has aided the stratigraphic subdivision of this otherwise monotonous sandstone sequence (Fig. 9). Notable are the sedimentary sulphide beds (mainly pyrrhotite and pyrite) in the Talisker Calc-siltstone, formerly exploited at Bruckunga for their sulphide content, and black carbonaceous and sulphide-bearing phyllites confined to the Brown Hill Sub-group. These originated as organic-rich muds and silts which due to stagnant bottom conditions permitted carbon and sulphides to form. Carbonates are rare within the Kanmantoo Group as their build-up was impaired by the continuous influx of terrestrial detritus swept into the rapidly subsiding basin.

Fig. 8. Disconformable contact between Precambrian and Cambrian strata, road-cutting old Sellick Hill road. Hammer (length 46 cm) on whitened contact marking eroded top of Precambrian. Precambrian (PЄ) siltstones and fine grained quartzite to left of contact; Cambrian (Є) arkose with thin siltstone interbeds to right of contact.

the growth of black biotite and other dark minerals in the already consolidated rocks due to elevated temperatures and pressures.

The Kanmantoo Group metasediments comprise the bulk of the Cambrian of the Adelaide region. In some areas their highly metamorphosed nature and complex geology suggested to some early investigators a Precambrian rather than the Cambrian age now accepted on the basis of geological mapping, radiometric dating and chance finds of trace fossils in the less metamorphosed areas. Surprisingly, a Silurian age (Cambrian in modern terms) assigned to rocks containing trilobite burrows in the Nuriootpa district (Woods 1862) has been ignored, overlooked or dismissed as having little stratigraphic significance (Tate 1879). Confirmation that trilobite trackways do occur in the Kanmantoo Group in that area is given in Fig. 11f.

The Delamerian Orogeny

The rocks of the Adelaide region were metamorphosed and in places folded at least three times (Offler & Fleming 1968) during the Early Palaeozoic Delamerian Orogeny (Thomson 1969). In the Late Cambrian the rocks were weakly folded and variously heated depending on their depth of burial and geographical location. Where temperatures were high enough, metamorphic minerals grew in the solid rock due to the re-distribution of ions. It was during this predominantly thermal event (Fleming & Offler 1968) that the Encounter Bay Granites, about 506 m.y. old, readily seen at Rosetta Head (The Bluff), on Granite Island and at Port Elliot, were intruded into the Petrel Cove Formation and Middleton Sandstone (Milnes, Compston & Daily, in press). Calculations based on pressures necessary to have

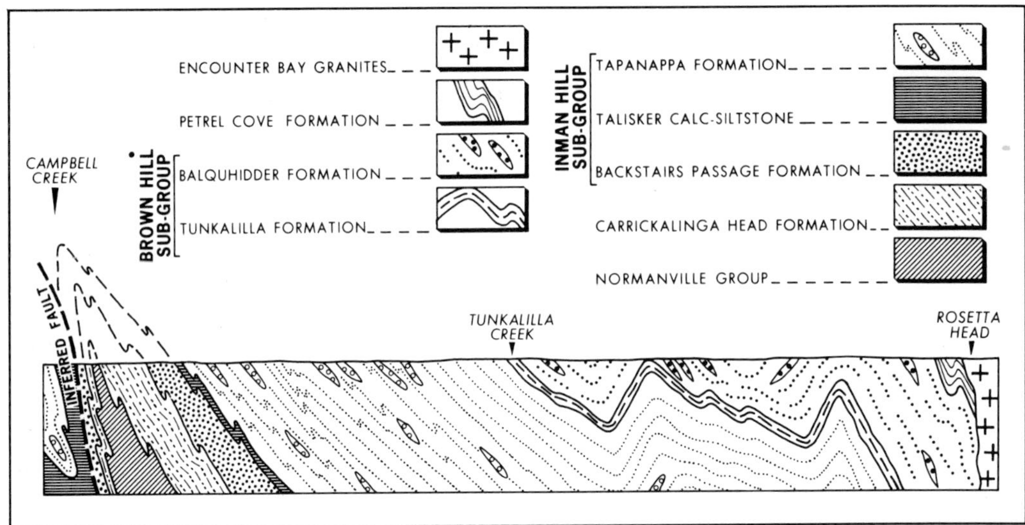

Fig. 9. Diagrammatic cross-section illustrating the stratigraphy and style of folds seen in the type section of the Kanmantoo Group between Campbell Creek and Rosetta Head (The Bluff). Data from Daily and Milnes (1973).

formed cordierite and andalusite adjacent to the granites imply that between 5 km and 10 km of younger Cambrian sediments, now eroded, covered these two youngest formations of the Kanmantoo Group when the granite magma crystallised (Daily & Milnes 1973).

Following this, the main phase of folding and metamorphism took place throughout the region. Pressures directed towards the west and northwest produced tight asymmetrical folds, often overturned to the west and northwest and more open upright folds further east, as for example along the south coast of Fleurieu Peninsula (Fig. 9). Late in this folding phase westerly directed thrust faults frequently developed on the under limbs of the folds. The combined earth pressures and elevated temperatures regionally metamorphosed the rocks of the Mt Lofty Ranges such that there is a progressive decrease in metamorphic grade away from the Palmer district (Offler & Fleming 1968, Fig. 6). In response to this metamorphism the rocks developed a slaty cleavage in low-grade areas and a schistosity where grades were higher. The lineated Palmer Granite (age beween 514 ± 33 m.y. and 489 ± 15 m.y., Milnes et al., in press) and other schistose granitic rocks in the eastern Mt Lofty Ranges, such as the Monarto Granite and Mt Crawford Granite Gneiss (Mills 1963), were

emplaced either prior to this phase of folding and were metamorphosed during that event, or were crystallised in the stress field created at the time of folding.

In the Encounter Bay area and probably elsewhere in the Mt Lofty Ranges the main effects of metamorphism waned in the Early Ordovician about 459 million years ago (Milnes et al., in press). Following this a weaker phase of folding (two phases have been recorded in some areas) produced crenulated schists. Radiometric age data, although inconclusive (Milnes et al., in press), suggest that this deformation took place possibly as late as 400 million years ago in the Late Silurian.

The Delamerian Orogeny culminated with the folded and metamorphosed rocks being thrust upwards along steeply dipping reverse faults (Sprigg 1946) to form part of an arcuate mountain chain, herein named the *Delamerides*, which stretched to the north beyond the Flinders Ranges and southwards into East Antarctica which lay juxtaposed to South Australia at that time.

C. Late Palaeozoic*

Little is known of the geological history of the Adelaide region between the end of the Delamerian Orogeny and the Early Permian, for no sediments covering that time interval are known for the region. During the interven-

---

* Original draft prepared by B. Daily.

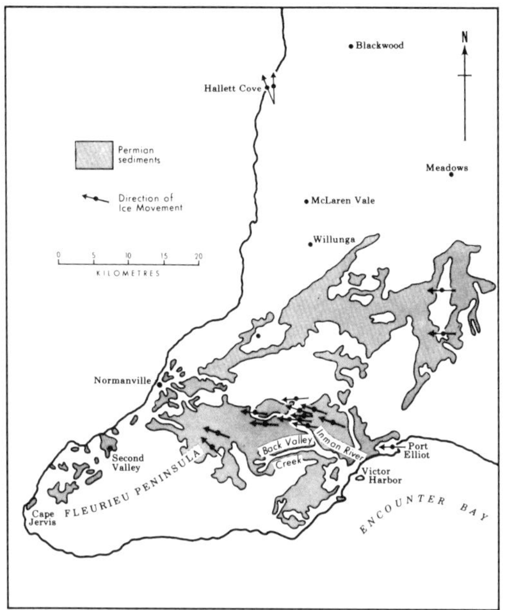

Fig. 10. Distribution of Permian sediments and Permo-Carboniferous glaciated pavements in the Adelaide region. Arrows show ice-movement directions recorded on exhumed glaciated surfaces cut across Precambrian and Cambrian rocks.

ing 120 m.y. erosion prevailed and by the end of the Permo-Carboniferous glaciation, the next major event in the history of the Adelaide region, the mountain chain had a thickness of many kilometres of rock removed from it. In the Encounter Bay region an estimated 10 km had been stripped to expose granite, and between Normanville and Second Valley erosion exposed the older Precambrian crystalline basement.

The earliest observed evidence of glaciation in Australia was the grooved, striated and polished glacial pavement recorded by Selwyn (1859) in Inman Valley, and now called "Selwyn Rock" or "Glacier Rock". In 1877 Ralph Tate discovered a glaciated pavement at Hallett Cove. Originally a Cainozoic age was assigned to the overlying sediments. The remarkable freshness and relatively unconsolidated nature of the deposits and their close association with fossiliferous Cainozoic strata argued a recent age. The forms and sediments were reminiscent of and were correlated with

the Pleistocene glacial features of Britain, Europe and North America. It was only when Permian plants were found in similar deposits at Bacchus Marsh in Victoria that a Permian age was considered likely for the South Australian occurrences. The Permian age of the Fleurieu Peninsula glacial sediments was subsequently confirmed by the discovery of arenaceous foraminifera in them at Second Valley and in clay shale 13.6 m above the base of the section at Cape Jervis (Ludbrook 1967).

It seems highly likely that the glaciation, of which evidence is recorded in the Permian section, commenced in the Late Carboniferous and continued into the Early Permian. Studies by Howchin (1910a, 1910b, 1926), Campana & Wilson (1955), Campana, Wilson & Whittle (1955), Ludbrook (1967), Crowell & Frakes (1971a, 1971b) and Milnes & Bourman (1972) suggest that during the Late Palaeozoic glaciation Fleurieu Peninsula was mountainous with overdeepened valleys such as Back Valley, having a minimum relief of over 500 m (probably much more) and along which ice moved in an east to west direction (Fig. 10). The valley bottoms must have been close to sea-level, for after glacial retreat the sea briefly invaded some of them. Crowell & Frakes (1971a) favour the former existence of a thick ice sheet rather than the mountain glaciers suggested by Campana, Wilson & Whittle (1955). The ice sheet was thick enough and presumably had adequate head to over-ride most topographic obstacles in its path.

At Hallett Cove, the glacial ice moved in a north to northwest direction (Howchin 1926; Sprigg 1942; Talbot & Nesbitt 1968; Oliver & Daily 1969; Nesbitt 1969; Cooper et al. 1970). The rock debris carried in the ice cut a highly irregular surface across the folded Late Precambrian quartzites and slates and produced the polished, grooved and striated glaciated pavements best seen in the Sandison Reserve.* The local direction of ice movement is given unambiguously by small *roches moutonnées* (Fig. 12c) and plucked crescentic fractures (crescentic gouges), both of which are steep-sided towards the down-stream direction.

With the retreat of the ice sheet in the Early Permian a thin veneer of ill-sorted, poorly- to non-stratified, pebbly and boulder-studded sandy clays called *till* was left blanketing the

---

* Establishment of the Sandison Reserve was initiated by A. R. Alderman and supported by the Sandison family. The best evidence in the world of Permian glaciation has been preserved through their contributions.

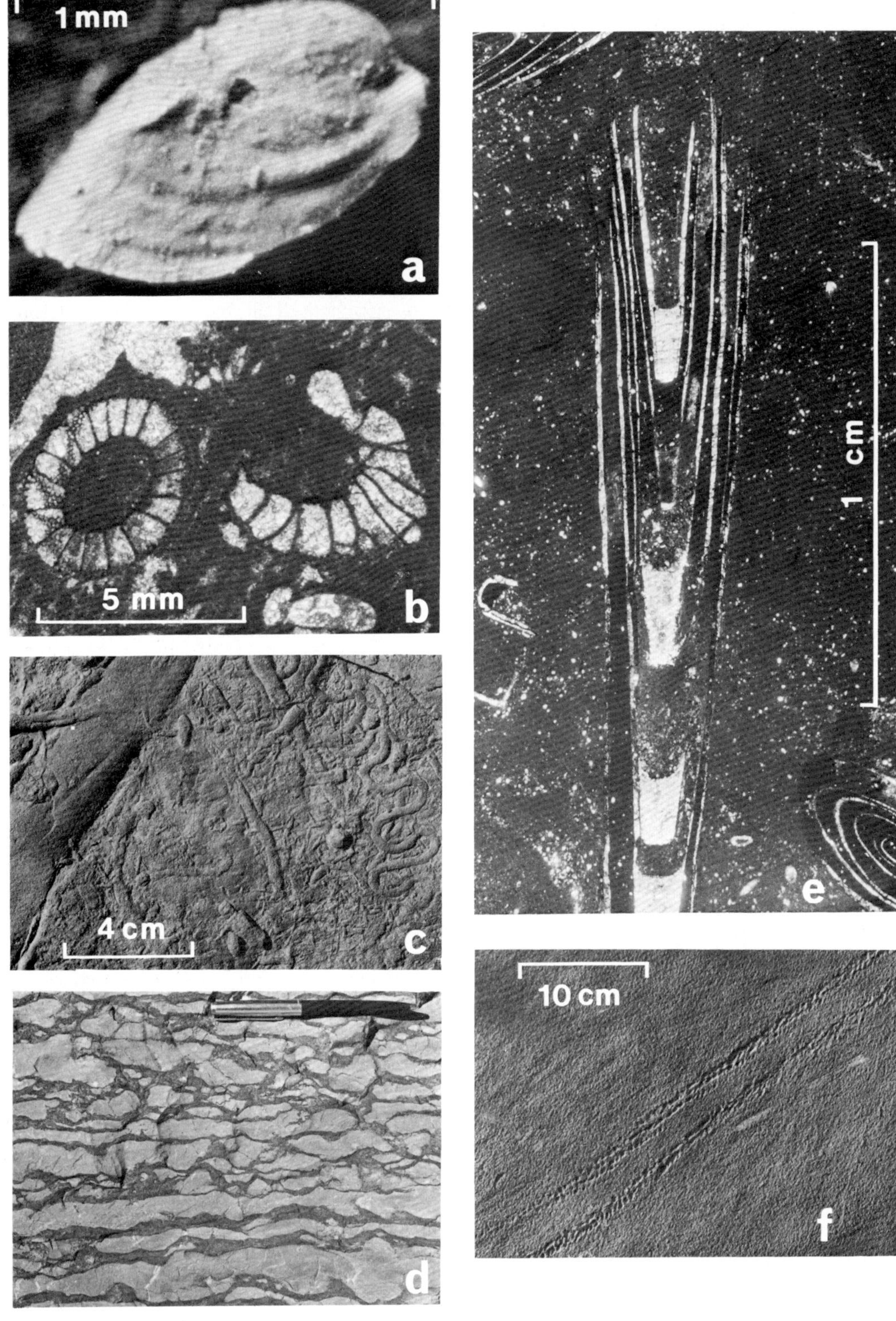

glaciated landscape. There are few exposures of Permian till in the Adelaide region as most of it was reworked and deposited elsewhere. Some patches up to 1.5 m thick are preserved at Hallett Cove resting unconformably on the Late Precambrian strata.

The preserved Permian sediments, collectively called the Cape Jervis Beds (Ludbrook 1967) were mainly deposited subaqueously. They reach a maximum thickness of about 300 m in Back Valley. The basal parts of the sequence everywhere drape over the non-planar glaciated bedrock often with substantial primary dips, but eventually the beds become horizontal upwards (Fig. 12a). Apart from the rare till, the relatively unconsolidated sediments are essentially bedded claystones and massive to cross-bedded sandstones. Some of the claystones and mudstones carry pebbles and boulders (erratics) rafted into the region by icebergs which on melting dropped their load. These dropstones are prominent in the purple varve-like beds below the Sugarloaf at Hallett Cove (Fig. 12d) and elsewhere (Mawson 1926).

Most of the erratics found on Fleurieu Peninsula were locally derived, especially the larger blocks, some of which are over 7 m in length (Howchin 1929). Erratics of Encounter Bay Granites transported as far north as Hallett Cove, and many Adelaidean and Cambrian lithologies such as Sturt Tillite, Tapley Hill Formation slates and Kanmantoo Group metasediments are well represented. The exotic erratic fraction includes many igneous rocks, especially porphyry and volcanics, and some sediments. Some lithologies are remarkably similar to ones occurring as far east as the Grampians in western Victoria. Many of the erratics are soled, faceted and striated.

Most of the Permian sediments were deposited under non-marine conditions, but some marine influence (brackish water facies) is indicated by the occurrence of the microscopic arenaceous foraminifera in clay shale described from Cape Jervis by Ludbrook (1967). Apart from faulting, the Permian deposits are undisturbed. Their preservation is due to their occurrence in over-deepened valleys which until uplift in the Tertiary were protected from erosion beneath the level of the Mesozoic peneplain.

D. Cainozoic*

The absence of any known sedimentary record in the Adelaide region representing the period of 200 m.y. between the Permian and the early Tertiary suggests that during this time the area was essentially a land mass undergoing weathering and erosion. The Permian glaciation had taken place on a land with accentuated local relief (Campana 1958). Preservation of Permian sediments, and the presence in the ranges of bleached profiles which may be as old as Mesozoic in places (Firman 1969a; Bourman 1973[1]; Daily, Twidale & Milnes 1974), indicate the slow rate of pre-Tertiary erosion; but by the early Tertiary the area had been bevelled down to a rather subdued and deeply weathered landscape. Thoroughly decomposed and bleached or mottled bedrock tens of metres in thickness is known to underlie Eocene sediments, e.g. beneath the Adelaide city area. Eocene clastic sediments deposited in the newly formed basins were derived from Proterozoic and Palaeozoic sedimentary rocks,

---

* Original draft prepared by J. B. Firman & J. M. Lindsay.
[1] Bourman, R. P. (1973).—Geomorphic evolution of southeastern Fleurieu Peninsula. M.A. thesis University of Adelaide (unpublished).

Fig. 11. (a) The enigmatic pseudobivalve *Heraultia* Cobbold previously classified both as a crustacean and a mollusc. *Heraultia* Cobbold is known from the early Cambrian of France and Siberia. Specimen from about 9 m below top of middle member, Mt Terrible Formation, new Sellick Hill road-cutting.
(b) Thin section of Fork Tree Limestone, Sellick Hill, showing cross-sections of Archaeocyatha (Howchin collection, University of Adelaide).
(c) Curved and meandering trace-fossils on a sandstone bedding surface, lower part of the Sellick Hill Formation, coastal section northwest of Myponga Beach.
(d) Mottled limestone, Sellick Hill Formation, old Sellick Hill Quarry. Pale areas are dark grey limestone, darker areas are carbonate-rich siltstones. The mottling is a secondary feature and formed probably prior to the lithification of the rock. Pen is 13 cm long.
(e) Thin section of limestone, Sellick Hill Formation, Myponga Beach, showing six septate hyolithid shells invaginated one into the other by current action. The calcitic shells, which have lost their pointed apices, have septal chambers filled with sparry calcite. The black sediment partly filling the shell cavities is phosphorite.
(f) Trackways ascribed to a trilobite on a Kanmantoo Group metasandstone bedding surface, approximately 6 km east of Truro.

ancient weathering zones and ?Mesozoic bleached zones.

Earth movements of Palaeocene to Middle Eocene age involving block faulting and tilting, led to the formation of a complex graben flanking the western margin of the ancestral Mt Lofty Ranges: the St Vincent Basin (Fig. 13). This graben exists today as part of a compound rift, a second-order structural feature which occupies much the same area as the present Gulf St Vincent and low-lying areas adjoining (Firman 1965). The main components of the Basin in the Adelaide region are the Adelaide Plains Sub-Basin (which includes the Adelaide city area on the Para Fault Block), the Hope Valley-Golden Grove Embayment, and the Noarlunga and Willunga Embayments. Other minor basins are the Myponga and Hindmarsh Tiers basins (both intermontane and connected at times during the past with the St Vincent Basin) and the intramontane Barossa Valley and Meadows Valley basins.

Delamerian structural trends strongly influenced the position and shape of these basins as they developed during the Tertiary (Sprigg & Stackler 1965). Indeed, the oscillation of fault blocks was probably an important factor in the structural development of the region as long ago as the Precambrian (Thomson 1965).

It is interesting to note that the earth movements which initiated the St Vincent Basin occurred at the same time as the separation of Australia from Antarctica along a line approximately the edge of the present continental shelf of South Australia, and the commencement of the Australian Plate's drift northwards

(McKenzie & Sclater 1971; Griffiths 1971; McGowran 1973; Deighton, Falvey & Taylor 1976).

The horsts, grabens and associated faults within and bounding the rift were discussed in a general way by Fenner (1930, 1931) and others; and in some detail with reference to specific areas by Miles (1952) and Campana & Wilson (1955). The position and style of the arcuate, steeply dipping faults is shown in Fig. 13, and the various cross-sections (Figs 16–19). These structural features persisted, controlling erosion and deposition throughout the remainder of the Cainozoic. Variations in thickness of strata and the progressive displacement which is known to occur across some of the faults, show that they have been active intermittently through much of the Cainozoic. Displacement along them is generally normal (northwest block down) but there is some local evidence of steep southeast dipping fault planes (i.e. reverse faults), e.g. on the Ochre Cove-Clarendon Fault. The fault blocks dip gently to the east and south, and displacement generally increases towards the south and southwest. The Adelaide Plains Sub-Basin and the Noarlunga and Willunga embayments are now asymmetric "half-grabens" or fault angle depressions, with a thickening of the sedimentary wedge to the southeast towards the boundary faults, partly, at least, due to post-depositional tilting and erosion. The representative cross-sections in Fig. 19 show the contrast between the size and structure of the sub-basin and the two embayments. Tertiary sediments in the latter are dragged up steeply adjacent to the faults, but there is little if any

---

Fig. 12. (a) Undeformed Permian pebbly clays and interbedded quartzose sands (pale coloured), which exhibit primary dips, rest unconformably on the irregular surface (marked by large arrows) glacially cut across an anticline of Late Precambrian dark coloured quartzites and slates, Waterfall Creek, Sandison Reserve. A thin horizontal capping of Pliocene to Recent marine and terrestrial strata rests unconformably on the Permian. Its base is marked by a thin outcropping band of calcareous sandstone designated by small arrows.
(b) A polished, grooved and striated Permian glacial pavement near the northern boundary of the Sandison Reserve. Chattermarks (small indentations within and transverse to length of grooves) caused by the vibratory movement of pebbles held in the base of the ice sheet, are visible in some of the deeper grooves. Note the 20° spread in the directions of ice movement which was from lower right to upper left. Pen is 13 cm long.
(c) Small *roches moutonnées* exhumed from beneath Permian strata, Sandison Reserve. Note the glacial striae parallel to length of hammer (32 cm long), large erratic (upper right) and the glacially plucked steep sides (leeward side) of the *roches moutonnées*. Glacial flow was from right to left.
(d) Permian boulder-bearing varve-like clays and sands at the base of the Sugar Loaf, Hallett Cove. Icebergs, which rafted the clasts into the area, on melting dropped their load into the finely bedded sediments to produce typical "dropstones" which punctured through earlier deposited layers. Subsequent sediments draped over the clasts to produce irregular bedding features (non-tectonic folds). Pen is 13 cm long.

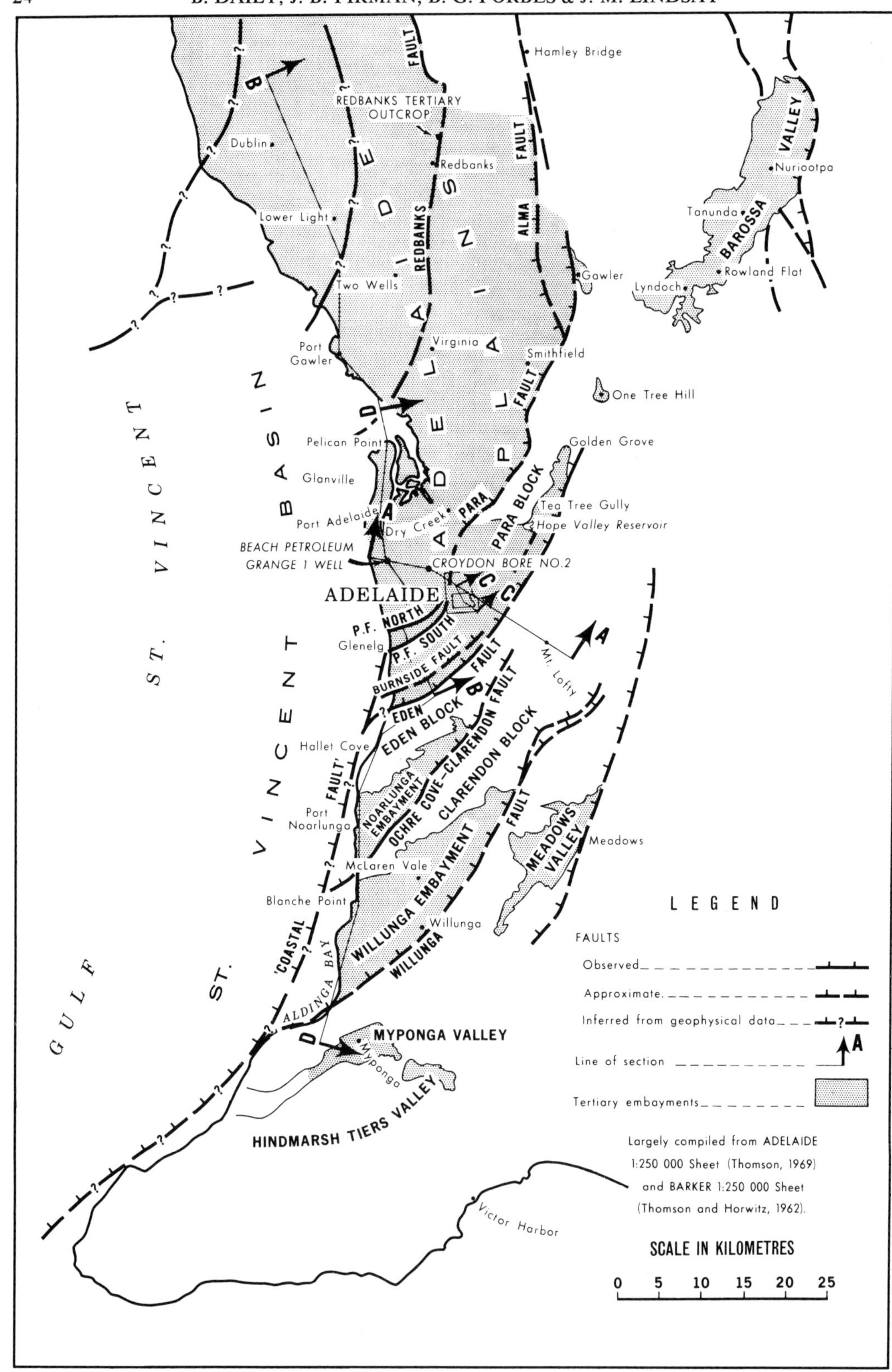
Hamley Bridge
REDBANKS TERTIARY OUTCROP
Dublin
Redbanks
Lower Light
VALLEY
Nuriootpa
Tanunda
BAROSSA
Gawler
Lyndoch
Rowland Flat
Two Wells
Virginia
Smithfield
Port Gawler
One Tree Hill
ADELAIDE SEDIMENTARY BASIN
ST. VINCENT BASIN
REDBANKS FAULT
ALMA FAULT
Pelican Point
Golden Grove
Glanville
PARA BLOCK
Tea Tree Gully
Port Adelaide
Hope Valley Reservoir
Dry Creek
PARA
ST. VINCENT
BEACH PETROLEUM
GRANGE 1 WELL
CROYDON BORE NO.2
ADELAIDE
C
A
P.F. NORTH
P.F. SOUTH
A
Glenelg
BURNSIDE FAULT
Mt. Lofty
FAULT
B
EDEN BLOCK
EDEN
Hallet Cove
OCHRE COVE–CLARENDON FAULT
CLARENDON BLOCK
FAULT
Port Noarlunga
NOARLUNGA EMBAYMENT
MEADOWS VALLEY
Meadows
McLaren Vale
WILLUNGA EMBAYMENT
Blanche Point
Willunga
GULF
WILLUNGA FAULT
'COASTAL
ST. VINCENT
ALDINGA BAY
D
MYPONGA VALLEY
Myponga
HINDMARSH TIERS VALLEY
Victor Harbor
LEGEND
FAULTS
Observed
Approximate
Inferred from geophysical data
Line of section
A
Tertiary embayments
Largely compiled from ADELAIDE
1:250 000 Sheet (Thomson, 1969)
and BARKER 1:250 000 Sheet
(Thomson and Horwitz, 1962).
SCALE IN KILOMETRES
0   5   10   15   20   25

direct evidence of this in the Adelaide Plains Sub-Basin.

Sedimentation was initially from freshwater streams draining the surrounding highlands. This was followed by several cycles of marine transgression and regression involving the deposition of marine sediments.

Except for excellent coastal exposures in the Noarlunga and Willunga Embayments, the older Cainozoic rocks of the Adelaide region are largely obscured due to their general lack of deformation and dissection, and to the ubiquitous cover of Pleistocene and Recent sediments and soils (Firman 1969b). However, much subsurface information concerning the Adelaide plains area is available from boreholes, most of which were drilled for water (Lindsay 1969).

The Tertiary succession exposed at Maslin and Aldinga bays (Fig. 20) was described as a series of type sections by Reynolds (1953), and later contributions by Glaessner & Wade (1958) and Lindsay (1967). This sequence, taken as the standard stratigraphic framework for the eastern St Vincent Basin, is developed in similar coastal exposures in the Noarlunga Embayment (Lindsay 1970). Younger Pleistocene-Recent sequences in this area (Fig. 23) have been described by Ward (1965), Twidale, Daily & Firman (1967), and Firman (1967a). The Tertiary succession at Maslin and Aldinga bays, with some variation, has been traced through the Adelaide Plains area in subsurface (Lindsay 1968, 1969; Ludbrook 1969) as shown in Figs 16–18. The Quaternary succession in the Adelaide Plains area has been described by Firman (1969a).

North Maslin Sands

These fluviatile sands are the oldest Tertiary sediments known in the Adelaide region. They range up to 30 m in thickness, and were deposited by strongly flowing rivers which drained the highlands marginal to the graben. Where exposed in the Maslin Bay sand quarries they rest with angular unconformity on the underlying weathered Adelaidean slates and Permian rocks (Reynolds 1953; Olliver & Weir 1967). They occur similarly at Christies Beach in the Noarlunga Embayment, and are known at depth beneath the Adelaide Plains (Fig. 17).

A lens of black carbonaceous clay in the sands at Maslin Bay contains plant microfossils of early Middle Eocene age (*i.e.* about 50 m.y.) (McGowran, Harris & Lindsay 1970) thus dating the commencement of the observed Tertiary sedimentation in this portion of the St Vincent Basin. The clay lens, representing a swamp deposit, has a flora compared by Lange (1970) with those of Queensland rain forests, indicating a climate considerably warmer and wetter than Adelaide's at present.

Various non-marine quartz sands marginal to the basin and its embayments are correlatives, at least in part. Examples are sands with white clay lenses, quarried near Highbury and Golden Grove; sands at Bakers Gully and Kangarilla; and sands quarried near Gawler.

A laterite profile is developed in fossiliferous Eocene gravelly sands near Bakers Gully. The cap on the laterite profile is the Yallunda Ferricrete (Firman 1967a).

South Maslin Sands

In the middle to late Eocene, the sea advanced into the St Vincent Basin from the widening seaway developing between Australia and Antarctica, resulting in the accumulation of the marginal marine South Maslin Sands which attain a maximum thickness of 42 m in the Croydon bore (Lindsay 1968, 1969). These display oxidised iron colours in outcrop at Maslin Bay and Port Noarlunga, but are dark grey-brown carbonaceous, and pyritic in the subsurface of the Adelaide Plains Sub-Basin (including the Adelaide city area) and the Willunga Embayment (Lindsay 1969). Ferruginisation of Tertiary basal sands (*e.g.* at Christies Beach) has been attributed to the formation of "laterite" (Glaessner & Wade 1958), but much of it can be explained by relatively recent oxidation following coastal erosion during the Holocene. Other comparable ferruginisation has been shown by recent stratigraphic studies (e.g. Firman 1973) to derive from separate events not related to the bleaching and mottling of the laterite profile of the sort described by Prescott & Pendleton (1952).

Clinton Formation

In swampy tracts marginal to the basin in the Middle to Late Eocene, lignites and carbonaceous clays, silts, and sands accumulated in lower energy environments. These deposits are correlative with the older, Eocene phase of Clinton Coal Measures; a formation named by Harris (1966) primarily for the Clinton and

Fig. 13. Map showing present distribution of Tertiary basinal sediments, related faults and features, lines of cross-sections and localities mentioned in text.

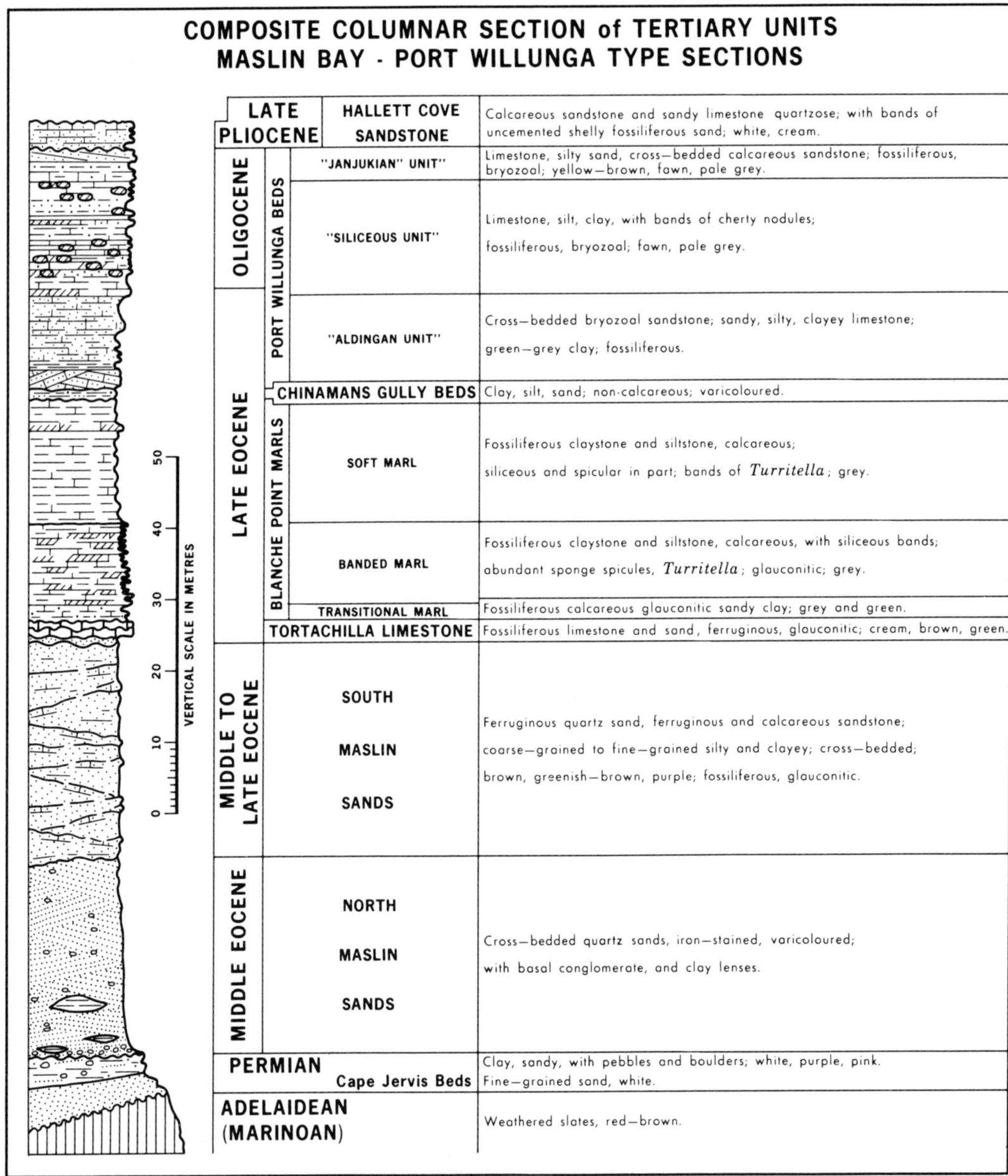

Fig. 14. Composite columnar section of Tertiary units Maslin Bay-Port Willunga Type Sections.

Inkerman-Balaklava coalfields, but also penetrated in boreholes and shafts near Noarlunga (Anon. 1921, 1924; Cornelius 1927), within the City of Adelaide, at Hope Valley (Cochrane 1953) and in the Adelaide Plains (Lindsay 1969). A small outcrop occurs on the south bank of the Onkaparinga River immediately south of Noarlunga township.

Tortachilla Limestone

A diachronous succession of sedimentary regimes from clastic to richly glauconitic to carbonate (limestone/marl) marked the progress of marine transgression through the southern Australian Tertiary basins (McGowran 1973). In the St Vincent Basin, the marginal-marine conditions represented by

Fig. 15. Sellicks Beach, southwest corner, looking southeast; bryozoal limestones of Port Willunga Beds (here of Early Miocene age) dragged up into a relatively steep dip adjacent to the Willunga Fault. Southern edge of Willunga Embayment, St Vincent Basin.

South Maslin Sands were followed by more pronounced marine influence marked first by deposition in a near-shore shelf environment of the distinctive Tortachilla Limestone which is up to 3 m thick. This richly fossiliferous and glauconitic formation dips gently south in picturesque cliffs at Maslin Bay near Blanche Point (Fig. 20a), and is also exposed at the foot of Witton Bluff, Port Noarlunga, and around the Onkaparinga River towards Noarlunga. Equivalents of the formation have been recognised in the subsurface of the Willunga Embayment and the Adelaide Plains. Faunal elements such as the foraminifera *Halkyardia* and *Linderina,* suggest warm water temperature.

Blanche Point Marls

The "Transitional Marl" Member (Reynolds 1953) was laid down on an irregular cemented surface of Tortachilla Limestone during a peak of warm marine transgression into the coastal embayments, as evidenced by the thin planktonic foraminiferal zone of *Hantkenina primitiva* (Fig. 22f) Late Eocene in age. The zone, important in inter-regional correlation, is better developed in similar sediments beneath the City of Adelaide (Lindsay 1969). Parr's finding of *Hantkenina* at Maslin Bay in 1948 (Glaessner 1951) provided a firm Eocene datum level for the stratigraphy of the whole succession.

The "Banded Marl" Member which is up to 20 m thick in the Adelaide Plains Sub-Basin (Lindsay 1969; Selby & Lindsay 1973) forms resistant headlands at Blanche Point (Fig. 20a) and Witton Bluff; the "Soft Marl" Member having a more subdued outcrop to the south at each locality and a thickness of 17 m at Aldinga Bay (Reynolds 1953). Both members are rich in siliceous sponge spicules and the gastropod *Turritella aldingae* (Fig. 22b). They were deposited on a low-energy muddy marine shelf with widespread sponge gardens. The dark grey carbonaceous lithology indicates poorly oxygenated bottom conditions, and the abundance of biogenic opaline silica as sponge spicules suggests the upwelling of cold deep waters over the continental shelf. A distinct cooling from the warm peak with *Hantkenina* can be inferred.

Chinamans Gully Beds

This thin clastic unit with very limited marine influence apparent, contrasts with the biogenic calcareous sediments which preceded and followed it, and marks a regressive episode of coastal marsh, swamp, or estuarine sedimentation. In the coastal section at Port Willunga and Port Noarlunga the Beds are only about 2 m thick and are weathered and oxidised to striking brown, red and green colours, but the equivalent interval inland in the subsurface is dark grey-brown, carbonaceous and lignitic. The age of the unit is Late Eocene (Lindsay 1967, 1969). Only occasionally has this regressive episode been recognised in the subsurface of the Adelaide Plains, e.g. in the Croydon bore (Lindsay 1968).

Port Willunga Beds

A further marine transgression into the Willunga and Noarlunga Embayments began, also in the Late Eocene, with deposition of basal Port Willunga Beds (Reynolds 1953; Lindsay 1967, 1970). Three members can be distinguished in the coastal cliffs at Port Willunga

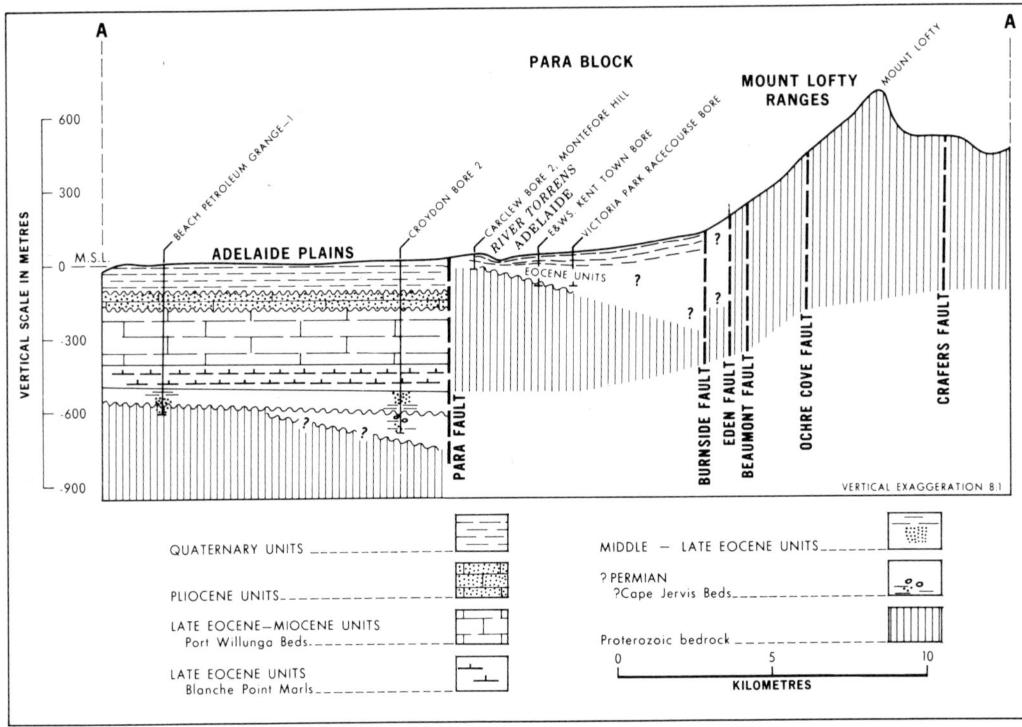

Fig. 16. Geological cross-section AA. Beach Petroleum Grange No. 1 Well to Mt Lofty.

(Fig. 20b) and at Port Noarlunga-Seaford. The basal fossiliferous sands, marls and limestones, contain the foraminifera *Halkyardia* and *Linderina* as evidence of warm-water deposition. The central unit, of Oligocene age, is marked by bands of cherty nodules and is prominent through the subsurface of the Adelaide Plains, where it attains a thickness of 70 m in the Croydon bore (Lindsay 1968, 1969). This silica-rich member of Port Willunga Beds is reminiscent of, but younger than and distinct from, Blanche Point Banded Marl. The abundance of opaline silica suggests further upwelling of cold, deep waters as the circum-Antarctic current was through the widening Oligocene seaway south of Australia (Kennett *et al* 1972, Kennett *et al.* 1975).

The cool conditions of the Early to Middle Oligocene ameliorated in the Late Oligocene, when a warmer transgressive period possibly also associated with tectonism left its mark in many localities. The "Janjukian Member" of Port Willunga Beds (Lindsay 1967), mostly Late Oligocene in age, was deposited widely through the Adelaide Plains Sub-Basin, and has a maximum thickness of 21 m in the Croydon bore (Lindsay 1968, 1969). At this time also, the sea entered the Hindmarsh Tiers Basin (Fig. 13), depositing fossiliferous limestones and sands which are quartzose and pebbly near the base (Furness 1975)[2]. Marine sedimentation had commenced slightly earlier in the adjacent Myponga Basin (Fig. 13) and Furness concludes that the two basins almost certainly had marine connection between them, and with the St Vincent Basin rather than with the Murray Basin.

At this time, in the intramontane Barossa Valley Basin, Oligocene-Miocene rivers, lakes and swamps accumulated non-marine carbonaceous sands, silts, clays and lignites, up to 130 m thick. These are the sands at Rowland Flat described by Harris & Olliver (1965); Olliver (1967); and Olliver & Weir (1967). Ironstone cappings on the terrestrial sands in the Barossa Valley were attributed by Glaessner & Wade (1958) to "lateritization" during the Eocene, but this ferruginisation is younger,

---

[2] Furness, L. J. (1975).—The hydrogeology of the Hindmarsh Tiers Basin. Flinders University of South Australia. B.Sc. Hons thesis, unpublished.

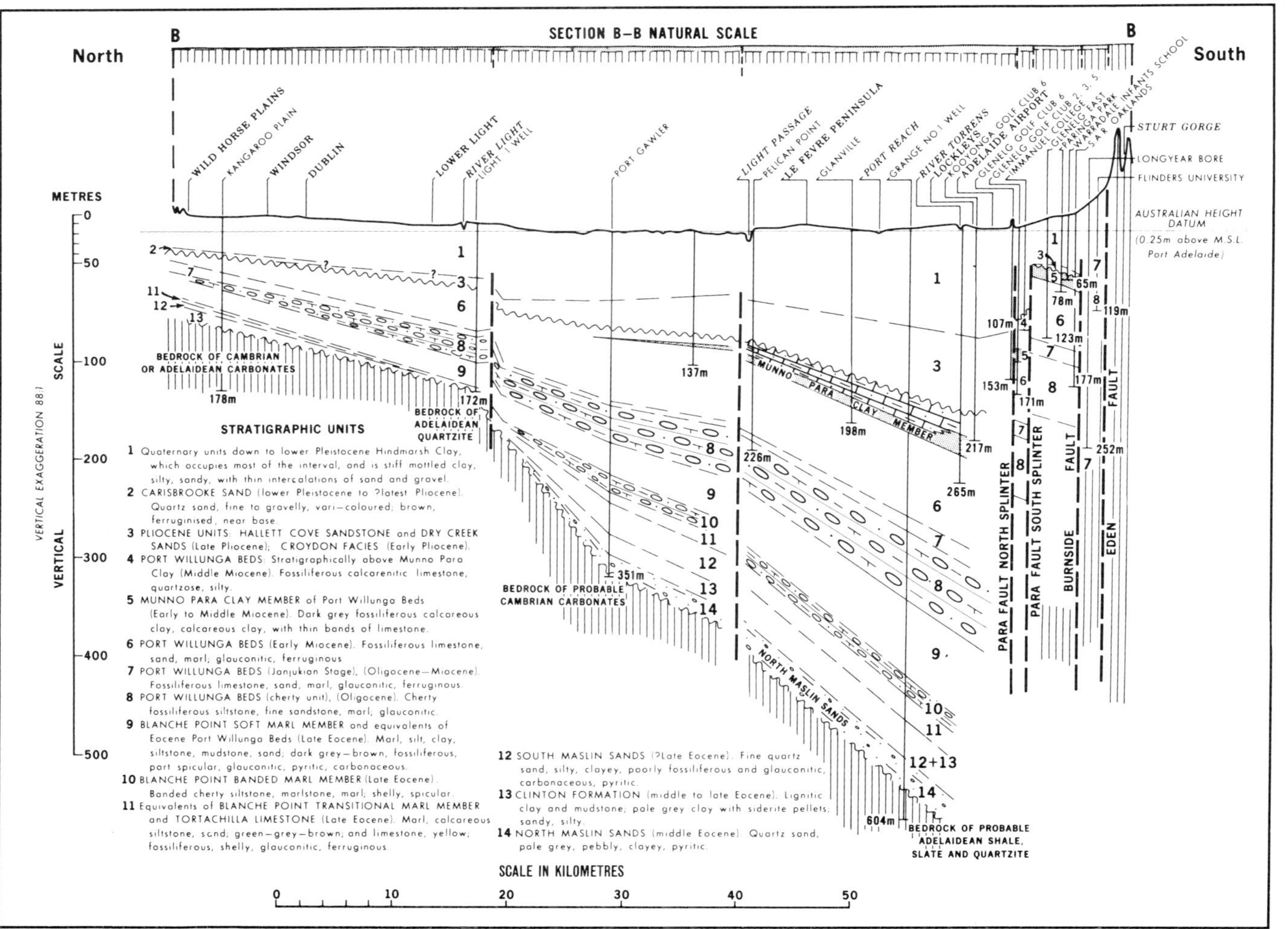

Fig. 17. Geological cross-section BB through Adelaide Plains.

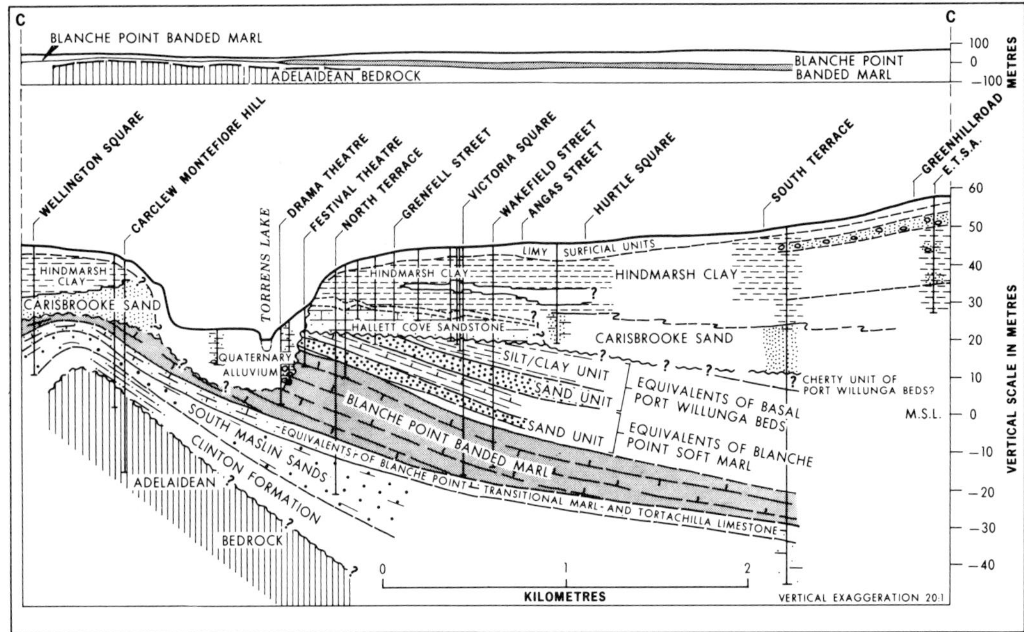

Fig. 18. Geological cross-section CC through City of Adelaide.

is not part of a laterite profile, and is not correlated with the full laterite profile developed elsewhere in the Adelaide region (Firman 1976b).

Deposition of fossiliferous limestones and sands (Port Willunga Beds) continued through the Early Miocene on the shallow, warm marine shelves of the Adelaide Plains Sub-Basin and the Willunga Embayment; and in the Myponga and Hindmarsh Tiers drowned valleys. However, no Port Willunga Beds younger than the Oligocene cherty member are known from the Adelaide city area (Fig. 18). At the beginning of the Middle Miocene, sea temperatures were warm enough for the large foraminifer *Lepidocyclina howchini* (Fig. 21d) with tropical affinities, to be established for a time in the Adelaide Plains, Willunga, and Myponga areas.

A marked change of environmental factors then led to deposition of the Munno Para Clay Member of the Port Willunga Beds (Lindsay & Shepherd 1966; Lindsay 1969) in the deepest parts of the Adelaide Plains Sub-Basin (Fig. 17). The presence there of another large foraminifera, *Flosculinella bontangensis* (Fig. 22c) in the succeeding marine limestones and sands (Lindsay 1969) suggests a further peak of warm Middle Miocene palaeotemperatures.

Marine sedimentation ceased during the Middle Miocene in the St Vincent Basin, as in other South Australian coastal Tertiary basins, due to widespread uplift of the continental margin and also related to a worldwide drop in sea level as water became locked into the rapidly developing Antarctic ice cap (Kennett *et al.* 1975; Shackleton & Kennett 1975; Savin *et al.* 1975).

Following the relatively quiet period of the Middle Tertiary, the mid-Tertiary land surface was disrupted by continued block faulting. These structural events accentuated the horsts and grabens and initiated further clastic sedimentation.

The region of greatest seismic activity in South Australia is parallel to and adjoins part of this persistent zone of block faulting, and stretches from Kangaroo Island in the south to near Leigh Creek in the northern Flinders Ranges (Sutton & White 1968). The Adelaide earthquake of 1st March, 1954, had its epicentre near the Eden Fault within this zone (Kerr-Grant 1956; Bolt 1957).

Low-angle unconformity between mid-Tertiary and Pliocene strata is displayed in the coastal cliffs at Port Willunga (Fig. 20b) where Late Pliocene Hallett Cove Sandstone dips southwards at less than ½°, and the underlying

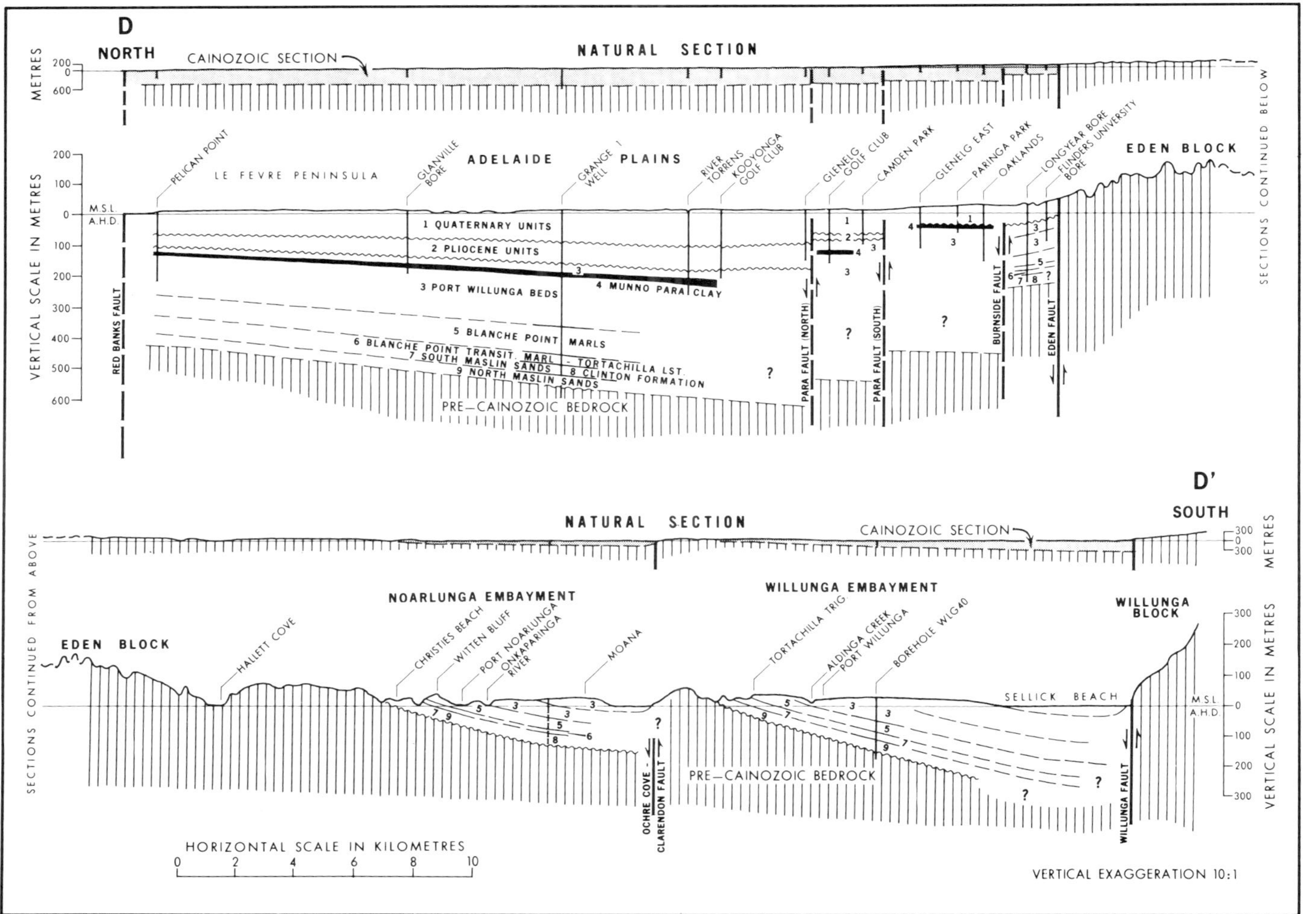

Fig. 19. Geological cross-section DD through southern Adelaide Plains, and Noarlunga and Willunga Embayments.

Eocene-Oligocene Port Willunga Beds dip at 2 or 3 degrees in a similar direction.

Pliocene units

After the late Miocene hiatus, a limited degree of marine sedimentation began again in the St Vincent Basin in the Early Pliocene (Ludbrook 1963) with deposition of the "Croydon facies", only known from bores in the Adelaide Plains near the present coast between Glenelg and Port Gawler (Lindsay 1969). These shallow marine-bay deposits are glauconitic bryozoal silts and clayey quartz sands with bands of calcareous cementation. The silts and sands are 46 m thick in Beach Petroleum Grange No. 1 Well.

In the Late Pliocene (Ludbrook 1963) a warm and shallow sea extended into the southern part of the Adelaide Plains Sub-Basin, possibly due to widespread eustatic rise of sea level. This transgression was associated with

Fig. 20. (a) Blanche Point, looking southwest across Gulf St Vincent. The cliff sequence comprises, from bottom to top: *Tortachilla Limestone* (dark, inter-tidal platform), Late Eocene; *Blanche Point Banded Marl* (banded appearance), Late Eocene; *Blanche Point Soft Marl* (at Blanche Point), Late Eocene; *Hallett Cove Sandstone* (Late Pliocene); *Seaford Formation* and *Burnham Limestone* (lower Pleistocene); dark-coloured slope-forming lower Pleistocene clayey sands and *Hindmarsh Clay;* calcrete profile (pale-coloured) middle Pleistocene. Willunga Embayment, St Vincent Basin.
(b) Port Willunga, fishermen's caves, looking south towards Snapper Point; remains of old jetty on right. The cliff sequence comprises, from bottom to top: *Port Willunga Beds,* part of type section (smooth, banded appearance), Late Eocene "Aldingan unit" in lower part of caves, and Oligocene "siliceous unit" from roof of caves upwards; *Hallett Cove Sandstone* (flaggy beds), Late Pliocene, resting with low-angle unconformity on south-dipping Port Willunga Beds; *Burnham Limestone* (rubbly appearance) and *Hindmarsh Clay* (dark, slope-forming), both lower Pleistocene; calcrete profile pale-coloured, top of cliff), middle Pleistocene. Willunga Embayment, St Vincent Basin.

Fig. 21. (a) Echinoid *Meoma decipiens* (Tate). Basal Port Willunga Beds. Onkaparinga estuary, near foot bridge. Port Noarlunga. Noarlunga Embayment. Late Eocene. Bar scale = 1 cm.
(b) Gastropod *Amoria grayi* Ludbrook. Dry Creek Sands, Hindmarsh bore, Adelaide Plains Sub-Basin. Late Pliocene. Bar scale = 1 cm.
(c) Hallett Cove Sandstone with several specimens of the large foraminifer *Marginopora vertebralis* Blainville. Aldinga Bay, Willunga Embayment. Late Pliocene. Bar scale = 1 cm.
(d) Large foraminifer *Lepidocyclina howchini* Chapman & Crespin. Port Willunga Beds, Adelaide Plains Sub-Basin. Early to Middle Miocene.
Dept Mines Observation Bore P3, Hd Munno Para, Sect. 2020, depth 164.6–166.1 m. Bar scale = 1 mm.
(e) Crinoid stems of *Pentacrinus* sp. Blanche Point Marls, Port Noarlunga, Noarlunga Embayment. Bar scale = 1 cm.
(f) Gastropod *Hexaplex (Murexsul) biconicus* (Tate). Dry Creek Sands: Abattoirs Bore, Hd Yatala, Sect. 97. Adelaide Plains Sub-Basin. Late Pliocene. Bar scale = 1 cm.
(g) Echinoid *Australanthus longianus* (Gregory). Tortachilla Limestone, Maslin Bay, Willunga Embayment. Late Eocene. Bar scale = 1 cm.
Photomicrograph (d) taken on University of Adelaide scanning electron microscope by B. J. Cooper (Department of Mines). Photographs a, b, c, e, f and g R. J. F. Jenkins, University of Adelaide.

the deposition of the *Dry Creek Sands* (Glaessner 1951) which are shelly sands with bands of fossiliferous calcareous sandstone and sandy limestone. Interbedded with the Dry Creek Sands and overlying them is the *Hallett Cove Sandstone* (Crespin 1954), the type section of which is at the Sandison Reserve, Hallett Cove.

This fossiliferous quartz sandstone to sandy limestone was also deposited in shallow bays. Where the transgressive sea flooded high coastal areas such as the Eden Block at Marino or Hallett Cove, the sandstone is only a metre or less thick. It is several metres thick at Port Willunga and on the Para Block beneath the City of Adelaide; and up to 12 m thick between the southern Adelaide Plains.

Carisbrooke Sand

Clayey quartz sand, mostly unfossiliferous, continued to be deposited in rivers, lakes and estuaries as the sea retreated at the end of the Pliocene due to climatic cooling and uplift. This sand, the Carisbrooke Sand (Lindsay 1969) is typically developed in the eastern parts of the Adelaide Plains Sub-Basin. In the Dry Creek bore (Tate 1890) it overlies the Dry Creek Sands with apparent conformity, and its age is thought to range from latest Pliocene to early Pleistocene. The relationship of this unit to other deposits of about the same age in the Adelaide region has been summarised by Firman (1976a).

At the close of the Pliocene the higher parts of the ancestral Mt Lofty Ranges stood above plains marking the sites of Tertiary seas and continental lowlands. Sands and gravels, now thoroughly ferruginised, and silicified, were deposited in the ancient river valleys (Firman 1976b). These deposits are the oldest of the alluvials attributed to ancient streams by How-

chin (1932) and are ". . . the alluvia consolidated by secondary silica" of that author. Deposits of this kind lie above the Late Pliocene Hallett Cove Sandstone and below the lower Pleistocene Burnham Limestone (Figs 20 & 23).

Pleistocene and Holocene units: During the last 2 m.y. or so there have been major climatic changes, causing waxing and waning of the polar ice and notable shifts in sea level. These events are all recorded directly or indirectly in the materials and forms of the waste mantle.

In the *lower Pleistocene,* earth movements of the so-called "Kosciuskan Orogeny" produced uplift and dissection of the Tertiary covermass and of weathered basement rocks in the Ranges. This led to deposition of clays, sands and gravels in adjoining basins and grabens. Near Adelaide, this sequence includes the *Hindmarsh Clay* (Firman 1966), a thick unit of clays and sandy clays with sand and gravel lenses. A thin shallow marine limestone, the *Burnham Limestone* (Firman 1976a) occurs at the base of the Hindmarsh Clay and is correlated with part of the Carisbrooke Sand (Lindsay 1969) and with sands reported from Lockleys (Ludbrook 1963). The Hindmarsh Clay is equivalent to units recognised by Ward (1965) in the Noarlunga and Willunga Embayments. Good exposures of this sequence are found in coastal cliffs between Kingston and Yankalilla.

A number of silicified and ferruginous zones within the Hindmarsh Clay sequence overlie the older bleached zones in the uplands. A mottled zone, younger than the Plio-Pleistocene ferruginised and silicified zone (see page —), is found within the Hindmarsh Clay. This zone, the Ardrossan Soil (Fig. 23) is a mottled zone

---

Fig. 22. (a) Bivalve mollusc *Spondylus spondyloides* (Tate). Hallett Cove Sandstone, Schnapper Point, Aldinga Bay. Willunga Embayment. Late Pliocene. Bar scale = 1 cm.
(b) Piece of Blanche Point Marls, with gastropod *Turritella aldingae* Tate. Blanche Point, Willunga Embayment. Late Eocene. Bar scale = 1 cm.
(c) Large foraminifer *Flosculinella bontangensis* (Rutten). Port Willunga Beds, Adelaide Plains Sub-Basin. Middle Miocene. Elizabeth Oval bore 1, Hd Munno Para, Sect. 3128, depth 112.8–114.3 m. Bar scale = 0.5 mm.
(d) Bryozoan. Maslin Bay, Willunga Embayment. Late Eocene. Bar scale = 1 cm.
(e) Coral *Flabellum distinctum* Ed. & Haime. Blanche Point Marls, Aldinga Bay, Willunga Embayment. Late Eocene. Bar scale = 1 cm.
(f) Planktonic foraminifer *Hantkenina primitiva* Cushman & Jarvis. Blanche Point Transitional Marl; Adelaide city area, Morphett St and Victoria Bridges foundation test bore 14, core at 17.4–17.7 m. Adelaide Plains Sub-Basin. Late Eocene. Bar scale = 0.1 mm.
(g) Nautiloid mollusc *Aturia clarkei* Teichert. Maslin Bay. Willunga Embayment. Late Eocene. Bar scale = 1 cm.
Photomigrographs (c) and (f) taken on University of Adelaide scanning electron microscope by B. J. Cooper (Department of Mines). Photographs a, b, d, e & g: R. J. F. Jenkins, University of Adelaide.

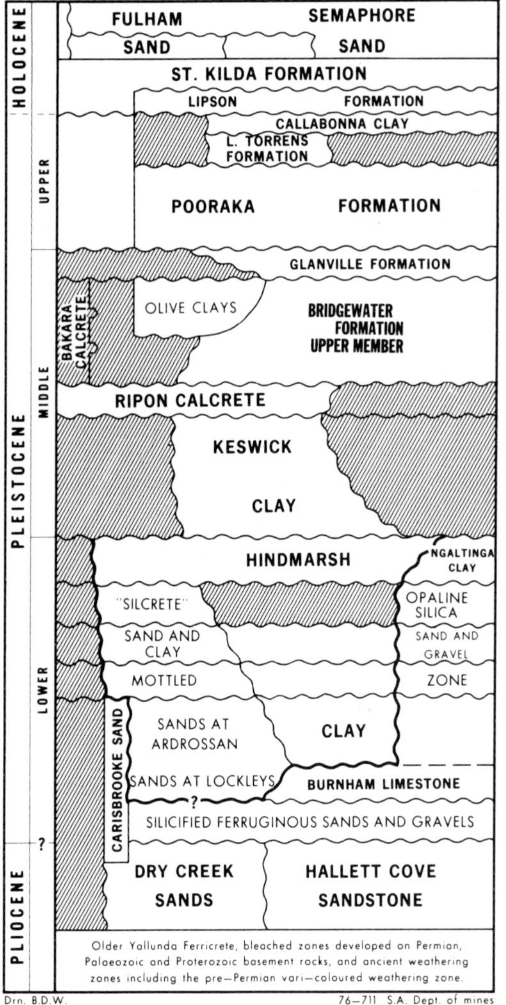

Fig. 23. Stratigraphic table late Cainozoic units.

in laterite profiles in the adjoining highlands (Firman 1976b). A further silicified and ferruginous zone showing vertical structures occurs above this mottled zone both in the sedimentary sequence and in the uplands, but still within the Hindmarsh Clay time-interval. This layer can be seen as a ferruginous cap-rock at the top of the laterite profile, e.g. at "The Gun Emplacement", Anstey Hill, and is a marker unit identifying the Karoonda Surface here and elsewhere in South Australia (Firman 1975).

In the *middle Pleistocene,* faulting, dated as post-Hindmarsh Clay and pre-Glanville Formation displaced the Hindmarsh Clay across the Para Fault. This faulting may have occurred at the same time as the faulting which displaced the younger clays at the top of the Hindmarsh Clay across the Eden Fault to the south. Middle Pleistocene deposits record strong shifts in sea level and the influence of a peri-glacial climate. The shifts in sea level have been of sufficient magnitude to change the rate and extent of fluvial processes, but high sea levels have not affected landforms east of the estuarine plain.

Aeolian sand with intercalated beds of shelly calcarenite is found in many places on the coastal margin, south of Adelaide, e.g. at the end of the Port Noarlunga jetty. This unit is the *Bridgewater Formation* (Boutakoff 1963; Firman 1969b). A lime-rock zone called the *Ripon Calcrete* separates the unit into a lower and an upper member. The Ripon Calcrete is a thick and extensive zone of cemented carbonate concretions, carbonate breccia and clastic material. In the upper member dune shapes are well preserved, as are root-like nodules, thin layers of loess and moderately hard calcrete as randomly arranged pedogenic layers.

Above the Hindmarsh Clay near Adelaide are remnants of various deposits including loess, sand and gravel cemented by the Ripon Calcrete and the younger *Bakara Calcrete* (Fig 23). The Ripon Calcrete occurs only near Dublin, North Adelaide, and at Noarlunga. Elsewhere it is reworked and cemented by the younger Bakara Calcrete (named from the Bakara Soil of Firman 1964). On the seaward margin, shallow-marine *Anadara*-bearing *Glanville Formation* is also cemented by Bakara Calcrete. Similar lime-cemented materials are known from the floor of Gulf St Vincent at least 20 m below modern sea level, suggesting that this cementation marks a major regression. The development of upper sandy layers in brown soils containing Bakara Calcrete is due to later sedimentary layering and soil differentiation.

These deposits record fluctuations in sea level, aeolian deposition, soil formation, faulting, and extensive subaerial erosion during the middle Pleistocene. C^{14} dating of shell from the Glanville Formation suggests that the unit has an age in excess of 45,000 years B.P. Recent uranium-series dating elsewhere on the coastal margin (Gill 1974) suggests that the true age of these deposits may be much greater than the C^{14} dates suggest.

At the end of the middle Pleistocene, Gulf St Vincent was ". . . a great plain extending

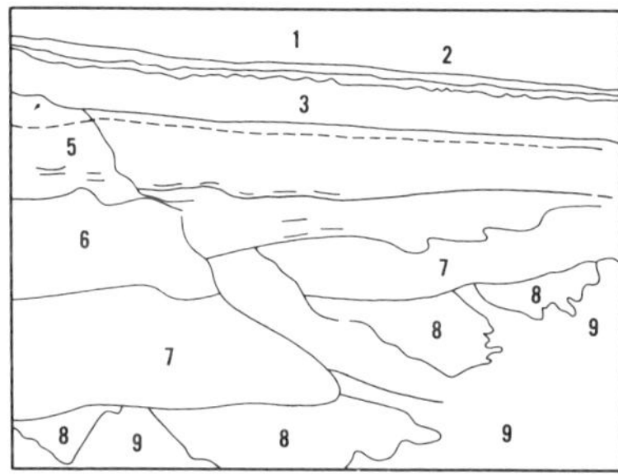

Fig. 24. The Amphitheatre on the seaward margin of the Eden Block at Hallett Cove. *Holocene (Recent)*: 1, Overlay of light brown silty fine sand. *Middle Pleistocene*: 2, Pale grey, off-white and pink Bakara Calcrete containing gravel derived from Ripon Calcrete. 3, Upper Member Bridgewater Formation; off-white gravelly sandy carbonate silt (gravel of reworked Ripon Calcrete), grading downwards into calcareous fine-grained quartz sand containing small soft carbonate rhizonodules and large carbonate lumps. This overlies a light reddish-brown fine-grained quartz sand with large irregular lumps of carbonate cemented quartz sand. *Lower Pleistocene* Hindmarsh Clay comprised of the following units: 4, ?Keswick Clay; brown and olive structured clay impregnated with lime at the top. 5, Light greyish brown clay with red mottles overlying unconformably 6, the Ardrossan Soil of vertical indurated ferruginous mottles up to 30 cm in length developed in weakly cemented pale grey clayey sand. This unit overlies red mottled sandy clay with olive clay at the base. The Holocene-Pleistocene sequence is about 20 m thick and overlies 7, the thin Late Pliocene Hallett Cove Sandstone resting disconformably upon 8, a yellow sand. 9, talus.

down towards Kangaroo Island", according to Tindale (1962)[3].

Renewed faulting at the beginning of the *upper Pleistocene* is suggested by the distribution of reworked older material within the extensive deposits of a younger alluvial sequence, the *Pooraka Formation*. This formation was deposited under the influence of a milder climate as valley-fill and as valley-flat deposits on the plains. At this time, the alluvial plain extended out into Gulf St Vincent following the earlier withdrawal of the sea at the end of the middle Pleistocene. Pooraka Formation contains sedimentary layers with distinctive carbonate nodules restricted to those layers. These nodules, which are most common on the western margin of the lower outwash plain, were probably formed in a valley-flat

---

[3] Tindale, N. (1962).—"Advent of Man in Australia." *In* Symposium on Geochronology and Land Surfaces in Relation to Soils in Australasia. Unpublished.

environment, following breaks in deposition of the sediments. Later carbonates which occur as patches and nodules in horizons cross-cutting sedimentary layers, mark an important period of soil development and identify the soil stratigraphic unit called *Loveday Soil.*

Inland of the deposits fringing the modern coast, a layer of red clay, the Callabonna Clay (Firman 1970), is found above Loveday Soil. This clay is well developed on the lower outwash plain. Fossil stream courses at the top of Pooraka Formation are marked by poorly sorted deposits of angular quartz sand.

During the Holocene, in the Mt Lofty Ranges, successively younger materials were formed on lower and increasingly steep slopes. These Recent slope deposits and alluvial sediments, and the bed load of the modern stream channels, all form part of an ancestral Pleistocene drainage pattern.

During the post-glacial rise in sea-level, i.e. the Flandrian transgression, a tidal-flat environment was established on the Gulf margin, leading first to deposition of the *Lipson Formation,* a silty clay with abundant plant fibres. This early phase of the transgression was succeeded by a high stand of the Flandrian sea, when shelly sand—the *St Kilda Formation*—was laid

down. This high sea level (about 1.5 m above modern sea level) is marked by shell beds along the gulf margins. C^{14} age determinations[4] on shell from St Kilda Formation north of the Adelaide region near Sandy Point, range from 3800 $\pm$ 500 years B.P. for shell from the base of the unit to 1120 $\pm$ 75 years B.P. for shell from standard beach ridges at the top of the unit.

The Recent marine gulf, estuarine and littoral sediments overlie the eroded seaward margin of the Pooraka Formation (Firman 1963, 1966).

Red-brown dune sand, the *Fulham Sand,* which was derived ultimately from Pooraka Formation on the eastern margin of the plain, was laid down upon St Kilda Formation following a retreat of the sea. Modern alluvial and estuarine deposits are found on the western margin of the plains. Beach ridges and dunes —the *Semaphore Sand*—fringe the modern coast.

Acknowledgments

Members of the Geological Survey of South Australia, and in particular W. K. Harris and B. P. Webb, greatly facilitated preparation of this manuscript.

References

Abele, C. & McGowran, B. (1959).—The geology of the Cambrian south of Adelaide (Sellick Hill to Yankalilla). *Trans. R. Soc. S. Aust.,* **82**, 301-320.

Alderman, A. R. (1973).—"Southern Aspect. An Introductory View of South Australian Geology." (S. Australian Museum: Adelaide).

Anon. (1921).—Boring operations for lignite at Hope Valley. *Min. Rev., Adelaide* **33**, 25-37.

Anon. (1924).—Boring operations for lignite. *Min. Rev., Adelaide* **39**, 33-45.

Barnes, T. A. & Kleeman, A. W. (1934).—The Blue Metal Limestone and its associated beds. *Trans. R. Soc. S. Aust.* **58**, 80-85.

Benson, W. N. (1909).—Petrographical notes on certain Precambrian rocks of the Mount Lofty Ranges with special reference to the geology of the Houghton district. *Trans. R. Soc. S. Aust.* **33**, 101-140.

Blissett, A. H. & Callen, R. A. (1969).— Myponga phosphate deposit. *Mineral Resour. Rev. S. Aust.* **127**, 93-100.

Bolt, B. A. (1957).—The epicentre of the Adelaide earthquake of March 1, 1954. *J. Proc. R. Soc. N.S.W.* **90**, 50-52.

Boutakoff, N. (1963).—The geology and geomorphology of the Portland area. *Mem. geol. Surv. Vict.* **22**.

Callen, R. A. (1970).—Phosphate deposits in the Myponga area. *Mineral Resour. Rev. S. Aust.* **129**, 42-53.

Callen, R. A. (1971).—Sedimentary phosphate exploration. Part 2—The Cambrian and Proterozoic of Fleurieu Peninsula, the Mount Lofty Ranges, and Yorke Peninsula. *Mineral Resour. Rev. S. Aust.,* **130**, 80-94.

Campana, B. (1958).—The Mt Lofty-Olary Region and Kangaroo Island. *In* Glaessner, M. F. & Parkin, L. W. (Eds), "The Geology of South Australia." *J. geol. Soc. Aust.* **5**(2), 3-27.

Campana, B. & Wilson, B. (1954).—*Yankalilla* map sheet, Geological Atlas of South Australia, 1:63,360 series (Geol. Surv. S. Aust.: Adelaide.)

Campana, B. & Wilson, R. B. (1955).—Tillites and related topography of South Australia. *Eclog. geol. Helv.* **48**, 1-30.

Campana, B., Wilson, B. & Whittle, A. W. G. (1955).—The geology of the Jervis and Yankalilla military sheets. Explanation of the geological maps. *Rep. Invest., geol. Surv. S. Aust.* **3**, 1-26.

Coats, R. P. (1965).—Tent Hill Formation correlations—Port Augusta and Lake Torrens. *Quart. geol. notes, geol. Surv. S. Aust.* **16**, 9-11.

---

[4] Scripps Institution of Oceanography, 1968, written communication, DM.1746/74.

COATS, R. P. (1967).—The "Lower Glacial Sequence"—Sturtian type area. *Quart. geol. Notes, geol. Surv. S. Aust.* **23**, 1-3.

COATS, R. P. & THOMSON, B. P. (1959).—*Truro* map sheet, Geological Atlas of South Australia, 1:63,360 series (Geol. Surv. S. Aust.: Adelaide.)

COCHRANE, G. W. (1953).—Brown coal prospects in the Paradise-Hope Valley-Golden Grove area. *Min. Rev. Adelaide* **94**, 78-81.

COMPSTON, W. & TAYLOR, S. R. (1969).—Rb-Sr study of impact glass and country rocks from the Henbury meteorite crater field. *Geochim. Cosmochim. Acta* **33**, 1,037-1,043.

COOPER, H. M., KENNY, M. & SCRYMGOUR, J. M. (1970).—*Hallett Cove*—a field guide. (South Australian Museum: Adelaide.)

COOPER, J. A. (1975).—Isotopic datings of the basement-cover boundaries within the Adelaide "Geosyncline". *Proterozoic Geology geol. Soc. Aust. First Australian Geological Convention Abstracts*, 12.

COOPER, J. A. & COMPTSON, W. (1971).—Rb-Sr dating within the Houghton Inlier, South Australia. *J. geol. Soc. Aust.* **17**, 213-219.

COOPER, J. A., WELLS, A. T. & NICHOLAS, T. (1971).—Dating of glauconite from the Ngalia Basin, Northern Territory, Australia. *J. geol. Soc. Aust.* **18**, 97-106.

CORNELIUS, H. S. (1927).—The Onkaparinga Coal Mining Syndicate. *Min. Rev. Adelaide* **45**, 102-103.

CRESPIN, I. (1954).—Stratigraphy and micro-palaeontology of the marine Tertiary rocks between Adelaide and Aldinga, South Australia. *Rep. Bur. Min. Resour. geol. geophys. Aust.* **12**.

CROWELL, J. C. & FRAKES, L. A. (1971a).—Late Palaeozoic glaciation of Australia. *J. geol. Soc. Aust.* **17**, 115-155.

CROWELL, J. C. & FRAKES, L. A. (1971b).—Late Palaeozoic glaciation: Part IV, Australia. *Bull. Geol. Soc. America* **82**, 2,515-2,540.

DAILY, B. (1963).—The fossiliferous Cambrian succession on Fleurieu Peninsula, South Australia. *Rec. S. Aust. Mus.* **14**, 579-601.

DAILY, B. (1969).—Fossiliferous Cambrian sediments and low-grade metamorphics, Fleurieu Peninsula, South Australia. *In* Daily, B. (Ed.), "Geology Excursions Handbook," 41st ANZAAS Congress, Section 3, Adelaide, 1969, 49-54.

DAILY, B. (1976).—New data on the base of the Cambrian in South Australia. *Izv. Akad. Nauk. Ser. geol.* **3**, 45-52 [in Russian].

DAILY, B. & MILNES, A. R. (1971).—Stratigraphic notes on Lower Cambrian fossiliferous metasediments between Campbell Creek and Tunkalilla Beach in the type section of the Kanmantoo Group, Fleurieu Peninsula, South Australia. *Trans. R. Soc. S. Aust.* **95**, 199-214.

DAILY, B. & MILNES, A. R. (1972).—Revision of the stratigraphic nomenclature of the Cambrian Kanmantoo Group, South Australia. *J. geol. Soc. Aust.* **19**, 197-202.

DAILY, B. & MILNES, A. R. (1973).—Stratigraphy, structure and metamorphism of the Kanmantoo Group (Cambrian) in its type section east of Tunkalilla Beach, South Australia. *Trans. R. Soc. S. Aust.* **97**, 213-251.

DAILY, B., TWIDALE, C. R. & MILNES, A. R. (1974).—The age of the lateritised summit surface on Kangaroo Island and adjacent areas of South Australia. *J. geol. Soc. Aust.* **21**, 387-392.

DEIGHTON, I., FALVEY, D. A. & TAYLOR, D. J. (1976).—Depositional Environments and Geotectonic Framework: Southern Australian Continental Margin. *J. Aust. Petrol. Explor. Ass.* **15**, 25-36.

DICKINSON, S. B., SPRIGG, R. C., KING, D., WADE, M. L., WEBB, B. P., WHITTLE, A. W. G., STILWELL, F. L. & EDWARDS, A. B. (1954).— Uranium deposits in South Australia. *Bull. geol. Surv. S. Aust.* **30**.

FENNER, C. (1930).—The major structural and physiographic features of South Australia. *Trans. R. Soc. S. Aust.* **54**, 1-36.

FENNER, C. (1931).—"South Australia—A Geographic Study." (Whitcombe & Tombs: Melbourne.)

FIRMAN, J. B. (1963).—Quaternary Geological Events near Port Adelaide. *Quart. geol. Notes, geol. Surv. S. Aust.* **7**.

FIRMAN, J. B. & CHUGG, R. I. (1963).—Marine Erosion and Stranded Beach Deposits near Port Wakefield. *Quart. geol. Notes, geol. Surv. S. Aust.* **8**, 4-5.

FIRMAN, J. B. (1964).—The Bakara Soil and other stratigraphic units of late Cainozoic age in the Murray Basin, South Australia. *Quart. geol. Notes, geol. Surv. S. Aust.* **10**, 2-5.

FIRMAN, J. B. (1965).—Late Cainozoic Sedimentation in Northern Spencer Gulf, South Australia. *Trans. R. Soc. S. Aust.* **89**, 125-131.

FIRMAN, J. B. (1966).—Stratigraphic Units of Late Cainozoic Age in the Adelaide Plains Basin, South Australia. *Quart. geol. Notes, geol. Surv. S. Aust.* **17**, 6-9.

FIRMAN, J. B. (1967a).—Late Cainozoic stratigraphic units in South Australia. *Quart. geol. Notes, Geol. Surv. S. Aust.* **22**, 4-8.

FIRMAN, J. B. (1967b).—Stratigraphy of late Cainozoic deposits in South Australia. *Trans. R. Soc. S. Aust.* **91**, 165-178.

FIRMAN, J. B. (1969a).—Stratigraphy and landscape relations of soil materials near Adelaide, South Australia. *Trans. R. Soc. S. Aust.* **93**, 39-54.

FIRMAN, J. B. (1969b).—The Quaternary Period, *In* L. W. Parkin (Ed.), "Handbook of South Australian Geology." (Geol. Surv. S. Aust.: Adelaide.)

FIRMAN, J. B. (1971).—Paleosols—a stratigraphic definition. *In* Études sur la Quarternaire dans le Monde. Paris: Union Internat. pour l'etude due Quart. 8th INQUA Cong. 1969.

FIRMAN, J. B. (1973).—Regional stratigraphy of surficial deposits in the Murray Basin and Gambier Embayment. *Rep. Invest., geol. Surv. S. Aust.* **39**.

FIRMAN, J. B. (1975).—South Australia. *In* Rhodes, W. Fairbridge (Ed.), "The Encyclopedia of World Regional Geology. Part I. Western Hemisphere". (Dowden, Hutchinson & Ross: Stroudsburg.)

FIRMAN, J. B. (1976a).—Limestone at the base of the Pleistocene sequence in South Australia. *Quart. geol. Notes, geol. Surv. S. Aust.* **58**, 2-5.

FIRMAN, J. B. (1976b).—Laterite: Paleosols in the laterite profile. In prep.

FLEMING, P. D. & OFFLER, R. (1968).—Pretectonic crystallisation in the Mt Lofty Ranges, South Australia. *Geol. Mag.* **105**, 356-359.

FORBES, B. G. (1961).—Magnesite of the Adelaide System: a discussion of its origin. *Trans. R. Soc. S. Aust.* **85**, 217-222.

FORBES, B. G., COATS, R. P. & DAILY, B. (1972).—Truro Volcanics. *Quart. geol. Notes, geol. Surv. S. Aust.* **44**, 1.

GILL, E. D. (1974).—Carbon-14 and Uranium-Thorium Check on Suggested Interstadial High Sea level around 30,000 B.P. *Search* **5**(5), 211.

GLAESSNER, M. F. & PARKIN, L. W. (Eds) (1958).—"The geology of South Australia". *J. geol. Soc. Aust.*, 5(2).

GLAESSNER, M. F. (1951).—Three foraminiferal zones in the Tertiary of Australia. *Geol. Mag.* **88**, 273-283.

GLAESSNER, M. F. & WADE, M. (1958).—The St Vincent Basin *In* M. F. Glaessner & L. W. Parkin (Eds), "The Geology of South Australia". *J. geol. Soc. Aust.* **5**, 115-126.

GRIFFITHS, J. R. (1971).—Continental margin tectonics and the evolution of Southeast Australia. *J. Aust. Petrol. Explor. Ass.* **11**, 75-79.

HARRIS, W. K. (1966).—New and redefined names in South Australian lower Tertiary stratigraphy. *Quart. Notes, geol. Surv. S. Aust.* **20**, 1-3.

HARRIS, W. K. & OLLIVER, J. G. (1965).—The age of the Tertiary sands at Rowlands Flat, Barossa Valley. *Quart. geol. Notes, geol. Surv. S. Aust.* **13**, 1-2.

HEATH, G. R. (1963).—Stonyfell Quartzite: descriptive stratigraphy and petrography of the type section. *Trans. R. Soc. S. Aust.* **87**, 159-

HORWITZ, R. C. (1960).—Géologie de la région de Mt Compass (feuille Milang), Australie Méridionale. *Eclog. géol. Helv.* **53**, 211-263.

HOWCHIN, W. (1897).—On the occurrence of Lower Cambrian fossils in the Mount Lofty Ranges. *Trans. R. Soc. S. Aust.* **21**, 74-86.

HOWCHIN, W. (1904).—The geology of the Mount Lofty Ranges. Part I. The coastal district. *Trans. R. Soc. S. Aust.* **28**, 253-280.

HOWCHIN, W. (1906).—The geology of the Mount Lofty Ranges—Part II. (The lower and basal beds of the Cambrian). *Trans. R. Soc. S. Aust.* **30**, 227-262.

HOWCHIN, W. (1910a).—The glacial (Permo-Carboniferous) moraines of Rosetta Head and Kings Point. *Trans. R. Soc. S. Aust.* **10**, 1-12.

HOWCHIN, W. (1910b).—Description of a new and extensive area of Permo-Carboniferous glacial deposits in South Australia. *Trans. R. Soc. S. Aust.* **10**, 231-247.

HOWCHIN, W. (1915).—A geological sketch-map, with descriptive notes on the Upper and Lower Torrens-limestones in the type district. *Trans. R. Soc. S. Aust.* **39**, 1-15.

HOWCHIN, W. (1926).—The geology of Victor Harbor, Inman Valley, and Yankalilla districts, with special reference to the great Inman Valley glacier of Permo-Carboniferous age. *Trans. R. Soc. S. Aust.* **50**, 89-119.

HOWCHIN, W. (1929).—"The geology of South Australia". (Adelaide.)

HOWCHIN, W. (1933).—The dead rivers of South Australia. Part II. The eastern group. *Trans. R. Soc. S. Aust.* **57**, 1-41.

JOHNS, R. K. (1967).—Cambrian phosphorite, St Johns. *Quart. geol. Notes, geol. Surv. S. Aust.* **21**, 3.

KENNETT, J. P., BURNS, R. E., ANDREWS, J. E., CHURKIN, M. Jun., DAVIES, T. A., DUMITRICA, P., EDWARDS, A. R., GALEHOUSE, J. S., PACKHAM, G. H. & VAN DER LINGEN, J. J. (1972). —Australian-Antarctic continental drift, palaeocirculation changes and Oligocene deep-sea erosion. *Nature Phys. Sci.* **239**, 51-55.

KENNETT, J. P., HOUTZ, R. E., ANDREWS, P. B., EDWARDS, A. R., GOSTIN, V. A., HAJOS, M., HAMPTON, M., JENKINS, D. G., MARGOLIS, S. V., OVENSHINE, A. T. & PERCH-NIELSEN, K. (1975).—Cenozoic palaeoceanography in the southwest Pacific Ocean, Antarctic glaciation, and the development of the Circum-Antarctic Current. *In* Kennett, J. P., Houtz, R. E. *et al.,* "Initial Reports of the Deep Sea Drilling Project", **29**, 1,155-1,169. (U.S. Govt Printing Office: Washington).

KERR-GRANT, C. (1956).—The Adelaide earthquake of 1st March, 1954. *Trans. R. Soc. S. Aust.* **79**, 177-185.

LANGE, R. T. (1970).—The Maslin Bay flora, South Australia. 2. The assemblage of fossils. *N. Jb. geol. Palaont. Mh.* **8**, 486-490.

LINDSAY, J. M. (1967).—Foraminifera and stratigraphy of the type section of Port Willunga Beds, Aldinga Bay, South Australia. *Trans. R. Soc. S. Aust.* **91**, 93-110.

LINDSAY, J. M. (1968).—Notes on foraminifera and stratigraphy of the Grange and Croydon bores. *Quart. geol. Notes, geol. Surv. S. Aust.* **26**, 1-3.

LINDSAY, J. M. (1969).—Cainozoic foraminifera and stratigraphy of the Adelaide Plains Sub-Basin, South Australia. *Bull. geol. Surv. S. Aust.* **42**.

LINDSAY, J. M. (1970).—Port Willunga Beds in the Port Noarlunga-Seaford area. *Quart. geol. Notes geol. Surv. S. Aust.* **36**, 4-10.

LINDSAY, J. M. & SHEPHERD, R. G. (1966).— Munno Para Clay Member. *Quart. geol. Notes, geol. Surv. S. Aust.* **19**, 7-11.

LUDBROOK, N. H. (1963).—Correlation of the Tertiary rocks of South Australia. *Trans. R. Soc. S. Aust.* **87**, 5-15.

LUDBROOK, N. H. (1967).—Permian deposits of South Australia and their fauna. *Trans. R. Soc. S. Aust.* **91**, 65-92.

LUDBROOK, N. H. (1969).—Tertiary Period. *In* L. W. Parkin (Ed.), "Handbook of South Australian Geology", 172-203. (Geol. Surv. S. Aust.: Adelaide.)

MADIGAN, C. T. (1926).—Organic remains from below the Archaeocyathinae Limestone at Myponga Jetty, South Australia. *Trans. R. Soc. S. Aust.* **50**, 31-35.

MAWSON, D. (1925).—Evidence and indications of algal contributions in the Cambrian and Pre-Cambrian limestones of South Australia. *Trans. R. Soc. S. Aust.* **50**, 186-190.

MAWSON, D. (1926).—Varve shales associated with the Permo-Carboniferous glacial strata of South Australia. *Trans. R. Soc. S. Aust.* **50**, 160-162.

MAWSON, D. (1939).—The first stage of the Adelaide Series: as illustrated at Mount Magnificent. *Trans. R. Soc. S. Aust.* **63**, 69-78.

MAWSON, D. & SPRIGG, R. C. (1950).—Subdivision of the Adelaide System. *Aust. J. Sci.* **13**, 69-72.

McGOWRAN, B. (1973).—Rifting and drift of Australia and the migration of mammals. *Science* **180**, 759-761.

McGOWRAN, B., HARRIS, W. K. & LINDSAY, J. M. (1970).—The Maslin Bay flora, South Australia. 1. Evidence for an early Middle Eocene age *N. Jb. geol. Palaont. Mh.* **8**, 481-485.

McKENZIE, D. & SCLATER, J. G. (1971).—The evolution of the Indian Ocean since the Late Cretaceous. *J. Roy. Astr. Soc.* **25**, 437-528.

MILES, K. R. (1950).—The geology of the South Para Dam project. *Bull. geol. Surv. S. Aust.* **24**.

MILES, K. R. (1952).—Geology and underground water resources of the Adelaide Plains area. *Bull. geol. Surv. S. Aust.* **27**.

MILLS, K. J. (1963).—The geology of the Mt Crawford Granite Gneiss and adjacent metasediments. *Trans. R. Soc. S. Aust.* **87**, 167-183.

MILLS, K. J. (1973).—The structural geology of the Warren National Park and the western portion of the Mount Crawford State Forest, South Australia. *Trans. R. Soc. S. Aust.* **97**, 281-315.

MILNES, A. R. & BOURMAN, R. P. (1972).—A Late Palaeozoic glaciated granite surface at Port Elliot, South Australia. *Trans. R. Soc. S. Aust.* **96**, 149-155.

MILNES, A. R., COMPSTON, W. & DAILY, B. (in press).—Pre-to-syn-tectonic emplacement of Lower Palaeozoic granites in southeastern South Australia. *J. geol. Soc. Aust.*

NESBITT, R. W. (1969).—Notes on the geology of the Hallett Cove area and Sandison Reserve. *In* Daily, B. (Ed.) "Geology Excursion Handbook," 41st ANZAAS Congress, Section 3, Adelaide, 1969, 67-69.

NIXON, L. G. B. (1963).—Blue dolomite deposit —Pinkerton Gully. *Min. Rev. Adelaide* **115**, 142-159.

OFFLER, R. & FLEMING, P. D. (1968).—A synthesis of folding and metamorphism in the Mount Lofty Ranges, South Australia. *J. geol. Soc. Aust.* **15**, 245-266.

OLIVER, R. L. & DAILY, B. (1969).—Proterozoic and Permian glacigene deposits near Adelaide, South Australia. In Daily, B. (Ed.) "Geology Excursions Handbook." 41st ANZAAS Congress, Section 3, Adelaide, 1969, 63-66.

OLLIVER, J. G. (1967).—Test-boring—Rowland Flat Sand Deposit. *Min. Rev. Adelaide* **121**, 144-150.

OLLIVER, J. G. & WEIR, L. J. (1967).—The construction sand industry in the Adelaide Metropolitan Area. *Rep. Invest., geol. Surv. S. Aust.* **30**.

PERRY, W. J. & ROBERTS, H. G. (1968).—Late Precambrian glaciated pavements in the Kimberley region, Western Australia. *J. geol. Soc. Aust.* **15**, 51-56.

PRESCOTT, J. A. & PENDLETON, R. L. (1952).—Laterite and lateritic soils. *Commonwealth Bur. Soil Sci., Tech, Commun.,* 47.

PREISS, W. V. (1973).—Palaeocological interpretations of South Australian Precambrian stromatolites. *J. geol. Soc. Aust.* **19**, 501-532.

REYNOLDS, M. A. (1953).—The Cainozoic succession of Maslin and Aldinga Bays, South Australia. *Trans. R. Soc. S. Aust.* **76**, 114-140.

SAVIN, S. M., DOUGLAS, R. G. & STEHLI, F. G. (1975).—Tertiary marine paleotemperatures. *Bull. geol. Soc. Am.* **86**, 1,499-1,510.

SCHOPF, J. W. & BARGHOORN, E. S. (1969).—Micro-organisms from the Late Precambrian of South Australia. *J. Paleont.* **43**, 111-118.

SELBY, J. & LINDSAY, J. M. (1973).—Blanche Point Banded Marl: an alternative bearing horizon beneath the City of Adelaide. *Quart. geol. Notes, geol. Surv. S. Aust.* **48**, 7-11.

SELWYN, A. R. C. (1859).—Geological notes of a journey in South Australia from Cape Jervis to Mount Serle. *Parl. Pap. S. Aust.* **20**, 4.

SHACKLETON, N. J. & KENNETT, J. P. (1975).—Paleotemperature history of the Cenozoic and the initiation of Antarctic glaciation: oxygen and carbon isotope analyses in DSDP Sites 277, 279 and 281. *In* Kennett, J. P., Houtz, R. E., *et al.,* "Initial Reports of the Deep Sea Drilling Project", **29**, 743-755 (U.S. Govt Printing Office: Washington).

SPRIGG, R. C. (1942).—The geology of the Eden-Moana Fault Block. *Trans. R. Soc. S. Aust.* **66**, 185-214.

SPRIGG, R. C. (1946).—Reconnaissance geological survey of portion of the western escarpment of the Mount Lofty Ranges. *Trans. R. Soc. S. Aust.* **70**, 313-347.

SPRIGG, R. C. & CAMPANA, B. (1953).—The age and facies of the Kanmantoo Group. *Aust. J. Sci.* **16**, 12-14.

SPRIGG, R. C. & STACKLER, W. F. (1965).—Submarine gravity surveys in St Vincent Gulf and Investigator Strait, South Australia, in relation to oil search. *J. Aust. Petrol. Explor. Ass.* **5**, 168-178.

SPRY, A. H. (1951).—The Archaean Complex at Houghton, South Australia. *Trans. R. Soc. S. Aust.* **74**, 115-134.

SUMARTOJO, J. (1975).—The depositional environments of the Tindelpina Shale Member of the Tapley Hill Formation (Late Proterozoic), Adelaide Geosyncline—a geochemical study. *Proterozoic Geology geol. Soc. Aust. First Australian Geological Convention. Abstracts,* 25.

SUTTON, D. J. & WHITE, R. E. (1968).—The seismicity of South Australia. *J. geol. Soc. Aust.* **15**(1), 25-32.

TALBOT, J. L. (1963).—Retrograde metamorphism of the Houghton Complex, South Australia. *Trans. R. Soc. S. Aust.* **87**, 185-196.

TALBOT, J. L. & NESBITT, R. W. (1968).—*Geological Excursions in the Mount Lofty Ranges and the Fleurieu Peninsula.* (Angus & Robertson: Melbourne.)

TATE, R. (1879).—The Anniversary Address of the President. *Trans. R. Soc. S. Aust.* **2**, XXXIX-LXXV.

TATE, R. (1890).—On the discovery of marine deposits of Pliocene age in Australia. *Trans. R. Soc. S. Aust.* **13**, 172-180.

TAYLOR, J. K., THOMSON, B. P. & SHEPHERD, R. G. (1974).—The soils and geology of the Adelaide area. *Bull. geol. Surv. S. Aust.* **46**.

THOMSON, B. P. (1965).—Geology and mineralisation of South Australia. *In* J. McAndrew (Ed.), "Geology of Australian Ore Deposits". 2nd Edtn, 270-284. (Eighth Commonwealth Mining and Metallurgical Congress: Melbourne.)

THOMSON, B. P. (1966).—Stratigraphic relationships between sediments of Marinoan age—Adelaide region. *Quart. geol. Notes, geol. Surv. S. Aust.* **20**, 7-9.

THOMSON, B. P. (1969a).—ADELAIDE Map Sheet, Geological Atlas of South Australia, 1:250,000 series. (Geol. Surv. S. Aust.: Adelaide.)

THOMSON, B. P. (1969b).—Precambrian crystalline basement. *In* L. W. Parkin (Ed.), "Handbook of South Australian Geology", 21-48. (Geol. Surv. S. Aust.: Adelaide.)

THOMSON, B. P. (1969c).—Precambrian basement cover, the Adelaide System. *In* L. W. Parkin (Ed.), "Handbook of South Australian Geology", 49-83. (Geol. Surv. S. Aust.: Adelaide.)

THOMSON, B. P. (1976).—Kanmantoo Trough—regional geology and comments on mineralisation. In C. L. Knight (Ed.). "Economic geology of Australia and Papua New Guinea." *Aust. Inst. Min. Met.* **1**, 555-560.

THOMSON, B. P., COATS, R. P., MIRAMS, R. C., FORBES, B. G., DALGARNO, C. R. & JOHNSON, J. E. (1964).—Precambrian rock groups in the Adelaide Geosyncline: a new sub-division. *Quart. geol. Notes, geol. Surv. S. Aust.* **9**.

THOMSON, B. P., DAILY, B., COATS, R. P. & FORBES, B. G. (in press).—Late Precambrian and Cambrian geology of the Adelaide "Geosyncline" and Stuart Shelf, South Australia. *25th International Geological Congress, Australia, Excursion 33A guidebook.*

THOMSON, B. P. & HORWITZ, R. C. (1961).—Cambrian-Precambrian unconformity in Sellick Hill-Normanville area of South Australia. *Aust. J. Sci.* **24**, 40.

TRUDINGER, J. P. (1973).—Engineering geology of the Kangaroo Creek Dam. *Bull. geol. Surv. S. Aust.* **44**.

TWIDALE, C. R., DAILY, B. & FIRMAN, J. B. (1967).—Eustatic and climatic history of the Adelaide area, South Australia: A Discussion. *J. Geol.* **75**(2), 237-242.

WARD, W. T. (1965).—Eustatic and climatic history of the Adelaide area, South Australia. *J. Geol.* **73**, 592-602.

WHITE, A. J. R., COMPSTON, W. & KLEEMAN, A. W. (1967).—The Palmer Granite—A study of a granite within a regional metamorphic environment. *J. Petrology* **8**, 29-50.

WILSON, A. F. (1952).—The Adelaide System as developed in the Riverton-Clare region, northern Mount Lofty Ranges, South Australia. *Trans. R. Soc. S. Aust.* **75**, 131-149.

WOODS, J. E. (1862).—"Geological observations in South Australia." (Longman: London.)

CHAPTER 2

GEOMORPHOLOGICAL EVOLUTION

by C. R. TWIDALE*

Introduction

In the early days of the Colony of South Australia the Mt Lofty Ranges were known as the Tiers (Hardy 1939), presumably on account of the step-like appearance so typical of the uplands immediately to the east and southeast of the present city of Adelaide (Fig. 1). Thus the presence of the two major topographic features of the Mt Lofty Ranges, namely the steep scarps, or risers, and the several levels of benches and high plains, or treads, were tacitly acknowledged by the early settlers. And though the scientific community took rather longer to appreciate these characteristics, our predecessors in the Royal Society subjected the scarps and surfaces of the Mt Lofty Ranges to careful scrutiny and extended debate. Fully sixty-five years ago Benson (1911) concluded that the major relief of the Ranges is consequent upon a series of fault blocks, and that the prominent linear or gently arcuate scarps are of fault origin. And half a century ago Hossfeld (1926)[1] launched what is a continuing debate on the age and origin of the summit high plain that dominates the Mt Lofty Ranges.

Despite their broad simplicity the present uplands and adjacent plains are in detail complex and the landform assemblages varied. That the contemporary physiographic landscape is the result of the interplay between diastrophism, principally faulting, on the one hand, and weathering, erosion and deposition on the other over a vast period of geological time (Campana 1958; Twidale 1976) is generally agreed. But there harmony ends and discord begins, for controversy surrounds the age and origin of several important aspects of the geo-morphological evolution of the Adelaide district. Reference is made to the various arguments as they arise, and though it would be more than remiss not to indicate the present writer's leanings, it is emphasised that, as Read (1957, p. xi) has pointed out, "the interpretation of field-work is often a strictly personal affair, depending on the observer's character, training and experience". The account that follows is, it is hoped, justified by a reasonable evaluation of the field evidence, but it is in many respects only an interpretation, subject to review and modification. The perceptive reader will note in other chapters different views concerning several aspects of the geological and geomorphological evolution of the region under consideration, but controversy is an integral part of all science, and so of geomorphology.

The present geomorphological pattern can only be understood in terms of past events, so that the approach adopted here is chronological, beginning with the earliest landforms that have survived and working forward to the present day.

Older landform elements: pre-Tertiary history

Exhumed late Palaeozoic forms

The oldest landforms of the region under discussion comprise glacial pavements and other landforms of possible glacial origin which are of late Palaeozoic age and which have been exhumed from beneath or are associated with Permian glacigene sediments. The glacial pavements are most numerous in the Inman Valley and adjacent areas (Selwyn 1859; Howchin 1898; Bourman 1969[2]; Milnes & Bourman 1972) and in the Strathalbyn-Finniss area of

---

* Department of Geography, University of Adelaide, Adelaide, S. Aust. 5000.

[1] Hossfeld, P. S. (1926).—"The geology of portions of the counties of Light, Eyre, Sturt and Adelaide." M.Sc thesis, University of Adelaide (unpubl.).

[2] Bourman, R. P. (1969).—"Landform studies near Victor Harbor." B.A. Hons. thesis, University of Adelaide (unpubl.).

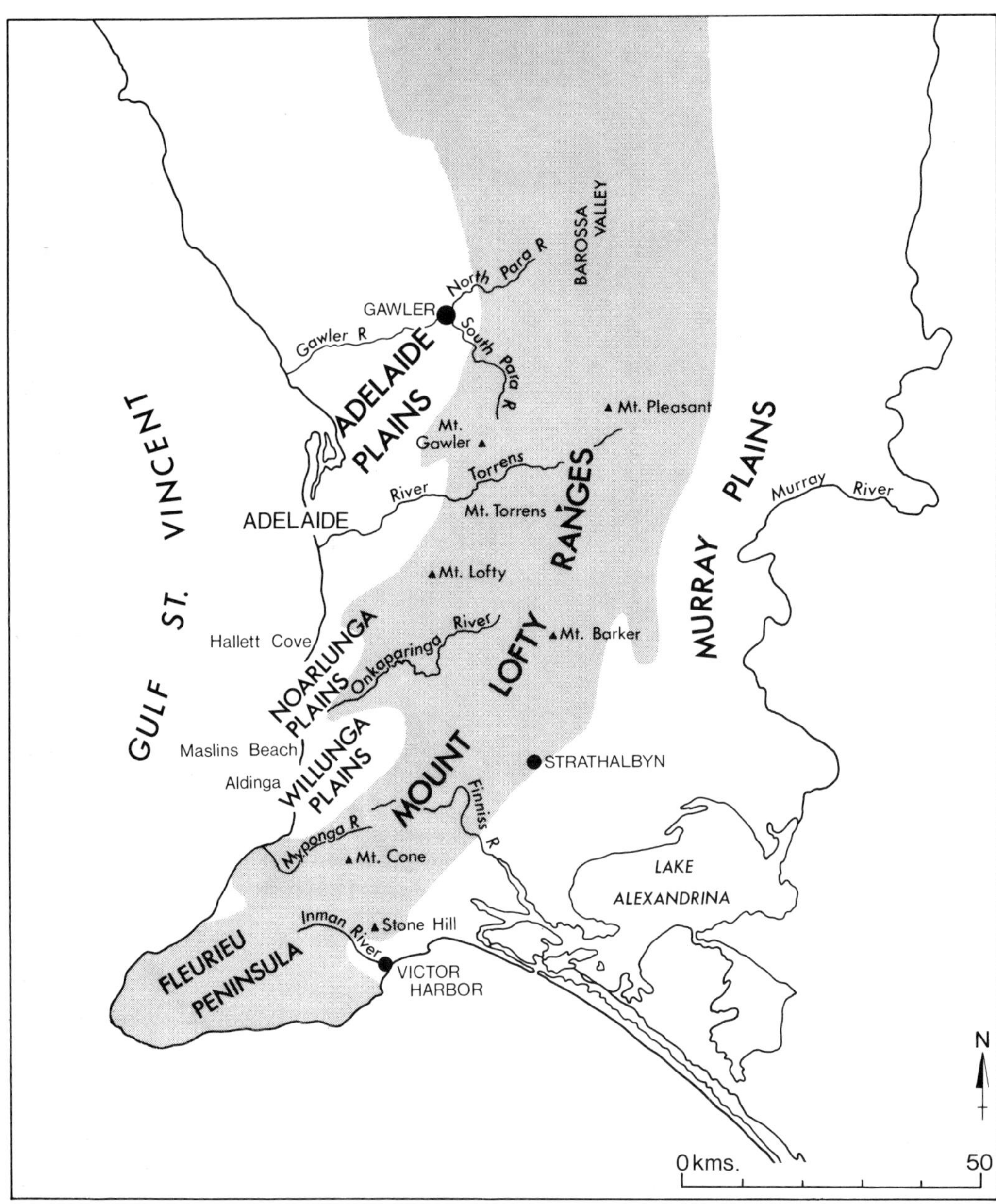

Fig. 1. The Adelaide region.

the southwestern Ranges whence a varied assemblage of Permian glacial forms has been described (Maud 1972). The glacial pavements and deposits of Hallett Cove (Fig. 2a), however, are historically significant, for it was there that the glaciation to which they are related was first shown to predate the Pleistocene (Tate 1886, 1889; Howchin 1895).

A glacial origin has been mooted for several other features in the Victor Harbor area, Crozier Hill, for example, being cited as a large *roche moutonnée* or crag and tail feature, but the residual could equally well be an expression of local structure and here, as elsewhere, the evidence and argument are equivocal. On the other hand, Stone Hill, on the

Fig. 2. (a) Permian glaciated pavement at Hallett Cove. The pavement is eroded in Precambrian shales. (C. R. Twidale.)

(b) Granite boulders near Palmer, eastern Mt Lofty Ranges. (C. R. Twidale.)

(c) Penitent rocks or "tombstones" in granite gneiss near Tungkillo, eastern Mt Lofty Ranges. (C. R. Twidale.)

divide between the Inman and Hindmarsh rivers, has been shown to be an exhumed *roche moutonnée* (Bourman & Milnes 1976). And though great care has to be taken in correctly identifying some glacial forms, few uncertainties attach to the nature and origin of the many splendid glacial pavements of the southern Ranges (but see Daily, Gostin & Nelson 1973).

The Inman Valley glacial forms are exposed in the floor of a valley that stands 250-300 m below the level of the adjacent summit surface that is the result of long-continued weathering and erosion during the Mesozoic (see below) so that the glacial valleys must once have been much deeper than they now appear. However, this does not necessarily imply that the late Palaeozoic glaciation was of Alpine type, with rivers of ice flowing in deep valleys and separated by angular mountainous divides. The present Antarctic ice sheet all but buries a land surface which in places has enormous relief amplitude but which elsewhere, and like much of subglacial Greenland, is relatively subdued. The Permian glaciation in the Adelaide region could have been of Alpine type but it could equally have involved an ice sheet which scoured out the deeply weathered valley floors. What is certain is that whatever the nature of the ice mass it moved from southeast to northwest in terms of the present orientation of the Australian continent.

The summit high plain

The summit surface is a plain of low relief located at high elevations and eroded across rocks of various types, structures and ages (Fig. 3). It stands at 300–350 m in the south and 350–400 m in the north, though residual remnants like Mt Lofty (727 m), Mt Gawler (543 m), Mt Torrens (584 m), Mt Barker (517 m) and Mt Cone (415 m) rise above the general level. Commonly, and especially on Fleurieu Peninsula, the surface carries a capping of laterite which is underlain by weathered Precambrian and Cambrian bedrock. The regolith is up to 10 m thick. Though deeply dissected, the surface remains intact over wide areas.

The nature and age of this deeply weathered land surface has long engaged the interest of local scientists, various interpretations having been presented by Hossfeld (1926[1], 1952[3]), Fenner (1930, 1931), Sprigg (1945, 1961), Campana (1958), Campana & Wilson (1954, 1955), Glaessner (1953), Glaessner & Wade (1958), Horwitz & Daily (1958), Horwitz (1960), Firman (1967), Brock (1971) and Daily, Twidale & Milnes (1974).

In general terms it is fair to state that many geologists regard the summit carapace and sur-

---

[3] Hossfeld, P. S. (1952).—"Geological publications on South Australia and the Northern Territory." Ph.D. thesis, University of Adelaide (unpubl.).

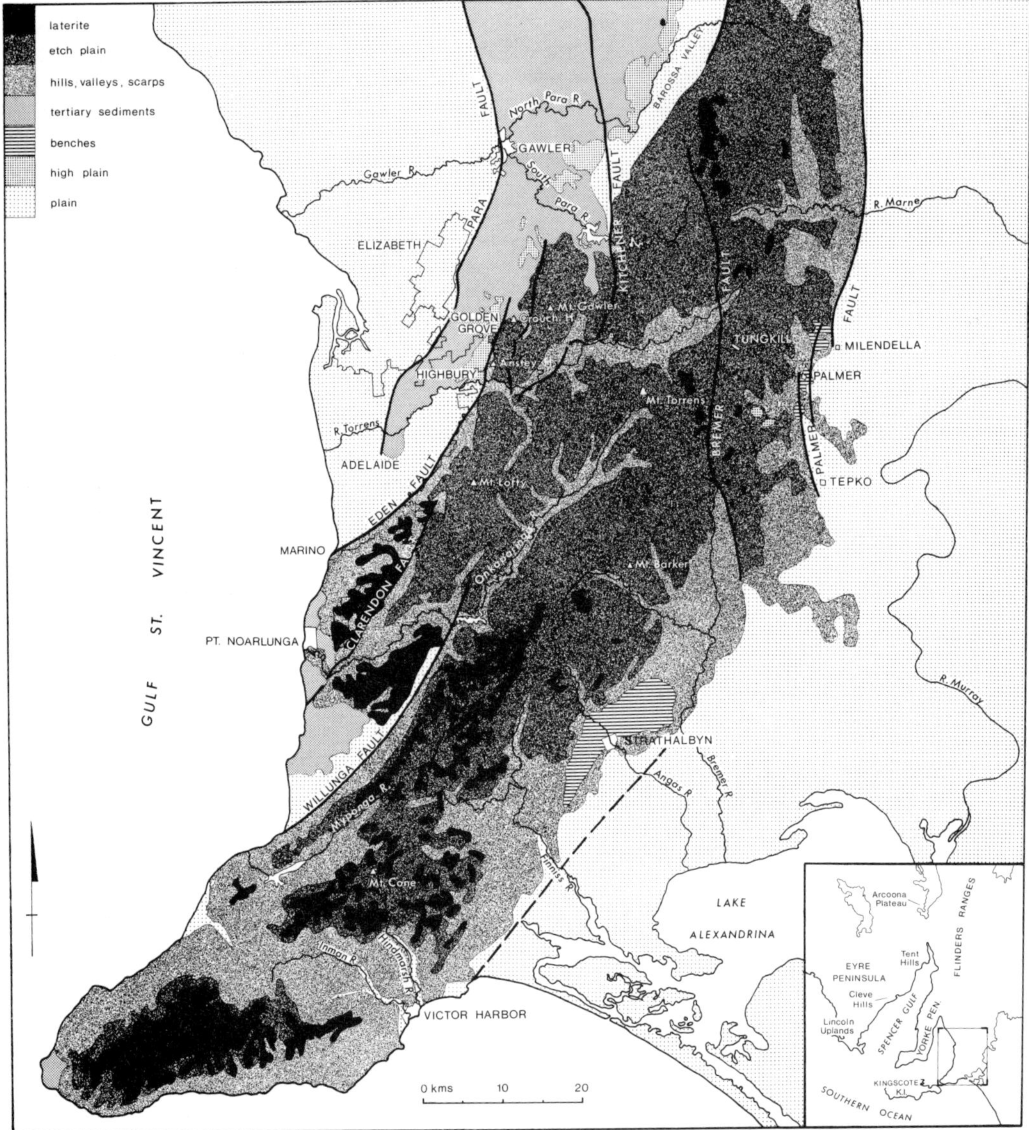

Fig. 3. The summit surface on the Mt Lofty Ranges showing the lateritised remnants and the etch surface, plus major faults.

face as of later Tertiary age (see Chapter 1) largely because the sediments deposited in the basins marginal to the upland (Willunga, Noarlunga, Adelaide, Murray) during the later Tertiary indicate only areas of low relief nearby. Furthermore, they suggest that the basin deposits were formerly more extensive, though how much more extensive is not clear.

Against this interpretation may be marshalled the following arguments:

(1) There is no evidence directly to correlate the laterite of the summit surface with the later Tertiary basin deposits.

(2) Arguments concerning the character of the land from which the sediments were derived are largely meaningless until the old shorelines and the old sea levels have been established, and until palaeocurrent analyses have been completed to show the provenance of the detritus. But if the sum-

mit laterite were developing in the later Tertiary the basin sediments should surely indicate this? And the basin sediments cannot have extended far beyond their limits, because laterite is preserved close to the edges of several of the fault blocks.

(3) Though faulting probably occurred recurrently and continuously along the old-established fracture zones there were periods of particularly pronounced activity. One such phase of marked faulting took place toward the end of the Mesozoic or early Cainozoic, for it was then that the gulfs, the Highbury fault angle depression and the Barossa Valley, came into being as tectonic lows. Sedimentation began in these depressions early in the Tertiary and the summit surface stands well above, and therefore probably predates, this depositional phase. Indeed, in the Angaston and Truro areas the dislocation of the palaeo-surface by the hinge faulting that gave rise to the present Barossa Valley is very apparent.

(4) The occurrence of marine Eocene (Bourman & Lindsay 1973) and Miocene (Campana & Wilson 1954: Horwitz 1960) strata in deep valleys and basins eroded well below the level of the lateritised summit surface in the southern Mt Lofty Ranges is suggestive of the pre-Tertiary age of the latter.

(5) A laterite summit surface occurs in southern Eyre Peninsula and an equivalent, possibly slightly younger high plain is preserved in the Cleve Hills, Tent Hill region and in the southern Arcoona Plateau (Twidale, Shepherd & Thomson 1970; Twidale, Bourne & Smith 1976). In all areas this summit surface (or surfaces) stands high above silcrete capped remnants of presumed early-middle Tertiary age (Wopfner 1960; Wopfner & Twidale 1967: Wopfner, Callen & Harris, 1973) and is therefore older — and probably much older.

(6) There is no evidence of Mesozoic marine sedimentation in the Adelaide region. The sediments of the Triassic lacustrine basins of the Flinders Ranges (see Parkin 1953; Shepherd & Thatcher 1959; Johnson 1960) indicate subdued relief in the surrounding areas, and with a humid tropical climate.

There was marine sedimentation through the later Mesozoic at least in the Great Artesian Basin and its several cognate basins, as well as in the Duntroon Basin (Smith & Kamerling 1969) beneath the Southern Ocean south of South Australia, so that during the later Mesozoic the present upland of the State, including the Adelaide district, could have been suffering erosion and have been a source area for the sedimentary basins mentioned.

(7) The most recent evidence concerning the age of the laterite and the associated land surface derives from Kangaroo Island where the duricrust occupies a high plain similar in all essential respects to the summit surface of the nearby Mt Lofty Ranges. West of the present site of Kingscote the laterite was eroded prior to a Middle Jurassic extrusion of basalt (Wellman 1971)[4] over weathered Permian glacigene strata exposed in the lowland formed by the dissection of the duricrust. Thus the laterite is younger than the Permian rocks on which it is weakly developed, but is older than the basalts. The same conclusion concerning the relative ages of the laterite and basalt was reached by Sprigg, Campana & King, 1954; but they had no absolute datings of the basalt to guide them. Palaeoclimatic and palaeontological considerations, though not palaeomagnetic restitutions, suggest that the most probable age of the laterite, and for the surface on which it evolved, is Triassic (Daily, Twidale & Milnes 1974).

How such an ancient land surface, remnants of which are preserved on Eyre Peninsula as well as Kangaroo Island and the Mt Lofty Ranges, and which has apparently always been exposed to the elements, could survive some 200 m.y. of weathering and erosion poses a considerable problem. Its preservation would be comprehensible had it been buried and recently exhumed as was suggested by Fenner (1930, 1931) but there is no evidence of this; on the contrary the identification of old shorelines of middle Tertiary age at both Gawler and Strathalbyn (Glaessner 1953) and the general distribution of Cainozoic marine strata show that the Mt Lofty Ranges were uplands throughout the Tertiary and were never buried by the sea and related sedimentary deposits.

---

[4] Wellman, P. W. (1971).—"The age and palaeomagnetism of the Australian Cainozoic volcanic rocks." Ph.D. thesis, Australian National University (unpubl.).

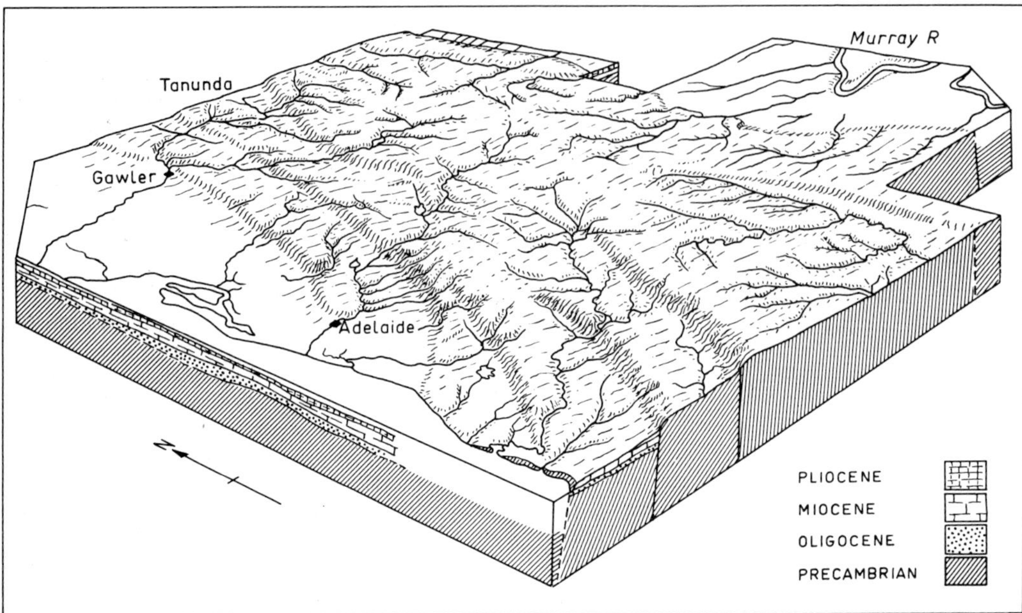

Fig. 4. Block diagram of the Mt Lofty Ranges in the vicinity of Adelaide showing the essential horst structure and the fault-angle valleys developed on the western side. (After Sprigg, 1945.)

The preservation of the summit surface in the Ranges is possibly due to a combination of their structure, the protection afforded by the lateritic capping, and the inherent characteristics of river erosion (Crickmay 1971, 1975; Twidale 1976).

For these reasons the summit surface and its laterite capping are both regarded as of Mesozoic, probably early Mesozoic, age. If subsequently the laterite in question and the associated land surface are shown to be more youthful then the whole sequence of geomorphological development must needs be compressed.

The occurrence of long ribbons of rounded quartzite pebbles and boulders on the lateritised land surface in the eastern Mt Lofty Ranges, and the obvious former creek channels exposed in road cuttings and quarries in the area south and southeast of Adelaide, indicate that the summit surface was eroded and reduced to low relief as a result of stream action. Whether this surface of low relief was formed close to regional baselevel as an orthodox peneplain (Davis 1909) is open to question. Laterite is preserved on the horst blocks and its distribution suggests it originally extended to and beyond the margins of the raised areas, so that it should be possible to locate laterite on the downfaulted, as well as the upfaulted, blocks.

But no buried laterite profiles have yet been found. There are several possible explanations for this (Twidale & Bourne 1975), and in addition there has been propounded an alternative model of landscape development that limits the development of laterite to the upfaulted blocks, with the summit surface a perched peneplain graded to raised local baselevels (see Kennedy 1962; Twidale 1968).

However, the antiquity of the laterite; its distribution, particularly its occurrence at the very margins of the fault blocks on the tilted surface of, for example, the Clarendon Block; its extension beyond the horst structure, for instance, to the north of the Barossa Valley, where it is preserved on the crests of the ridges; and the fact that major faulting did not occur in the early Mesozoic but rather toward the end of that era, and the stratigraphy of the adjacent basin deposits, favour the conventional interpretation of a lateritised peneplain developed by weathering and erosion under epigene conditions and subsequently disrupted by faulting.

The making of the present landscape

Faulting, stream rejuvenation and landscape revival

The Mesozoic land surface was disrupted by pronounced faulting along old-established lines in either late Mesozoic or the earliest Tertiary.

Fig. 5. (a) The summit etch surface east of Mt Pleasant; a—remnant with ferruginous capping, b—etch surface, c—gneiss blocks. (C. R. Twidale.)
(b) The summit etch surface in the southern Mt Lofty Ranges, with a lateritic relict preserved in Mt Cone. (C. R. Twidale.)
(c) The 8 m platform developed on Pleistocene calcarenite (aeolianite) at Cape Jervis. (E. J. Brock.)

The fault zones had been established during the early Palaeozoic orogeny and the recurrent movements caused the development of a clearly defined though complex horst (the main upland region), and irregularly shaped graben and fault angle depressions (Figs 1 and 4). The seas invaded some of the depressed blocks,

which became the ancestral gulfs and basins, and the raised blocks became the uplands we know today. Thus, the earth movements that gave rise to the framework of the modern physiographic landscape are of great antiquity. By the beginning of the Tertiary, the Mt Lofty Ranges had been resurrected, and the outlines of the present gulfs, of Yorke and Eyre peninsulas, and of the Lake Torrens Basin, had been determined.

As in the Flinders Ranges (Twidale 1966), the broad outlines of the contemporary scene are of considerable antiquity.

The topographic depression we call the Barossa Valley, probably came into being at this time, though sedimentation did not take place on the lowland until the Oligocene and Miocene. However, it is not clear whether the Valley is a tectonic or a structural feature, whether the topographic low is a fault-angle depression or a large fault-line valley scoured out over the ages by the precursor of the North Para drainage system. But faults have indubitably played an important part in shaping the Barossa Valley and other major features of the Adelaide district.

The uplifted blocks were raised well above sealevel or regional baselevel, and various rivers and streams which drained from them were rejuvenated. The many gorges, large and small, that are characteristic of the Ranges were then initiated. The major rivers and their tributaries also undermined the duricrust, causing its collapse. As the Cainozoic progressed the laterite was gradually stripped, revealing the weathering front or limit of significant weathering of intrinsically fresh rock (Twidale & Bourne 1975). In contrast to the duricrusted surface where weathering has subdued all but the more pronounced differences, and notably the several quartzite residuals already mentioned, local structure is clearly discernible in the landforms of the etch surface. Thus west of Palmer, corestones have been revealed to give the boulder-strewn landscape typical of many granite outcrops (Fig. 2b). The gneissic slabs or tombstones of the Tungkillo area (Fig. 2c) and of much of the eastern Mt Lofty Ranges, where the Kanmantoo Series is exposed, originated in similar fashion, by differential joint-controlled subsurface weathering, followed by stripping of the weathered mantle.

In more general terms the various folds involving sedimentary and metasedimentary sequences were exposed and suffered differen-

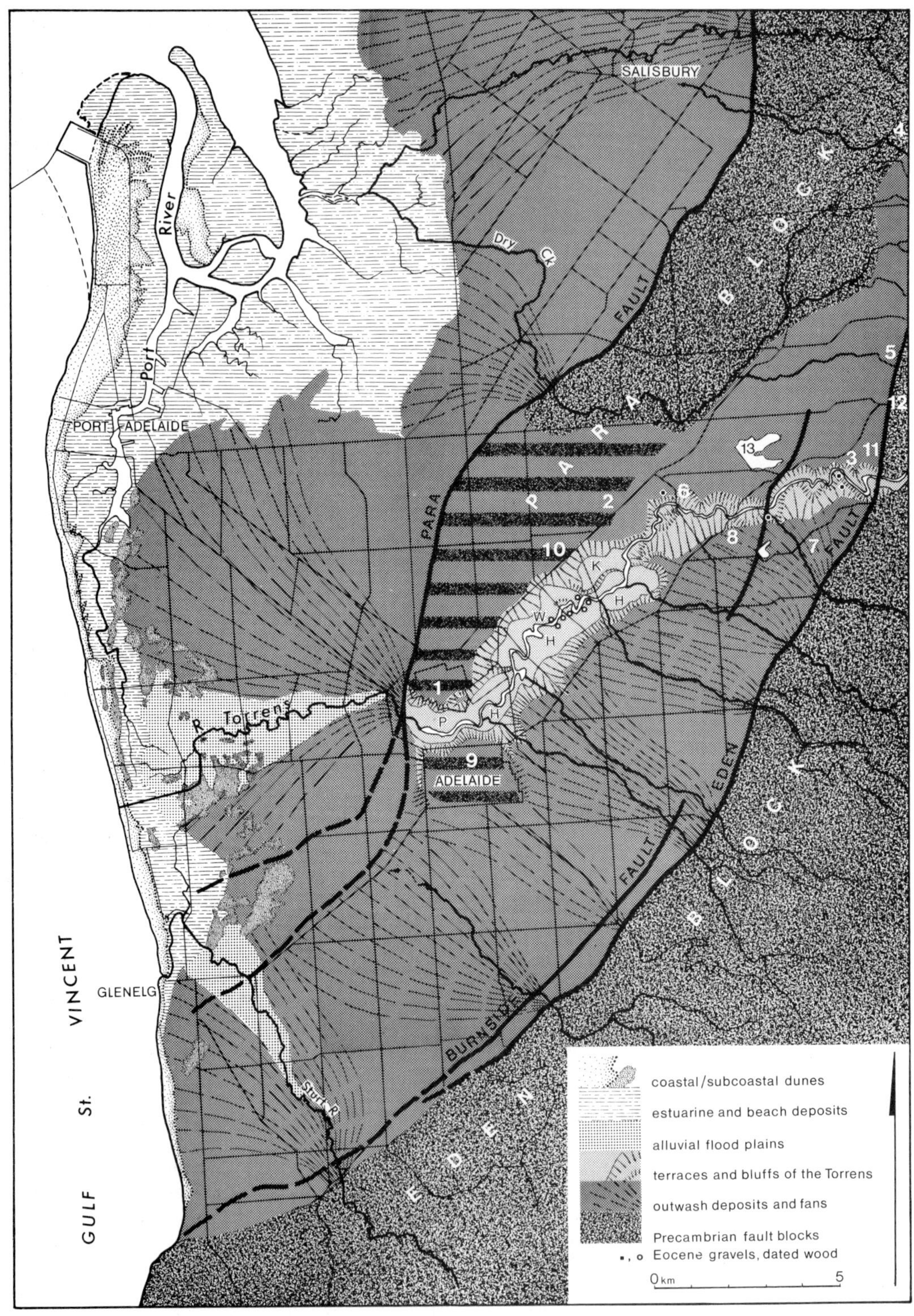
GULF
St.
VINCENT
SALISBURY
Dry Ck
PARA FAULT
PARA BLOCK
PORT ADELAIDE
Port River
River
PARA
13
6
2
10
K
W
H
H
H
3 11
12
5
4
8
7
EDEN FAULT
1
P
H
9
ADELAIDE
Torrens
BURNSIDE FAULT
EDEN BLOCK
GLENELG
Sturt R.
coastal/subcoastal dunes
estuarine and beach deposits
alluvial flood plains
terraces and bluffs of the Torrens
outwash deposits and fans
Precambrian fault blocks
Eocene gravels, dated wood
0 km   5

tial weathering and erosion so that not only river courses but also the residual ridges left upstanding display strong structural guidance. Thus the River Torrens in its gorge section follows the strike of the local bedrock over long sectors, and sandstone ridges and valleys eroded in slates and phyllites are prominent in the dissected areas.

But more significant than these interesting but local features, was the initiation of the gently rolling etch plain (Fig. 5a) which is such a prominent feature of the modern landscape in the eastern and southern Mt Lofty Ranges (Twidale & Bourne 1975), but which is also present elsewhere as for instance at Mt Cone (Fig. 5b). Rejuvenated streams began to re-excavate deep valleys in the southern Ranges, eroding the Permian glacial strata and exposing the glacial pavements of the Inman Valley and environs.

Thus, dissection and stripping of the Mesozoic high plain began in the early Tertiary as a result of a renewal of faulting along the old-established lines. The major rivers began to cut through the uplifted and duricrusted high plain and as they incised into intrinsically fresh bedrock they adapted to the new and changing structures they exposed. Gradually, through the Cainozoic, the rejuvenated streams also eroded headwards, breached divides and effected captures. The Torrens, Sturt, Onkaparinga, Myponga and Finniss all display angular courses due to the capture of strike sectors by rivers flowing either down the local dip or against it as does the smaller Dry Creek, north of Adelaide (Fenner 1931). The regional drainage pattern is radial off the horst block but in detail is determined by faulting, stream rejuvenation and piracy (Sprigg 1945) and the deep incision of streams into structures to which the drainage is constantly adjusting (Twidale 1972). Over long sectors the rivers run parallel to structures in the local bedrock. The various drainage anomalies referred to in the literature (see e.g. Fenner 1930, 1931) are largely illusory.

The implication of these remarks is that the major rivers are of considerable antiquity, but it is difficult to establish the age of any of them. High level alluvial deposits occur in association with several of the present river valleys, as for instance just below the abattoir at Noarlunga, but their age is debatable. The lower Torrens valley has been closely studied, but even there the evidence is equivocal, though there is strong suggestion that the ancestral Torrens was already in existence during the early Tertiary. As is described below, the sediments of the Highbury-Golden Grove area are generally fine or medium grained, but at the southern edge of the outcrop there are three small deposits of rounded quartz gravels, cobbles and boulders (Fig. 6). The gravels are unconsolidated and they include lenses of mottled sand and clay. At Beefacres they display a rudimentary cross-bedding, and calcrete is developed in the upper horizons. The gravels are some 4–5 m thick at Beefacres, 3 m at Paradise and 2 m at Athelstone. At each of the three sites there is a considerable range of particle size but the mode, mean and maximum all increase from west to east, so that at Athelstone there are some boulders fully 80 cm in diameter.

The base of the gravel rests on Precambrian phyllites at all three sites and stands at about 69 m above present sealevel at Beefacres, 76 m at Paradise and 91 m at Athelstone so that the unconformity rises in an easterly direction. Except for locally derived phyllites all the gravels could have originated in the Mt Lofty Ranges to the east. Thus the provenance of the materials, the rudimentary bedding displayed at Beefacres, the particle size gradation and the location of the deposits are all consistent with derivation from the east, and the limited lateral distribution of the boulders and gravels suggests that they are remnants of a former valley train.

Their age is uncertain. Before subdivision of the land at Beefacres they could be traced laterally both as scattered surface remnants as well as in shallow excavations to the east and northeast. The gravels close to the valley graded into sandstone which in general appearance can be correlated with the poorly lithified Hope Valley Sands, which include interbedded lignite of Eocene age (Miles 1952). If this interpretation is valid then the valley train gravels are also of Eocene age.

Fig. 6. Landforms of the lower Torrens Valley. K—Klemzig Terrace; W—Walkerville Terrace; H—Hackney-Felixstowe Terrace; P—Parklands Terrace; 1—North Adelaide; 2—Hillcrest; 3—Highbury; 4—Golden Grove; 5—Tea Tree Gully; 6—Beefacres; 7—Athelstone; 8—Paradise; 9—City Plateau; 10—Hampstead Plateau; 11—Pearce Quarry; 12—Gun Emplacement; 13—Hope Valley Reservoir.

Another possibility, however, is that the gravels are much younger, that they were laid down possibly in later Tertiary or even Pleistocene times long after the lake had disappeared. Certainly in view of the demonstrable presence of the sea in the Adelaide embayment at times during the Tertiary the survival of the gravels is more acceptable if they are taken to be of Cainozoic age. On the other hand there are only a few relics extant and the only evidence to hand points to an Eocene age.

Thus the question of the antiquity of the Torrens remains unresolved.

Landform development during the Tertiary

At the beginning of the Tertiary, the downfaulted blocks were invaded by the sea, and marine embayments rather more extensive than the present gulfs came into being. Thus with a huge embayment of the ocean occupying the present Murray Basin through much of the Tertiary the Mt Lofty Ranges of Eocene-Miocene times formed a peninsula bounded on three sides by the sea. Furthermore, considerable terrestrial deposition took place in the major fault angle depressions, around Baker Gully, at the eastern extremity of the Willunga embayment, and on the lower slopes of the Para Fault Block as far north as Golden Grove and Crouch Hill. For much of middle Tertiary time the sands and clays were laid down in the Barossa Valley.

Tertiary strata occur below the floor of Gulf St Vincent, and extend into the Willunga, Noarlunga and Adelaide embayments. These sequences are beautifully exposed in coastal cliffs south of Adelaide. But the terrestrial strata are not so well known, though quarrying has revealed a great deal in recent years. The sand and clay of the Highbury area are of particular interest in the present context. They grade into the fossiliferous Eocene beds of the Hope Valley region. The maximum thickness of the beds so far discovered is about 30 m but there has been some erosion (see below) and their original thickness may have been as much as 60 m.

What is the origin of these strata? They have for many years been regarded as lacustrine deposits (Miles 1952; Aitchison, Sprigg & Cochrane 1954; Dalgarno 1961[5]), partly because of the amplitude of the cross-bedding displayed by the sands; partly because of the overall lack of coarse debris and the undisturbed character of the sediments (cf e.g. Tickell 1976); partly because of the local derivation of the sediments which are consistent with the erosion of a land surface deeply weathered under humid tropical conditions so that little coarse debris survives as a basal conglomerate—the gravel swept off the old land surface by rejuvenated streams immediately following uplift of the land; and partly because of the similarities they display with deposits elsewhere that are attributed to lacustrine sedimentation.

However, this interpretation has now been abandoned in some quarters, and these terrestrial sediments, and those of the Barossa Valley, are explained as fluviatile deposits laid down by broad braided rivers. This suggestion, which is reminiscent of the "Dead Rivers" concept of Howchin (1931, 1933), is difficult to sustain with respect to the Highbury-Golden Grove sequence. If they are of riverine origin, whence did the drainage system originate? Why are there no exotic materials? Where are the coarse lenses typical of cut and fill characteristic of riverine deposits (and so clearly revealed, for instance, in the alluvial apron fronting the Willunga Scarp (see Twidale 1968, 1969)? Where are the stream channels filled with coarse bed load cut into the underlying bedrock? Such are displayed beneath pediment mantles, for instance, but the unconformity beneath the early Tertiary sands, wherever it has been observed is remarkably even. The only sign of coarse deposits occurred, before quarrying, at Crouch Hill, where there was a bench with numerous scattered quartz and quartzite cobbles, and interpreted as a shoreline marking the former northern limit of the lake deposits. There are quartzite gravels in the fault-disturbed exposure at Pearce Quarry which are taken to be colluvial or slope deposits comparable to those mantling steep slopes in the vicinity at present and indicating that hills occurred nearby at the time the Tertiary strata were laid down.

Thus the riverine concept is not wholly satisfactory. However, regardless of the origin of the early Tertiary strata, it is clear that the Eocene sand and clay once extended to the west of the present continuous outcrop for there are outliers of the sand high up on the gentle eastern slope of the Para Block. The latter is the pre-Eocene surface cut in Pre-

---

[5] Dalgarno, C. R. (1961).—"Geology of the Barossa Valley." M.Sc. thesis, University of Adelaide (unpubl.).

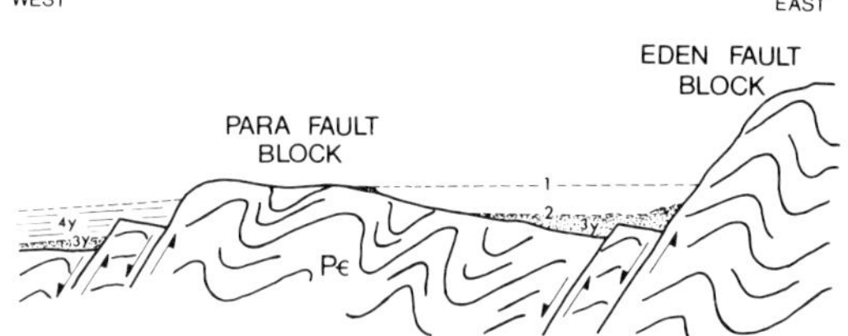

Fig. 7. Diagrammatic cross-section through the Para and Eden scarps showing the exhumed pre-Eocene surface exposed on the dip slope of the Para Block. 1—minimum upper level of Eocene beds; 2—present surface cut in Eocene strata; 3y—Tertiary beds; 4y—Quaternary strata.

cambrian rocks from which the Tertiary strata have been stripped (Fig. 7). It is an exhumed land surface of pre-Tertiary age (Twidale 1968). There are no indications of the easterly limit of the lake and associated deposits. The Tertiary outcrop terminates against the Eden Fault, and though well exposed there are no beach deposits. However, the angular quartz gravel comparable with contemporary colluvial or slope deposits indicates that though the edge of the lake cannot be precisely determined, the hilly shore was not far distant from the present margin of the outcrop.

At the end of the Eocene the seas still occupied the three embayments mentioned as well as the present Adelaide Plains and the Barossa Valley was still an area of deposition. During the middle Tertiary the lake sediments and some of the broad valley floors to the north were weathered and in particular were heavily silicified. Moreover, during the Oligocene and lower Miocene the sea occupied the Myponga and other nearby basins and valleys which had been eroded as a result of uplift and stream incision.

Later Cainozoic events and forms
Erosion and weathering

Following the draining of the Highbury area the exposed sands and clays were weathered and eroded. The beds covering the higher parts of the Para Fault Block were stripped away in later Cainozoic times, re-exposing the pre-Eocene surface eroded in Precambrian phyllites and slates. A weakly developed ferricrete was formed on the Eocene sands at least in the piedmont zone for relicts survive in the so-called "Gun Emplacement" (Fenner 1939), at the foot of Anstey Hill, and elsewhere north of Tea Tree Gully. Similar broadly convex spurs eroded in Precambrian rocks carrying a fer-

ruginised mantle and jutting out from the main line of the hill front occur at intervals along the Eden Scarp from Golden Grove in the north to the sea at Marino. They have been interpreted as remnants of old marine shorelines of later Tertiary (Sprigg 1961) or early Pleistocene (Ward 1965) age. But no evidence has been adduced in support of this contention and it is more likely that they mark a former scarp foot zone or zones, where, because of the concentration of moisture (Twidale 1967), there tends to be precipitation of salts, particularly iron oxides, and which have been lifted as a result of recurrent fault dislocation subsequent to the main elevation of the Ranges in the late Mesozoic and possibly as part of the late Tertiary-early Quaternary movements that are described below.

Fenner (1939) interpreted the deposits of the Gun Emplacement as alluvia of Plio-Pleistocene age. They are neither alluvial nor transported but the result of weathering. However, the character of the weak duricrust suggests analogies with the ferruginisation that occurred on Yorke Peninsula and Eyre Peninsula during the late Pliocene and early Pleistocene (Horwitz & Daily 1958; Twidale, Bourne & Smith 1976). Other ferruginised surfaces occur in valley floors in the One Tree Hill area, and in the southern Mt Lofty Ranges, near Myponga (Horwitz 1960). Some of this ferruginous and siliceous material is derived from the disintegration of the summit surface laterite. Iron deposition is also pronounced on perched plateau remnants isolated by faulting and erosion, on the eastern side of the Ranges, in the Milendella, Palmer and Tepko regions (Twidale & Bourne 1975), as well as to the west of Strathalbyn, where there is a platform of considerable extent located at an elevation intermediate between the summit high surface and the Murray Plains.

The Cainozoic sediments of the Barossa were also dissected by the North Para and its various tributaries and alluvial fans were laid down at the base of the fault or fault-line scarp.

Recurrence of faulting

Most of the faults of the Adelaide region were significantly active during the later Cainozoic (Fenner 1930, 1931; Steel 1962; Firman 1969). The only possible exception is the Clarendon Fault (Ward 1965) but even there modern activity is indicated by the seismic record and though no dislocation can be demonstrated in coastal sections this is no guarantee of stability inland. However, most of the others have moved and all generate earth tremors at present (Sutton & White 1968).

Post Eocene movement of the Eden Fault is demonstrated by the development of fault drag displayed in the former Pearce Quarry just north of the Torrens Gorge mouth and in coastal sections near Sellick Hill. The complex Para faults have been very active in post Pliocene times with a proven total displacement of almost 400 m, the westerly blocks subsiding relative to those of the east (Steel 1962). Brock (1971) has adduced evidence which indicates a late Tertiary uplift of Fleurieu Peninsula of at least 230 m and possibly as much as 460 m. The reality of continued activity was shown forcibly by the Adelaide earthquake of 1954 (Grant 1956).

Thus the patterns of faulting and of consequent topography established by the early Tertiary were reinforced by the later Cainozoic movements and these two phases of diastrophism together account for the major feature of the local relief. The faulting was accompanied by renewed stream incision, and by a continuation of the dissection and stripping of the raised blocks that had begun earlier, and by deposition on the lower. The discontinuous nature of the faulting is suggested by the valley-in-valley forms seen, for example, in the Torrens Gorge (Walker 1962[6]; Trudinger 1973).

There was, however, another potent reason for changes in stream regime, namely movements of their ultimate baselevel, sealevel.

Quaternary changes of sealevel

At the end of the Tertiary the sea withdrew from the Willunga, Noarlunga and Adelaide embayments and plains, and indeed from the gulfs generally. The sea floor was exposed in the City and Hampstead plateaus (Fig. 6). The shore retreated far to the south of the edge of the continental shelf as a result of a lowering of sealevel of about 100 m consequent on the development of ice sheets in high latitudes in both hemispheres. Ice sheets had begun to develop on Antarctica and elsewhere much earlier, probably in the Miocene (see for example Hollin 1962, 1969) but it was only during the late Pliocene that the withdrawal of water from the hydrologic cycle began to cause significant eustatic or world-wide changes in sealevel.

There is no doubt that the waxing and waning of the icesheets caused major lowering of sealevel several times during the Quaternary but the question of high sealevels is a vexed one. Some workers, following Depéret (1918) and Zeuner (1940), have identified many high stands of the sea dating from the later Cainozoic. Thus Ward (1965) and Ward & Jessup (1965) have claimed evidence for stands of the sea at approximately 183, 165, 113, 32, 11, 8, 3 and 2.5 metres above present sealevel during the Quaternary in the Adelaide district. But quite apart from serious doubts about the validity of some of the evidence (Twidale, Daily & Firman 1967) and the lack of any independent dating of the features in question, extension of the sea to the higher levels Ward postulates would imply the inundation of much of South Australia, and of this there is no evidence. Most of the forms and deposits which suggested the possibility of high stands of the sea to Ward are either marine or scarp foot features, some possibly of pre-Pleistocene age, that have been raised to their present positions by faulting. However, there is good evidence in the form of raised beaches (Figs 5c and 8a) and of river terraces graded to raised beaches, of a stand of the sea about 8 m higher than present at many sites around the coast of Fleurieu Peninsula (Brock 1971). Stands at about +5 m and +3 m are also widely cited, though evidence from Eyre and Yorke peninsulas suggests that some may be related to modern sealevel, and be due either to spray and pool weathering or to storm tides, or to a combination of these.

The sea has stood at essentially its present level for only a few thousand years—some 6000 years according to Fairbridge (1960, 1961), and at most 2000 years according to Shepard & Curray (1967). During that time the sea has eroded some forms and built up others, using sediment derived from the erosion of the land and brought to the coast by rivers as well as the debris derived by marine weathering and erosion. During this brief period shore platforms have been eroded in some places, e.g. Hallett Cove, and several zones of coastal dunes developed, e.g. between Glenelg and the Port River (Fig. 6). And of course these coastal changes continue.

The distribution of coastal forms is determined by structure. Where the fault blocks

---

[6] Walker, G. T. (1962).—"Physiography of the Torrens Gorge." B.A. Hons. thesis, University of Adelaide (unpubl.).

a

b

c

Fig. 8. (a) 8 m platform cut in Cambrian Kanmantoo schists at Cape Jervis. (E. J. Brock.)

(b) The serrated shore platform cut in steeply dipping Precambrian shales at Hallett Cove. Note the curved limb of the fold expressed in the cliff morphology. (C. R. Twidale.)

(c) Smooth platform in flat-lying early Tertiary limestone at Blanche Point. (C. R. Twidale.)

stand above sea level their coastal margins have suffered differential erosion; resistant strata survive as cliffed headlands bordered by shore platforms (Fig. 8b & c), but weaker formations have been worn down to give embayments with small bayhead beaches composed of shingle or sand depending on the character of the local bedrock. In detail structure strongly influences the morphology of both cliffs and platforms (Fig. 8b & c).

Thus the coast of Fleurieu Peninsula consists of a succession of rugged cliffs and promontories with intervening coves and beaches. Only at Cape Jervis where there is a coastal flat standing some 8 m above sealevel—a raised beach—is this rugged coastline absent. Similar alternations of headland and bay, of rocky coast and sandy beach, occur where the Clarendon and Eden fault blocks reach the coast, as for example at Hallett Cove; and again there are raised coastal plains, at about 30 m elevation, in the Seaford area south of Noarlunga.

Standing in strong contrast are the tectonic lows, particularly the Adelaide Plains, with their long sandy beaches backed by modern and older linear zones of coastal foredunes, and with mangrove swamps around and to the north of the Port River; and the fault angle depressions (Willunga plains) where the horizontal and subhorizontal Tertiary sequences are exposed in steep cliffs (Fig. 8c) which back and separate spectacular sandy beaches such as those at Maslin and Aldinga.

River regimes: cut and fill

As a result of the several lowerings of sealevel during the Quaternary, river activity changed in several ways. Major rivers extended their courses across the exposed continental shelf, and though it has been shown that the River Murray ran far to the south of the present mouths and plunged over the continental slope where it formed a series of enormous submarine canyons well over 1.5 km deep (Von der Borch 1968), there has been insufficient work as yet to demonstrate the prior courses of other streams such as the ancestral Torrens.

But the Torrens, and many of the other rivers of the Mt Lofty Ranges experienced alternations of erosion and deposition, phases of cut and fill. Between the Gorge mouth and Hillcrest the river runs across the strike of the tilted and lithologically varied Precambrian strata, and so flows through alternations of defile and basin, the latter displaying local terrace developments. Between Hillcrest and the Weir, following erosion of the valley below the level of the Pliocene sediments of the City Plateau and North Adelaide, there was first a period of deposition, followed by a phase of incision accompanied by lateral corrasion during which prominent terraces were eroded (Fig. 6) during the late Pleistocene and early

Fig. 9. (a) Evidence of cut and fill exposed in gully near Sellick Hill. The alluvial apron (A) was deposited in late Pleistocene times. A narrow valley was cut in this material, but this was partially filled with alluvium (B). x–y is the unconformity separating the older and younger alluvium, and is in fact the former valley side. The terrace (T) is the former valley floor. Recently this younger alluvium has in turn been gullied; and the evidence suggests that this has taken place in the last century, since European settlement of the area. (C. R. Twidale.)

(b) Furrows dating from an earlier phase of agriculture in the Yankalilla area have been exploited by run-off to form linear, evenly-spaced, gullies. (C. R. Twidale.)

(c) Earthflows developed in Permian glacigenes near Yankalilla. (C. R. Twidale.)

Recent. There followed an infilling of the narrow but deep river trench about 5,800 years ago and finally the most recent incision. During most of this time the river debouched on to the Adelaide Plains near the southern extremity of the Para Block and then ran in a north-westerly direction toward what is now the Port River, depositing an alluvial fan as it did so. It was only relatively late in the Pleistocene or early Recent that the river choked its own course and diverted itself to the south and its present position.

Many of the other rivers of the region display a similar chronology. The valleys of the Gawler River, the lower Onkaparinga (between Noarlunga and the sea) and many minor valleys like those cut into the alluvial apron fronting the Willunga Scarp, display similar cut and fill (Fig. 9a), though it has not been established that the phases of erosion and deposition evidenced in the various valleys can be correlated. Nor has it been established why these

Fig. 10. Gullying of the alluvial apron fronting the Willunga Scarp. The presence of a well (X) close to the deep trench suggests that the gully was not in existence when the well was sunk in the early years of European settlement. (C. R. Twidale.)

phases of cut and fill occurred, though their wide distribution in a variety of topographic and geological settings strongly suggests that climatic changes are basically responsible (Twidale 1969). However, the most recent stream incision in some places at any rate (see below) is demonstrably consequent upon European settlement.

But the whole question is complicated by several factors, including the recurrence of dislocation along the Para Fault and elsewhere, so that it is imprudent at present to make any sweeping generalisations. Thus, though there has been widespread and marked deposition of alluvial fans on the western margin of the Mt Lofty Ranges in the late Pleistocene (Williams 1969, 1973) there are ·no corresponding features on the eastern or Murray Basin side. Part of the reason may be found in the lower rainfall from the eastern areas, but more significant is the incision of streams debouching from the hills on to the Murray Plains. They have cut shallow trenches and there is considerable scarp foot erosion (Twidale & Bourne, 1975). This has resulted in the evacuation of most of the debris deposited at the hill-plain junction in contrast to the western side where there is not only greater relief amplitude but open unimpeded plains where the streams can spread and deposit much of their loads.

Man

Finally and most recently the natural quasi-equilibrium has been disturbed by the advent of man. First, several thousands of years ago the aborigines settled in the Adelaide region, and though it is difficult to determine precisely what impact they had on their environment, though the depletion, through man-made fire, of the vegetation must have induced and accelerated erosion and deposition (but see Chapter 8). However, it is difficult to distinguish between these and natural lightning strikes. No such doubts can be entertained about European man, who began settling the area in 1836, who spread rapidly to the most distant parts of the State and who, because of his exponentially increasing numbers and advancing technology, has had very obvious effects on the environ-ment and on landform development. In a variety of ways, but, as elsewhere, principally by clearing the "natural" vegetation, man has rendered the land surface vulnerable to erosion, so that a recent minor episode of erosion and of course concomitant deposition is in train (Fig. 10). Similar effects have been achieved by the making of roads and tracks, by the introduction of exotic plants and animals, and by ploughing the land (Fig. 9b). The accelerated soil erosion that results from man's interference is manifested most dramatically in gullying, but sheet erosion or the stripping of the surface soil is perhaps more significant. How long this most recent phase of marked erosion will last is difficult to predict, but in western Europe a similar epicycle of pronounced soil erosion during the early Nineteenth Century was cut short as a result of improved farm practices and techniques (Vogt 1953).

Landslides and earth flows are another result of man's intereference with the natural eco-system (Fig. 9c). They have developed on steep slopes in many parts of the Mt Lofty Ranges, but are especially numerous in the southern areas where Permian glacigene rocks are exposed (Van Deur 1975)[7], and where in some localities up to 25 per cent of the land surface is affected by either mass movements or gullying.

Conclusion

The geomorphological landscape of the Adelaide region is basically one of great antiquity. The age of some facets of the landscape is to be measured in terms of several scores of millions of years, but in broad outline the framework of the modern topography was determined by the diastrophic disruption of a Mesozoic weathered land surface. The Cainozoic has seen the recurrence of faulting, and changes of sealevel consequent upon the development and dissipation of ice sheets. These have resulted in the perpetuation of the ancient framework, the stripping of the weathered mantle from the old land surface, the deep dissection of major rivers, and alternations of cut and fill in the valleys so formed.

References

AITCHISON, G. D., SPRIGG, R. C. & COCHRANE, G. W. (1954).—The soils and geology of Adelaide and suburbs. *Bull. Geol. Surv. S. Aust.* 32.

BENSON, W. N. (1911).—Note descriptive of a stereogram of the Mt Lofty Ranges, South Australia. *Trans. R. Soc. S. Aust.* 35, 108-111.

---

[7] Van Deur, W. J. (1975).—"Earthflows in the Yankalilla district of South Australia." B.A. Hons. thesis, University of Adelaide (unpubl.).

BOURMAN, R. P. & LINDSAY, M. F. (1973).—Implications of fossiliferous Eocene marine sediments underlying part of the Waitpinga drainage basin, Fleurieu Peninsula, South Australia. *Search* **4**, 44.

BOURMAN, R. P. & MILNES, A. R. (1976).—Exhumed *roche moutonnée*. *Aust. Geogr.* **13**, 214-216.

BROCK, E. J. (1971).—The denudation chronology of the Fleurieu Peninsula, South Australia. *Trans. R. Soc. S. Aust.* **95**, 85-94.

CAMPANA, B. (1958).—The Mt Lofty-Olary region and Kangaroo Island. Chapter 1 *In* M. F. Glaessner & L. W. Parkin (Eds), "The Geology of South Australia." (Melbourne University Press: Melbourne.)

CAMPANA, B. & WILSON, R. B. (1954).—The Geology of the Jervis and Yankalilla Military Sheets. *Rept. Invest. geol. Surv. S. Aust.* **3**.

CAMPANA, B. & WILSON, R. B. (1955).—Tillites and related glacial topography of South Australia. *Eclogae géol. Helv.* **48**, 1-30.

CRICKMAY, C. H. (1971).—"The Role of the River." (Crickmay: Calgary.)

CRICKMAY, C. H. (1975).—"The Interpretation of Scenery." (Crickmay: Calgary.)

DAILY, B., GOSTIN, V. A. & NELSON, C. A. (1973).—Tectonic origin for an assumed glacial pavement of Late Proterozoic age, South Australia. *J. geol. Soc. Aust.* **20**, 75-78.

DAILY, B., TWIDALE, C. R. & MILNES, A. R.. (1974).—The age of the lateritized summit surface on Kangaroo Island and adjacent areas of South Australia. *J. geol. Soc. Aust.* **21**, 387-392.

DAVIS, W. M. (1909).—"Geographical Essays." (Dover: Boston.)

DÉPERET, C. (1918).—Essai de co-ordination chronologique général de temps quaternaires. *Acad. Sci. (Paris) Comptes rendus* **167**, 418-422.

FAIRBRIDGE, R. W. (1960).—The changing level of the sea. *Sci. Am.* **202**(5), 70-79.

FAIRBRIDGE, R. W. (1961).—Eustatic changes in sea-level. *Physics & Chemistry of the Earth* **4**, 99-185.

FENNER, C. (1930).—The major structural and physiographic features of South Australia. *Trans. R. Soc. S. Aust.* **54**, 1-36.

FENNER, C. (1931).—"South Australia: A Geographical Study." (Whitcombe & Tombs: Melbourne.)

FENNER, C. (1939).—The significance of the topography of Anstey Hill, South Australia. *Trans. R. Soc. S. Aust.* **63**, 79-87.

FIRMAN, J. B. (1967).—Stratigraphy of late Cainozoic deposits in South Australia. *Trans. R. Soc. S. Aust.* **91**, 165-180.

FIRMAN, J. B. (1969).—Quaternary Period. Chapter 6. *In* L. W. Parkin (Ed.), "Handbook of South Australian Geology." (Geol. Surv. S. Aust.: Adelaide).

GLAESSNER, M. F. (1953).—Some problems of Tertiary geology in southern Australia. *J. R. Soc. N.S.W.* **87**(2), 31-45

GLAESSNER, M. F. & WADE, M. (1958).—The St Vincent Basin. Chapter 9. *In* M. F. Glaessner & L. W. Parkin (Eds), "The Geology of South Australia." (Melbourne University Press: Melbourne.)

GRANT, C. Kerr (1956).—The Adelaide earthquake of 1 March 1954. *Trans. R. Soc. S. Aust.* **79**, 177-185.

HARDY, M. (1939).—"A History of Crafers." (Crafers Centenary Committee: Adelaide.)

HOLLIN, J. T. (1962).—On the glacial history of Antarctica. *J. Glaciol.* **4**, 173-195.

HOLLIN, J. T. (1969).—The Antarctic ice sheet and the Quaternary history of Antarctica. *Palaeo ecol. Afr.* **5**, 109-138.

HORWITZ, R. C. (1960).—Géologie de la région de Mt Compass (feuille Milang), Australie méridionale. *Eclogae géol. Helv.* **53**, 211-263.

HORWITZ, R. C. & DAILY, B. (1958).—Yorke Peninsula. Chapter 3. *In* M. F. Glaessner & L. W. Parkin (Eds), "The Geology of South Australia." (Melbourne University Press: Melbourne.)

HOWCHIN, W. (1895).—New facts bearing on the glacial features of Hallett's Cove. *Trans. R. Soc. S. Aust.* **19**, 61-69.

HOWCHIN, W. (1898).—On the evidence of glacial action in the Port Victor and Inman Valley districts, South Australia. *Rep. Aust. Ass. Advanc. Sci.* **7**, 114-127.

HOWCHIN, W. (1931).—The dead rivers of South Australia Part I. Western group. *Trans. R. Soc. S. Aust.* **55**, 113-135.

HOWCHIN, W. (1933).—The dead rivers of South Australia Part II. The eastern group. *Trans. R. Soc. S. Aust.* **57**, 1-41.

KENNEDY, W. Q. (1962).—Some theoretical factors in geomorphological analysis. *Geol. Mag.* **99**, 304-312.

JOHNSON, W. (1960).—Exploration for coal, Springfield Basin. *Rept. Invest. geol. Surv. S. Aust.* **16**.

MAUD, R. R. (1972).—Geology, geomorphology, and soils of central County Hindmarsh (Mount Compass-Milang), South Australia. *CSIRO Soil Publ.* No. 29.

MILES, K. R. (1952).—The geology and underground water resources of the Adelaide Plains area. *Bull. Geol. Surv. S. Aust.* 27.

MILNES, A. R. & BOURMAN, R. P. (1972).—A late Palaeozoic glaciated granite surface at Port Elliott, South Australia. *Trans. R. Soc. S. Aust.* **96**, 149-155.

PARKIN, L. W. (1953).—The Leigh Creek coalfield. *Bull. Geol. Surv. S. Aust.* 31.

READ, H. H. (1957).—"The Granite Controversy." (Murby: London.)

REYNOLDS, M. (1953).—The Cainozoic succession of Maslin and Aldinga Bays, South Australia. *Trans. R. Soc. S. Aust.* **76**, 114-140.

SELWYN, A. R. C. (1859).—Preliminary report by A. R. C. Selwyn of a geological survey of portions of South Australia. *S. Aust. Parl. Pap.* **119**, 1.

SHEPARD, F. P. & CURRAY, J. R. (1967).—Carbon-14 determination of sea level changes in stable areas. *In* "Progress in oceanography—Vol. 4 The Quaternary history of the ocean basins," pp. 283-291. (Pergamon Press: London & New York.)

SHEPHERD, R. G. & THATCHER, D. (1959).—The geology of the Quorn Military Sheet. *Rept. Invest. geol. Surv. S. Aust.* **13**.

SMITH, R. & KAMERLING, P. (1969).—Geological framework of the Great Australian Bight. *Aust. Petrol. Explor. Assoc. J.* **9**, 60-66.

SPRIGG, R. C. (1945).—Some aspects of the geomorphology of portion of the Mount Lofty Ranges. *Trans. R. Soc. S. Aust.* **69**, 277-302.

SPRIGG, R. C. (1961).—The oil and gas prospects of the St Vincent Gulf Graben. *In Aust. Petrol. Explor. Assoc. Conf. Pap.* 71-88. (Melbourne: Victoria.)

SPRIGG, R. C., CAMPANA, B. & KING, D. (1954).— Kingscote, *Geological Atlas of South Australia Map. 1:253,440 series.* Ref. 1/53 16, Zone 5. (Geol. Surv. S. Aust.: Adelaide.)

STEEL, T. M. (1962).—Subsurface stratigraphy in the western suburbs of Adelaide. *Quart. geol. Notes geol. Surv. S. Aust.* 2.

SUTTON, D. J. & WHITE, R. E. (1968).—The seismicity of South Australia. *J. geol. Soc. Aust.* **15**, 25-32.

TATE, R. (1886).—Post-Miocene climate in South Australia. *Trans. R. Soc. S. Aust.* **8**, 49-59.

TATE, R. (1889).—Glacial phenomena in South Australia. *Rep. Aust. Ass. Advanc. Sci.* **1**, 231-232.

TICKELL, S. J. (1976).—Braided river deposits in the Waterberg Supergroup. *Trans. geol. Soc. S. Afr.* **78**, 83-88.

TRUDINGER, J. P. (1973).—Engineering geology of the Kangaroo Creek Dam. *Bull. Geol. Surv. S. Aust.* 44.

TWIDALE, C. R. (1966).—Chronology of denudation in the southern Flinders Ranges, South Australia. *Trans. R. Soc. S. Aust.* **90**, 3-28.

TWIDALE, C. R. (1967).—Origin of the piedmont angle as evidenced in South Australia. *J. Geol.* **75**, 393-411.

TWIDALE, C. R. (1968).—"Geomorphology, with Special Reference to Australia". (Nelson: Melbourne.)

TWIDALE, C. R. (1969)—A possible late Quaternary change of climate in South Australia, pp. 43-48. *In* H. E. Wright (Ed.). "Quaternary Geology and Climate", Vol. 16, VII Congress INQUA. (Nat. Acad. Sci.: Washington.)

TWIDALE, C. R. (1972).—The neglected third dimension. *Z. Geomorph.* **16**, 283-300.

TWIDALE, C. R. (1976).—On the survival of palaeoforms. *Am. J. Sci.* **276**, 77-95.

TWIDALE, C. R. & BOURNE, J. A. (1975).—Geomorphological evolution of part of the eastern Mount Lofty Ranges, South Australia. *Trans. R. Soc. S. Aust.* **99**, 197-210.

TWIDALE, C. R., BOURNE, J. A. & SMITH, D. M. (1976).—Age and origin of palaeosurfaces on Eyre Peninsula and in the southern Gawler Ranges, South Australia. *Z. geomorph.* **20**, 28-55.

TWIDALE, C. R., DAILY, B. & FIRMAN, J. B. (1967).—Eustatic and climatic history of the Adelaide area, South Australia. Discussion. *J. Geol.* **75**, 237-242.

TWIDALE, C. R., SHEPHERD, J. A. & THOMSON, R. M. (1970).—Geomorphology of the southern part of the Arcoona Plateau and of the Tent Hill Region west and north of Port Augusta, South Australia. *Trans. R. Soc. S. Aust.* **94**, 55-67.

VOGT, J. (1953).—Erosion des sols et techniques de culture en climat tempere maritime de transition (France et Allemagne). *Rev. Géomorph. Dynam.* **4**, 157-183.

VON DER BORCH, C. C. (1968).—Southern Australian submarine canyons: their distribution and ages. *Mar. Geol.* **6**, 267-279.

WARD, W. T. (1965).—Eustatic and climatic history of the Adelaide area, South Australia. *J. Geol.* **73**, 592-602.

WARD, W. T. & JESSUP, R. W. (1965).—Changes of sea level in southern Australia. *Nature* **205**(4,973), 791-792.

WILLIAMS, G. E. (1969).—Glacial age of piedmont alluvial deposits in the Adelaide area, South Australia. *Aust. J. Sci.* **32**, 257.

WILLIAMS, G. E. (1973).—Late Quaternary piedmont sedimentation, soil formation and palaeoclimates in arid South Australia. *Z. Geomorphy.* **17**, 102-125.

WOPFNER, H. (1960).—On some structural developments in the central part of the Great Australian Artesian Basin. *Trans. R. Soc. S. Aust.* **83**, 179-193.

WOPFNER, H. & TWIDALE, C. R. (1967).— Geomorphological history of the Lake Eyre basin, Chapter 7: *In* J. N. Jennings & J. A. Mabbutt (Eds), "Landform Studies from Australia and New Guinea." (Australian National University Press: Canberra.)

WOPFNER, H., CALLEN, R. & HARRIS, W. K. (1973).—The Lower Tertiary Eyre Formation of the southwestern Great Artesian Basin. *J. geol. Soc. Aust.* **21**, 17-51.

ZEUNER, F. E. (1940).—"Dating the Past." (Methuen: London.)

CHAPTER 3

SOILS

by K. H. NORTHCOTE[*]

Introduction

One of the first observations recorded in the *Transactions of the Royal Society of South Australia* regarding soils near the Adelaide region is that of Tepper (1880) who discussed the so-called "Bay of Biscay Soil of South Australia and its formation". It is now well known that these gilgai soils with their capacity for seasonal movement (swelling when moist and shrinking when dry) are troublesome and require special care in the construction of foundations for buildings, roads etc. In contrast, they are natural highly productive cereal growing soils. The first formal study of soils in the region was that of Teale (1918)[1] who produced soil and vegetation maps of the Government Forest areas in the Hundred of Kuitpo. Moreover, his description of laterite on the plateau between Meadows and Blackfellows creeks was the first such description in South Australia.

The Adelaide region may be considered to cover about 6,000 km², extending from the Barossa Valley in the north to Cape Jervis in the South, and from the Gulf St Vincent in the west to the slopes of the Mt Lofty Ranges where they descend to the Murray Plains in the east. It has a varied natural environment in which there is still room for thoughtful planning and development. Although the contribution of soil data in this regard is often overlooked, pedologists of both the South Australian Department of Agriculture and the Division of Soils, CSIRO, realize its importance. They are currently carrying out comprehensive studies to provide basic information for the future. It is fitting that this should be so, as soil scientists working from the Adelaide region have contributed much to soils research not only locally but throughout Australia, and indeed the world. Many of these scientists were, or are, members of the Royal Society of South Australia. In particular, J. A. Prescott, C. S. Piper and J. K. Taylor were responsible for the development of modern soil science in Australia, beginning in the period 1926-27. From their early efforts, soils research in the modern sense, spread throughout Australia, eventually earning Australian soil scientists world-wide recognition, as shown by the choice of Adelaide for the venue of the 9th International Congress of Soil Science 1968.

Early inspiration came from Prescott's classic study of "The soils of Australia in relation to vegetation and climate". This report included one of the earliest soil maps of a continental area ever drawn. It showed the Adelaide region as being dominated by one soil group, the Red-brown Earths, as described for the first time by Prescott (1931). However, it is clear from statements in his report that Prescott was well aware of the presence of Rendzina, Black Earth, and podzolized soils within the region. This first soil map of Australia was published at a scale of about 1:20 million. Prescott's second map of Australia published at a scale of 1:10 million (Prescott 1944) showed essentially two soil zones for the region: Podzols in the hills; and Red-brown Earths on the plains. The soil map of Australia compiled by Stephens (1961) at a scale of 1:5 million showed three soil landscapes for the region: Lateritic Podzolic Soils in the southern dissected tableland area; Podzolic Soils in the hills north of that; and Red-Brown Earths on the plains. Finally, sheet 1 of the Atlas of Australian Soils (Northcote 1960) published at a scale of 1:2 million showed 22 compound soil landscapes composed of more than 20 principal profile forms (PPFs). Some of these PPFs, or major soil groups, recur in the different compound soil landscapes, but

---

[*] Division of Soils, CSIRO, Glen Osmond, S. Aust. 5064.
[1] Teale, E. O. (1918).—Soil survey and forest physiography of Kuitpo, South Australia. *Bull. Univ. Adelaide Dept For.* No. 6.

always in comparable landscape situations. Small areas of other PPF[s], not included in the foregoing figures, occur in special locations. In comparison with areas of a similar size in Europe or North America, three to four times as many major soil groups have been mapped in the Adelaide region. Around Paris and near Sacramento, California six major soil groups have been distinguished. The number of major soil groups in the Adelaide region as compared to the European and North American areas, suggest some significant difference in the natural history of the three continents. In the Adelaide region, and indeed in most of Australia, while the Pleistocene is recorded as a period of active block uplift and high rainfall (Taylor, Thompson & Shepherd 1974) there were no extensive ice-sheets to strip off the pre-Pleistocene products of weathering (?soils) as happened in Europe and North America (Hallsworth, Stephens & Northcote 1966). The importance of Pleistocene glaciation in the latter lands is that soil formation could begin anew, after the glacial periods, on extensive, fresh, soil parent materials. Whereas in the Adelaide region, soil formation continued on land surfaces that had been continually exposed to weathering, possibly even before Tertiary times, thus providing a much larger range of locally variable soil parent materials. Such a range has been demonstrated to occur by Stephens (1946) in one of the most significant papers ever written regarding soil genesis in southern Australia. His paper sets out both the sedentary and transported parent materials available for the other soil-forming agencies (particularly climate and the biological factors) to act upon, following the high rainfall, erosive phases of the Pleistocene. However, his diagram does not (and indeed was not intended to) account for the late-to post-Pleistocene geological formations of the coastal strip that are also parent materials of some soils found in the region.

Having established a *raison d'etre* for a range of soil parent materials within the Adelaide region, it seems desirable briefly to set out some of the properties of the soils before further consideration is given to their genesis.

Soil-landscapes of the Adelaide region

The Adelaide region may be divided into two sub-regions: the Coastal (or "Plains") strip that flanks Gulf St Vincent along the west-facing slopes of the Ranges or Hills which make up the other, and larger, sub-region.

These sub-regions and their broad soil-landscapes are shown in Figure 1. The Coastal sub-region is, from north to south, made up of the Adelaide, Noarlunga and Willunga basins or embayments. These are orientated along the line of the generally southwest to northeast trending faults and fault scarps that have produced the tilted fault-blocks of the western and northern portions of the Hills sub-region. Generally, the southern part of the Ranges, Fleurieu Peninsula, has been more stable. Here, there are more extensive areas of lateritic materials (Td2, Fig. 1) which, however, are also present on ridge crests, remnants of the summit high plain (Twidale 1976) throughout much of the Hills sub-region. They may be especially prominent in those parts of the angled southwest to northeast trending fault-blocks where these merge into the basins of the coastal strip. The tectonic events leading to the tilted, block-faulted landscape caused extensive erosion and consequent stripping of older weathering products, and their redeposition both in local lower-lying situations and in the Coastal sub-region. Removal of the older weathering products exhumed underlying weathered materials, and fresh basement rocks, which then became available for renewed weathering and soil formation, as indicated by Stephens' diagram (1946). Thus it is not surprising to find repeated patterns of soils and soil-landscapes throughout the region (Fig. 1 and Table 1). However, to complicate the situation, there is also the series of hard laterites that seem to range in age from the Pliocene to the Recent found in the exhumed Permian glacial valleys of the Mt Compass area (Maud 1972).

Peneplanation of the region during the Mesozoic and early Tertiary would have provided suitable conditions for extensive laterite development during the Pliocene or even earlier, except in areas of marine inundation in the coastal sub-region. Probably, laterite development on the summit high plain which resulted from a prolonged and intense weathering produced a soil similar to the Eleanor Sand of Kangaroo Island (Northcote 1946). Soils like the Eleanor have profiles that in sequence from the surface are characterised by sand, laterite, yellow-brown mottled kaolinitic clays, red mottled kaolinitic clays, white kaolinitic clays passing to white weathered rock (with rock structure evident) and finally to fresh, unweathered rock at several metres depth. Such deeply weathered profiles (DWP) occur not

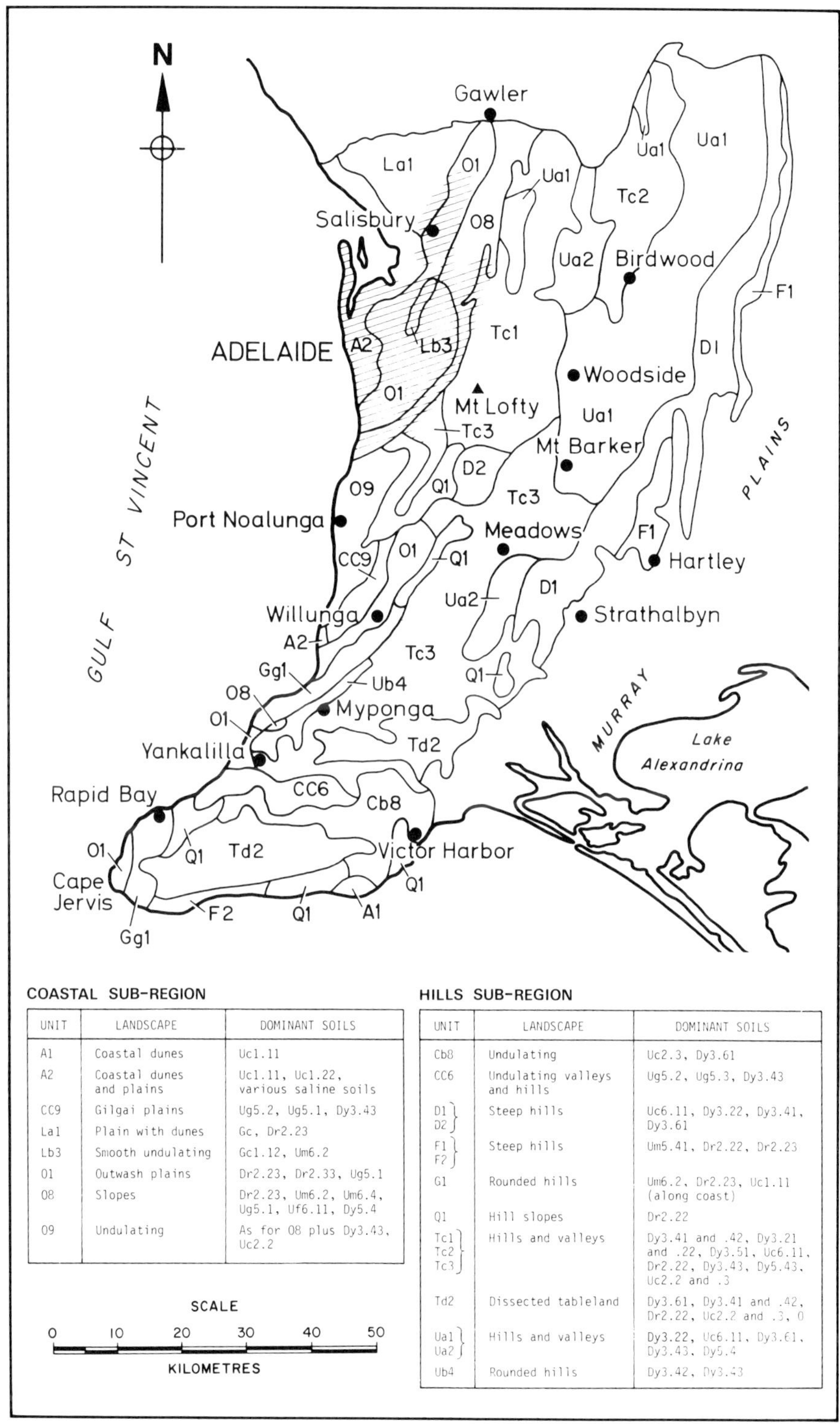

COASTAL SUB-REGION

UNIT	LANDSCAPE	DOMINANT SOILS
A1	Coastal dunes	Uc1.11
A2	Coastal dunes and plains	Uc1.11, Uc1.22, various saline soils
CC9	Gilgai plains	Ug5.2, Ug5.1, Dy3.43
La1	Plain with dunes	Gc, Dr2.23
Lb3	Smooth undulating	Gc1.12, Um6.2
O1	Outwash plains	Dr2.23, Dr2.33, Ug5.1
O8	Slopes	Dr2.23, Um6.2, Um6.4, Ug5.1, Uf6.11, Dy5.4
O9	Undulating	As for O8 plus Dy3.43, Uc2.2

HILLS SUB-REGION

UNIT	LANDSCAPE	DOMINANT SOILS
Cb8	Undulating	Uc2.3, Dy3.61
CC6	Undulating valleys and hills	Ug5.2, Ug5.3, Dy3.43
D1 D2	Steep hills	Uc6.11, Dy3.22, Dy3.41, Dy3.61
F1 F2	Steep hills	Um5.41, Dr2.22, Dr2.23
G1	Rounded hills	Um6.2, Dr2.23, Uc1.11 (along coast)
Q1	Hill slopes	Dr2.22
Tc1 Tc2 Tc3	Hills and valleys	Dy3.41 and .42, Dy3.21 and .22, Dy3.51, Uc6.11, Dr2.22, Dy3.43, Dy5.43, Uc2.2 and .3
Td2	Dissected tableland	Dy3.61, Dy3.41 and .42, Dr2.22, Uc2.2 and .3, O
Ua1 Ua2	Hills and valleys	Dy3.22, Uc6.11, Dy3.61, Dy3.43, Dy5.4
Ub4	Rounded hills	Dy3.42, Dy3.43

Fig. 1. Generalised soil-landscapes of the Adelaide region.

TABLE 1
Some properties of soils of the Adelaide region.

	Soils		Usual Occurrence			Morphological Properties	
PPF	GSG[1]	Location	Landscape	Generalized Parent Material	Profile	A horizon	B horizon
Uc1.11	CS	Coastal	Coastal Dunes	Calcareous Sands	U	A_1	—
Uc1.22	SS	Coastal	Inland Dunes	Siliceous Sands	U	A_1	—
Uc2.2	P	Hills	Dunes and Sand Sheets	Siliceous Sands	U	A_1, A_2	Colour B
Uc2.3	P	Hills	Sand Sheets	Siliceous Sands	U	A_1, A_2	Cemented colour B
Uc6.11	L	Hills	Slopes	Coarser-grained quartz-rich	U	A_1, pedal	
Um5.41	L	Hills	Slopes	Coarser-grained quartz-rich	U	A_1, firm	Firm, compact
Um6.2 Um6.4	T-R	Hills-Coastal	Slopes	Calcareous	U	Red, brown, black shallow pedal loams	
Uf6.11	R	Hills-Coastal	Slopes	Calcareous	U	Black, shallow pedal clays	
Ug5.1	BE	Coastal	Plains and Valleys	Clays	U	Black deep, cracking clays	
Ug5.2 Ug5.3	GB	Hills	Plains and Valleys	Clays	U	Grey and Brown, deep, cracking clays	
Gc1.12	SB	Coastal	Smooth, gentle slopes	Calcareous	G	Calcareous throughout but with horizon of $CaCO_3$ maxima	
Dr2.22	NB	Hills	Slopes	Finer-grained, felspar-rich	D	Hardsetting	Red pedal clays
Dr2.23	RB	Coastal	Slopes and Plains	Finer-grained, felspar-rich	D	Hardsetting	Red pedal clays calcareous
Dr2.33	RB	Coastal	Plains	Finer-grained, felspar-rich	D	Hardsetting	Red pedal clays calcareous
Dy3.21	YP	Hills	Slopes	Coarser-grained quartz-rich	D	Hardsetting	Mottled yellow pedal clays
Dy3.22	GP	Hills	Slopes	Coarser-grained quartz-rich	D	Hardsetting	Mottled yellow pedal clays
Dy3.41 Dy3.42	S	Hills	Slopes	Coarser-grained quartz-rich	D	Hardsetting	Mottled yellow pedal clays
Dy3.43	SS	Coastal	Slopes and Plains	Coarser-grained quartz-rich	D	Hardsetting	Mottled yellow pedal clays calcareous
Dy3.61	LP, YP	Hills	Ridge Crests	Deep weathering profile (DWP)	D	Hardsetting	Mottled yellow clays
Dy5.42 Dy5.43	SS	Hills	Slopes and Plains	Coarser-grained quartz-rich	D	Sandy	Mottled yellow clays

Notes:
1. Nearest equivalent Great soil Group (Stace et al. 1968): CS = Calcareous Sands; SS = Siliceous Sands; P = Podzols and Humus Podzols; L = Lithosols; R = Rendzina; T-R = Terra Rossa-Rendzina-like soils; BE = Black Earths; GB = Grey, Brown and Red Clays; SB = Solonized Brown Soils; NB = Non-calcic Brown Soils; RB = Red-brown Earth; YP = Yellow Podzolic Soils; GP = Grey-brown Podzolic Soils; S = Soloths; SS = Solodized Solonetz and Solodic Soils; SP = Lateritic Podzolic Soils.

TABLE 1
Some properties of soils of the Adelaide region.

	Physical Properties			Chemical Properties		
Below B horizon	Moisture Regime	Shrink-swell capacity	Bearing capacity	SRT	Ex. Cats[3]	Deficiencies Reported[5]
various	Highly permeable	nil	Stable after compaction	Alkaline	—	N, P, K, Cu, Co, Zn, B, Fe, Mn
various	Highly permeable	nil	as above	Neutral	—	N, P, K, Zn, Ca, Mg
various	Highly permeable	nil	as above	Acid	— 1[4]	N, P, K, Ca, Mg, Cu, Zn
various	Seasonal waterlogging	nil	Unstable when wet	Acid	— 8 in pan[4]	N, P, K, Ca, Mg, S, Cu, Co, Mo, Zn
Country rock	Permeable	nil	Stable	Acid	3-4	N, P, K, plus
Country rock	Permeable	nil	Stable	Acid	—	N, P, K, plus
Highly calcareous	Permeable	Small	Moderate[2]	Acid, neutral, alkaline	—	N, P, Zn, Cu, Mn
Highly calcareous	Permeable	Small	Moderate[2]	Alkaline	—	P
various	Low permeability	Large	High[2]	Alkaline	>30	P, N, Zn, Mo, S, Mn
various	Low permeability	Large	High[2]	Alkaline	>20[4]	P, N, Zn, Mo, S, Mn
various	Moderate permeability	nil	Low	Alkaline	8-15[4]	P, N, Zn, Mn, Fe
Phyllites, slates, shales	Permeable	Moderate in B	Moderate to High	Neutral	20-25	N, P, Zn
various	Permeable	Moderate in B	Moderate to High[2]	Alkaline	15-30	N, P, Zn
various	Less permeable than Dr 2.23	Moderate in B	as above	Alkaline	15-30	N, P, Zn
Sandstones, quartzites, schists	Permeable	Moderate in B	High	Acid	10-20	N, P,
as above	Permeable	Moderate in B	High	Neutral	10-20	N, P,
as above	Seasonal waterlogging	Moderate in B	High	Acid, neutral	5-20	N, P, K, Ca, Mo, S, Cu, Zn
various	Seasonal waterlogging	Moderate in B	High	Alkaline	15-30	N, P, Mo, S, Cu, Zn
Mottled and pallid clays	Seasonal waterlogging	Small in B	High	Acid	5-10	P, N, K, Mo, Cu, Zn
various	Seasonal waterlogging	Small to moderate in B	High	Neutral, alkaline	6-18	N, P, K, Zn, Cu, Mo, Ca

2. Calcareous layers where soft and powdery have low bearing capacity.
3. Total metal cations (Ca + Mg + K + Na) in m.e. per 100 g of soil for upper part B horizon.
4. Data from similar soils elsewhere.
5. On these soils, PPFs, but not necessarily in this Region.

only as present day soils, but their dissection products have provided parent materials for a wide range of new soils both on the truncated remnants and on transported materials of DWP. The ironstone gravelly Dy3.61 soils e.g. the Kuitpo, Hahndorf and Kangarilla soils of Rix & Hutton (1953) represent the first stage in truncation of DWP. They are found on remnants of the summit high plain throughout the Hills sub-region, thus forming the oldest soil-landscape in the region. Theoretically, it should be possible to list the soil-landscapes as they developed. However, data are generally insufficient to permit this.

Some properties of soils of the Adelaide region

Table 1 sets out briefly the usual occurrence of some common soils together with some of their morphological, physical and chemical properties. Generally, it is these and other properties that most directly affect soil use. These properties result from the natural history of soil formation. However soil properties can be observed, measured and used without knowledge of soil formation. Nevertheless, insight into natural history improves understanding and appreciation of why such soil properties exist and how they are likely to react to new circumstances such as those imposed by human activities.

Morphological properties

The four primary profile forms (Northcote 1971) are represented in the Adelaide region. The O soils (peats) that are a minor component in the Mt Compass area are not listed in Table 1. The U soils are those in which texture of the mineral soil remains nearly constant throughout the profile; G profiles show a gradual increase in apparent clay content with depth; while D profiles have a marked change in texture between their A and B horizons—sand, sandy loam or loam A horizons passing to clay B horizons. Table 1 shows that not only do the complex Dy3.61 soils derived from DWP occur in the region, but also the simple Uc1.11 and Uc1.22 soils, in which a slightly organic A_1 horizon is the main pedological feature. There are a variety of soils between these extremes.

Physical properties

Physical properties vary in accordance with morphological ones. They range from the highly permeable sand soils, Uc, through the well aerated and drained Dr soils to Dy soils that have seasonal (late winter to early summer) waterlogging, due to water perching on

the clay subsoil. Sand soils with subsoil hardpans, Uc2.3, are also waterlogged seasonally. Such conditions create land-use problems for the farmer and for the home-builder. Then there are the "Bay of Biscay" or Cracking Clay soils, Ug5, with their dramatic capacity to shrink and swell. The Dr2.23 and .33 soils also have clay subsoils that shrink and swell, often by as much as 15 cm in a 2 m thickness of clay, but their loamy surface soils, especially if 30 cm or more thick, form a buffer between such subsoil movements and surface structures. Another important property of some soils in the Coastal sub-region and also on the slopes to the Hills is their low bearing capacity caused by soft, powdery, calcareous layers which collapse when wet and under load e.g. from building foundations. This problem may be worse in those soils with attendant salinity or sodicity as is often the case in Gc1.12 soils.

Chemical properties

The soil reaction trend (SRT) defines the pH values for the solum (A + B horizons), and shows that the soils range from acid to neutral and alkaline. Moreover, the data for the exchangeable metal cations serves to illustrate, and at the same time extend the wide chemical variations found in the soils. Kaolin is the dominant clay mineral in Dy3.61, and is very common in the other Dy soils, apart from Dy3.22 which, like the Dr soils, is dominated by illite. The Ug5 and Uf6 soils have very mixed clay mineral assemblages in which montmorillonite and/or randomly interstratified materials predominate. These data indicate that considerable variations existed in the parent materials from which the soils developed.

Early in the history of nutrient element deficiencies in soils, Anderson (1946) found molybdenum to be beneficial to subterranean clover grown on Dy3.61 soils (Rix & Hutton 1953). Since that time knowledge of the nutritional requirements of plants has developed considerably. Table 1 shows deficiencies of nutrient elements reported for PPF[s] from records throughout Australia for a range of plants (Northcote, Hubble, Isbell, Thompson & Bettenay 1975), and serves to illustrate differences in soil plant relationships found in the Adelaide region.

Soil genesis

Not only does Table 1 demonstrate the range and variation in soil properties found in the region, but it also suggests the diversity of the natural history of the soils.

The development of soil is determined by the five soil forming factors: parent material, topography (landscape), time, climate and vegetation. Thus each science pertinent to the individual soil forming factors must establish sound data before pedological studies can be furthered. Geology and geomorphology are particularly important, especially under Australian conditions, for reasons already discussed. Often the pedologist has to make his own interpretation of these sciences where data are either incomplete or lacking. Unfortunately, this may mean that the pedologist can only study a limited area, and thus his data may not be sufficient to provide an interpretation for areas of the size of the Adelaide region. Indeed, it is a salutary fact that only now are sufficient geological and geomorphological data becoming available to allow pedological interpretations. Of course, this can be a reciprocal matter as both geologists and geomorphologists can benefit from the closer studies made by pedologists. However, the soil maps of the region as a whole are too broad, so that an integrated natural history of soil genesis must await the present work being undertaken to compile soil-landscape maps at a sale of 1:50,000. Nevertheless, some review of the effects of the soil-forming factors in the light of present knowledge is justified.

Parent materials

It is generally agreed, Taylor, Thomson & Shepherd (1974), Wright (1973), Maud (1972), Ward (1966), Lang (1965), Litchfield (1960, 1951), Jackson (1957), Aitchison, Sprigg & Cochrane (1954), and Rix & Hutton (1953) that parent material has exerted a dominant influence on the formation of the soils. This conclusion agrees with the hypothesis put forward by Brewer (1954) that parent material, not climate, controls the distribution of soils in southeastern Australia which, in turn, agrees with Stephens' (1946) earlier assessment of pedogenesis following the dissection of lateritic regions in southern Australia. A major difficulty is to define soil parent material. One approach is to deduce the general nature of the soil parent material from the soil profile as shown in Table 1. On this basis, five generalized soil parent materials are recognized for the region, and may be used as a means for grouping formations reported as soil parent materials by various workers in the

region, as shown in Table 2. Brewer (1954) pointed out the need, and indeed suggested a method, of more adequately classifying geological materials (formations) from the standpoint of their propensity to weather to soil. This opinion is supported by Ward (1966) who pointed out that the rate and kind of soil development depends on the lithology and mineralogical composition of rocks. Indeed, the diversity and complexity of entries in Table 2 clearly demonstrates the urgency for better definition of soil parent materials, if studies on soil genesis are to be furthered. At present the most useful designations for soil parent materials are lithological, Table 2. But even so, the stage of weathering of the rock itself is important. For example, Northcote & de Mooy (1957) found Dy3.22 soils on "fresh" mica schist but Dy3.61 soils on weathered schists in the Barossa Hills immediately north of the Adelaide region. Again, unconsolidated sediments are often more difficult to classify in a way that is meaningful for soil studies than are basement rocks. Special mention should be made of Firman's (1969) time-stratigraphic approach which shows six alternative layer assemblages for the different Red-brown Earths (Dr2.23 and .33) of the Coastal sub-region near Adelaide, Table 2. Their only common layers are his first two; that is, slope deposits and a thin clay layer. But as these are also the first two layers for his Podzolic, Black Earth, Solodic, Terra Rossa, Rendzina, Yellow Podzolic and Red Podzolic soils of the coastal sub-region, it is virtually meaningless in a soil parent material sense. While such time-stratigraphic units may be interesting, they are evidently not useful for determining the nature of soil parent materials, except perhaps for special cases like the Gc1.12 soils where Pooraka Formation is the necessary second layer, rather than the thin clay layer. But even here the pedologoist would be better served if the mineralogy of Pooraka Formation could be defined.

Aeolian accessions are claimed to have influenced soil parent materials in the region. Ward (1966) invokes an aeolian origin for his Ngaltinga Formation on which cracking clays, Ug5.1, formed; whereas Taylor, Thompson & Shepherd (1974) consider such soils were formed from clay alluvium (Keswick and/or Hindmarsh formations). However, Wells (1965) makes the reasonable suggestion that these soils formed on late Tertiary or early Quaternary clay plains. Such plains may have

TABLE 2.
Formations reported as parent materials for some soils listed in Table 1.

Generalized Soil Parent Materials	Soils	Formations Reported
Sands	Uc1.11	Semaphore[1,6]; Foredune[4]
	Uc1.22	Fulham[1]; Semaphore/Fulham[6]
	Uc2.2	Myponga[3]; Seaford[4]; Laterite on Oligocene(?) fluviatiles[5];
	Uc2.3	Myponga[3]; dune sands of Taringa[4]
Calcareous Materials	Um6.2-Um6.4	Calcrete over calcareous rocks e.g. dolomite, slates, limestones, calcareous shales and siltstones[1]; calcrete over limestone[3]; Ngaltinga, also slates, limestones[4]; calcareous shales[5]; slope deposits/thin clay layer/Bakara Calcrete/Loess/basement rocks[6]
	Gc1.12	Aeolian calcareous earth[1]; slope deposits/Pooraka/Bakara Calcrete/Loess/basement[6]; Loessial lime[7]
Clays	Ug5.1	Alluvial deposits—Keswick and/or Hindmarsh[1]; Alluvium from calcareous shales[3]; Aeolian deposit—Ngaltinga[4]; Alluvium with accession of lime[5]; slope deposits/thin clay layer/older carbonate horizons/Keswick[6]
Finer-grained felspar-rich	Dr2.22	Shale[3]; Shale[5]
	Dr2.23 Dr2.33	?Pooraka[1]; Calcareous bedrock[1]; Ochre Cove with aeolian lime[4]; Christies Beach (alluvium from basement rocks)[4]; basement rocks with aeolian lime[4]; colluvial-alluvial fan outwash[7]; slope deposits/thin clay layer/Loveday Carbonate/older carbonate/old red clay/Hindmarsh[6]; Slope deposits/thin clay layer/Loveday Carbonate/Pooraka/older carbonate/old red clay/Hindmarsh/basement rock[6]; slope deposits/thin clay layer/older carbonate/Keswick/Tertiary Sand/basement rock[6]; slope deposits/thin clay layer/Bakara Calcrete/basement rock[6]; slope deposits/thin clay layer/Loveday Carbonate/Pooraka[6]; slope deposits/thin clay layer/basement rocks[6]
Coarser-grained quartz-rich	Dy3.21, Dy3.22	Mica Schists[8]
	Dy3.41	Sandstones and siltstones[1]; Siliceous pre-Permian rocks[3]; Sandstones, quartzites and schists[5]; slope deposits/thin clay layer/Tertiary Sand/basement rocks[6]
	Dy3.43	Basement slates and quartzites[4]; Ochre Cove with aeolian lime[4]; Taringa[4]; Kurrajong;[4] Transported material with accession of lime[5]
	Dy3.61	Truncated lateritic profiles on pre-Permian, Permian and Tertiary rocks and sediments[3]; Pre-weathered (lateritized) basement slates and quartzites[4]; Truncated relict laterite on quartzites, sandstones and shales[5]; ?slope deposits/thin clay layer/ferricrete/Tertiary Sand/basement rocks[6]; Gneiss and schist affected by previous lateritic processes[8]

References:
1. Taylor, Thompson & Shepherd (1974); 2. Wright (1973); 3. Maud (1972); 4. Ward (1966); 5. Rix & Hutton (1953); 6. Firman (1969); 7. Aitchison, Sprigg & Cochrane (1954); 8. Northcote & de Mooy (1957).

resulted from the dessication of earlier alluvial or lacustrine lakes such as that of the Highbury-Golden Grove area (Twidale 1976).

Calcium-magnesium carbonates are the other material for which an aeolian origin is claimed. Crocker (1946) postulated wide exposures of the continental platform during Pleistocene low sea-levels, and that lime (carbonates) winnowed from exposed sub-coastal shelly material became airborne as calcareous dust that was then deposited widely. Taylor, Thompson & Shepherd (1974) quoting Firman (1969) agree that a blanket of wind blown carbonate was deposited to give the Bakara Calcrete which subsequently eroded to provide the Pooraka Formation of the North Adelaide (City-Hampstead) plateau. Largely on the basis of the location of the scarp of the Para block, they suggest that the carbonate dust was erratically effective for not more than about 16 km from the present coast. Certainly, the very low contents of metal cations, including calcium, (Dy3.61, Dy3.41, Table 1), in the soils of the Hills sub-region suggest that carbonate dust was not deposited at that distance from the coast, unless it was removed by rain almost as rapidly as it was deposited. Probably, an assessment of the full import of carbonate dust in soil formation of even the Coastal sub-region cannot be made with any certainty as the possible sources of carbonates are so varied: many of the exposed Tertiary and basement rocks are calcareous for example. Rix & Hutton (1953) identified the shallow Um6.43 soil, Clarendon loam, as being developed from calcareous Precambrian shales. Further evaluation of the parent materials of the calcareous soils in the region seems justified.

In conclusion, it seems that there is a range of soil parent materials of different ages, some having been deeply weathered over a long time. However, all the soils exist now, and must be accredited with being present day soils, even though their imperfectly defined parent materials are distinctly not so. Indeed, the full importance of DWP does not seem to have been appreciated, or investigated, by most workers; for example, the ultimate source of the parent material for the Uc2 soils could well be the transported former surface material (A horizon sand) from DWP which would agree with Northcote & Tuckers' (1948) finding for Soil Type 17, a Uc2 soil on Kangaroo Island. Again, Taylor & O'Donnell (1932) reported the Burbrook soil series (Dy3.41) as "the universally occurring soil over a wide range of conditions". The description, taken in conjunction with its low content of exchangeable cations and kaolinitic clay, suggests that these soils could have been formed from the pallid zone—the weathered (kaolinized) sandstone part of Stephens' (1946) pedogenic sequence rather than from "fresh" sandstone as seems to be implied by Rix & Hutton (1953). Further evaluation of soil and parent material relationships are warranted.

Topography

While all soil workers in the region have recognized with Maud (1972) that a close relationship exists between geomorphology and soil, no clear analysis has been given. It is not always certain what different authors include in landscape or topography. There are two main aspects of landscape that influence soil: one is the individual land form which is a structural entity, e.g. remnants of the summit high plain; and the other is the local relief, or slope elements, occurring on different parts of that landform. Thus Wright (1973) found Dr soils on

shale in mid-slope positions, but Dy soils on the wetter, lower, more gently sloping sites. Such soil differences on different slope elements of a topographic unit, or land form, are not always recognized, yet are important considerations in soil genesis and in land use. Again, springs and soaks that may be associated with some slope elements in the Ranges, bring dissolved salts to the surface and influence soils as pointed out by Rix & Hutton (1953).

It is also vital to determine the significance of the land form *in toto* regarding the geological history of the area and thus the soil parent materials. This leads to a further unfortunate conclusion: namely, that all too often the modern day soil is tacitly assumed to be the same soil as that formed when a particular landform first developed. Clearly this is not so as soils change due to chemical and biological influences. Surely then it is only the soil parent material that can be fully equated with a given landform. Otherwise a circular argument results, as epitomised, for example, by Ward's (1966) statement that "identification of groundsurfaces has particular value in that it permits both the correlation and age classification of all aggraded and eroded surfaces characterized by soils that owe their differences to differences in age". On the contrary, it is important for an understanding of soil genesis to evaluate land form in its own right, and to integrate this with the geological record so that soil parent materials can be defined. It is the clear definition of soil parent materials that will enable more useful pedological studies to be made.

There are many interesting and important soil-landscape relations in the region. That of ironstone gravelly Dy3.61 soils with remnants of the former summit high plain has already been mentioned. In the Hills sub-region, it is noteworthy that on "fresh" rock hillsides that were either above the general level of the summit high plain e.g. Mt Panorama, Mt Barker, or were exposed by subsequent erosion, similar soils occur depending on the mineralogy of the rock e.g. Kondoparinga Loam (Dr2.22) on shales and slates. Valleys have their distinctive soils e.g. Dy3.43, Dy5.42 and .43 in the Hundred of Kuitpo, whilst the contrasting, depositional land forms of the Coastal sub-region, soil-landscape relations are equally precise. For example, the Uc1 soils of the coastal dunes are distinctive. There is an excellent correlation between the presence of Dr2.23 and .33 soils (Red-brown Earths) and the outwash plains of the Adelaide Basin, whereas the outwash plains of the Willunga Basin have either Dr2.23 or Dy3.43 soils depending on the mineralogy of the rocks providing the material for the different colluvial-alluvial deposits. And there is a close relationship between the occurrence of Gc1.12 soils and the North Adelaide plateau. Other soil-landscape relationships seem less clear in the Coastal sub-region because Ward (1966) claims that higher stands of the sea above present sea level during the Quaternary are involved in the Noarlunga and Willunga basins; but this is denied (Twidale, Daily & Firman 1967), except for the 8 m and 3 m stands (Twidale 1976). Some re-evaluation seems warranted.

Time

Most soil-workers in the region bring in the time factor at least by implication through geological history, but this relates more precisely to soil parent material, and not necessarily to the present day soil. Ward (1966) has introduced stages of profile differentiation including the "degree of textural contrast" to support his claims. Unfortunately, Ward (1966) does not state how he determines this, nor does he present supporting analytical data. Particle-size analysis for the Barossa Valley soils (Bond 1955) show that for the Dr2.23 soils of the eastern inclined plain (Northcote, Russell & Wells, 1954) derived from marble and calc-silicate rocks, the ratio of clay-sized particles in A_2 horizons to those in the upper B horizons ranges from 1:4 to 1:8. Whereas similar data for the Dy5.43 soils of the contiguous southern inclined plain derived from mica schists and related rocks has ratios ranging from 1:12 to 1:46. Here, the "degree of textural contrast" seems to relate to differences in soil parent materials rather than soil age. Ward's approach assumes that the textural contrast found in duplex soils is due entirely to eluviation of clay from A horizons to B horizons. Following the work of Oertel (1961, 1974) and others (e.g. Brewer 1968), this cannot be considered indisputable, so questioning the validity of profile morphology of duplex soils being used as a determinant of soil age. Firman's (1969) stratigraphic approach to soil occurrence seems to conflict with Ward's ideas on soil development in the region. Yet, if Firman is right, then it is indeed strange that "Dr" A horizons always occur on "Dr" B horizons and never on "Dy" B horizons, except for areas of man-made erosion. Thus, the duplex

nature of many Australian soils remains a pedological conundrum that requires solution before the natural history of the soils of the Adelaide region, and indeed many other areas in Australia, can be resolved.

Climate

Taylor, Thompson & Shepherd (1974) state that the nature of the present soils is related to the climate within the Coastal sub-region in the vicinity of Adelaide. They further point out that the climate is generally considered to be essentially unchanged since the end of the last great arid period, 4–5,000 years ago, and therefore it seems reasonable to assume that the majority of soils with developed profiles are associated with the present climatic regime. This concept could well apply to the whole of the Adelaide region. However, the significance of the great fluctuations in rainfall typical of the region, e.g. Adelaide's average annual rainfall for the period 1839–1966 was 530 mm, but during that time ranged from a low 257 mm to a high 786 mm (Bureau of Meteorology 1971), has not been considered. It could well cause some re-evaluation of the concept of the "great arid period".

Few soil workers have considered the effect of climate on the soils of the region apart from Ward (1966) who concluded that differences in rainfall did not correlate with the distribution of his soil classes. Variation in rainfall (see Chapter 4) must have affected present soils, possibly, for example, by the relative intensities of local leaching (Northcote 1974) which depends here largely on the average annual rainfall. Most soil studies in the region have not analysed climate in relation to soils. However, Ward (1966) states that the boundary between calcareous and non-calcareous soils is determined by the 685 mm isohyet, and Taylor, Thompson & Shepherd (1974) state that the development of Podzolic soils (acid Dy soils) on all kinds of parent rock is tied approximately to the 675 mm isohyet as a minimum rainfall, apart from soils on calcareous rocks. They probably mean the same thing and infer the same isohyet. However, there is clearly more to be learnt from further study of the climate, including rainfall. For example, carbonate nodules occur in Dy3.43 soils in valleys near Meadows where the rainfall is over 800 mm per annum. Again, the occurrence of neutral Dr2.22 soils (Kondoparinga Loam) as recorded by Rix & Hutton (1953) on shales and slates where the average annual rainfall is about 800-1,000 mm as compared with acid Dr2.21 soils recorded on slates by Wright (1973) where the average annual rainfall is over 1,000 mm is worthy of note. Studies of climate involving the leaching factor may provide further insight into the great variety of Dy soils and their landscapes. However, the clear definition of parent materials would be a necessary prerequisite. Other climatic factors besides rainfall could be significant. In particular, temperature fluctuations are likely to influence not only the weathering of rocks and soil parent materials generally, but also the biological component in soil formation.

Vegetation

Rix & Hutton (1953) have commented on the complexity of the vegetation pattern in the Hundred of Kuitpo; a comment that could be repeated for the Adelaide region as a whole. Maud (1972) found an extremely well-marked ecological relationship between vegetation, soils and climate. Indeed, the available information strongly suggests that the soils and the climate have determined the distribution of the vegetation; that is, vegetation does not seem to have been a determinant of the soils. Specht & Perry (1948) and Rix & Hutton (1953) found that microclimatic conditions due to aspect and topography are vital, and give rise to wide-ranging variations in plant communities.

No detailed ecological studies in which vegetation, soils and climate are mapped and evaluated separately, and then compared to establish relationships, have been made in the region. However, a general indication of vegetation in relation to soils seems to include the following:

1. An extensive sclerophyll forest characterized by *Eucalyptus obliqua* on shallow Uc and Um soils and acid Dy soils from the more siliceous basement rocks and DWP where the rainfall is 635 mm or more. Occurrences on Dy soils derived from DWP often have a more open undergrowth.
2. More stunted sclerophyll communities of *E. baxteri* and/or *E. cosmophylla* on large residual plateau areas of Dy soils on DWP or on hard laterite (Maud 1972).
3. *E. fasciculosa* sclerophyll communities often associated with *Banksia* spp. occur on Uc2 soils as at Blewitt Springs and nearby areas.
4. Dwarf *E. baxteri* sclerophyll communities usually with associated *Banksia* spp. occur on Uc2 soils of the Mt Compass and adjacent areas.
5. *E. leucoxylon* savannah woodlands on slopes of Dr2.2 soils derived from shales, phyllites etc.
6. *E. leucoxylon* savannah woodlands merge into *E. camaldulensis* savannah woodlands in valleys and along drainage lines. Soils are mainly neutral and alkaline Dy soils formed on transported parent materials.
7. Savannah woodlands apparently characterized by *E. odorata* and/or *E. camaldulensis* were usual on the Dr2.23 soils of the Coastal sub-region.

This list is not intended to be precise or exhaustive, nor to include special habitats such as the peat swamps of Mt Compass. It is clear that future ecological studies would be worthwhile, and that time to carry these out has already passed in large areas, such as the City of Adelaide and its suburbs.

Conclusions

Within the Adelaide region soil distribution is complex and there is a wide range and variability in soil properties. It is generally appreciated that the basis for this situation resides in the great mineralogical variations of the basement rocks, and the superposed geological and geomorphological events: particularly Permian glaciation, Mesozoic peneplanation, Tertiary deep weathering with laterite formation and marine incursions followed by upward block faulting in the Ranges, marine sedimentation and changes in sea level together with the possible incorporation of aeolian material in the Coastal strip. All these events have contributed to the diversity of soil parent materials from which both the present day soils and their subsolum materials (or the materials below the soil) have been developed. Unfortunately, soil parent materials and subsola have not been clearly defined. Although Stephens' diagram (1946) and Rix & Hutton's extension (1953) provide a framework for unravelling soil relationships in the Ranges, no similar basis exists for the plains, where moreover, the studies of Ward (1966) and Firman (1969) are not in agreement. Furthermore Stephens' principles derived for the Ranges have not been widely applied to the soils of the lowlands. Thus the relationships between soils and the other factors of the environment have not been clearly formulated. Indeed, the only soil study that covers the whole region (Northcote 1960) does so at the very broad scale of 1:2 million. This is not nearly detailed enough to provide adequate soil data to demonstrate soil relationships. Nor is it detailed enough to provide the necessary basis for correct land use in a region under great population pressure, where possible uses range from rural to urban and recreational. Clearly the objective of providing soils data at a scale of 1:50,000 for the region has considerable practical merit as well as much scientific interest.

Acknowledgments

I wish to express grateful appreciation to my colleagues, Mr D. J. Chittleborough of the South Australian Department of Agriculture and Fisheries, and Messrs J. T. Hutton, M. J. Wright and N. B. Billing of the Division of Soils, CSIRO for their helpful comments and suggestions on the original draft of this paper. However the statements remain my responsibility. I also wish to express my gratitude to Mr N. B. Billing and Mr G. E. Rinder of the Division of Soils, CSIRO for the compilation and drafting of Figure 1.

References

AITCHISON, G. D., SPRIGG, R. C. & COCHRANE, G. W. (1954).—The soils and geology of Adelaide and suburbs. *Bull. geol. Surv. S. Aust.* No. 32.

ANDERSON, A. J. (1946).—Molybdenum in relation to pasture improvement in South Australia. *J. Coun. Sci. Ind. Res., Aust.* **19**, 1-15.

BOND, R. D. (1955).—The laboratory examination of soils from zone 1, Nuriootpa area, Barossa district, S.A. *CSIRO Aust. Divn Soils Divnl Rep.* No. 4/55.

BREWER, R. (1954).—Soil parent material. *Aust. J. Sci.* **16**(3), 134-138.

BREWER, R. (1968).—Clay illuviation as a factor in particle-size differentiation in soil profiles. *Trans. 9th Int. Congr. Soil Sci., Adelaide.* **4**, 489-499.

BUREAU OF METEOROLOGY (1971).—"Climatic Survey, Adelaide, Region 1—South Australia." (Cwlth Aust., Dept. Interior: Canberra).

CROCKER, R. L. (1946).—Post-Miocene climatic and geologic history and its significance in relation to the genesis of the major soil types of South Australia. *Bull. Coun. Sci. Ind. Res. Aust.* No. 193.

FIRMAN, J. B. (1969).—Stratigraphy and landscape relations of soil materials near Adelaide, South Australia. *Trans. R. Soc. S. Aust.* **93**, 39-54.

HALLSWORTH, E. G., STEPHENS, C. G. & NORTHCOTE, K. H. (1966).—The soils of Australia. Yearbook of the Commonwealth of Australia. (52), 873-879.

JACKSON, E. A. (1957).—Soils of Portion of the Mt Lofty Ranges, S.A. *CSIRO Divn Soils, Soils and Land Use Ser.* No. 21.

LANG, D. (1965).—Data of the geomorphology and soils of the Yundi area, S.A. *CSIRO Divn Soils Divnl Rep.* No. 3/65.

LITCHFIELD, W. H. (1951).—Soil survey of the Waite Agricultural Research Institute, Glen Osmond, S.A. *CSIRO Divn Soils Divnl Rep.* No. 2/51.

LITCHFIELD, W. H. (1960).—Soils on the western slopes of the Mt Lofty range near Adelaide and Elizabeth, South Australia. *CSIRO Divn· Soils Divnl Rep.* No. 8/59.

MAUD, R. R. (1972).—Geology, geomorphology, and soils of central County Hindmarsh (Mount Compass-Milang), South Australia. *CSIRO Soil Publ.* No. 29.

NORTHCOTE, K. H. (1946).—A fossil soil from Kangaroo Island, South Australia. *Trans. R. Soc. S. Aust.* **70**(2), 294-296.

NORTHCOTE, K. H. (1960).—"Atlas of Australian Soils, Sheet 1 Port Augusta, Adelaide, Hamilton area." (CSIRO and Melbourne University Press: Melbourne).

NORTHCOTE, K. H. (1971).—"A factual key for the recognition of Australian Soils." 3rd Edn (Rellim: Glenside, S. Aust.).

NORTHCOTE, K. H. (1974).—Relationships of some hardsetting loamy soils with red clay subsoils in south-eastern Australia. *Trans. 10th Int. Congr. Soil Sci., Moscow* **6**(1), 257-263.

NORTHCOTE, K. H., HUBBLE, G. D., ISBELL, R. F., THOMPSON, C. H. & BETTENAY, E. (1975).—"A Description of Australian Soils." *CSIRO.*

NORTHCOTE, K. H. & DE MOOY, C. J. (1957).—Soils and land use in the Barossa district, South Australia. *CSIRO Divn Soils, Soils and Land Use Ser.* No. 22.

NORTHCOTE, K. H., RUSSELL, J. S. & WELLS, G. B. (1954).—Soils and land use in the Barossa district, South Australia zone 1. The Nuriootpa Area. *CSIRO Divn Soils, Soils and Land Use Ser.* No. 13.

NORTHCOTE, K. H. & TUCKER, B. M. (1948).—A soil survey of the Hundred of Seddon and part of the Hundred of MacGillivray, Kangaroo Island, South Australia. *Bull. Coun. Sci. Industr. Res. Aust.* 233.

OERTEL, A. C. (1961).—Pedogenesis of some red-brown earths based on trace-element profiles. *J. Soil Sci.* **12**, 242-258.

OERTEL, A. C. (1974).—The development of a typical red-brown earth. *Aust. J. Soil Res.* **12**(2), 97-105.

PRESCOTT, J. A. (1931).—The soils of Australia in relation to vegetation and climate. *Bull. Coun. Sci. Ind. Res. Aust.* 52.

PRESCOTT, J. A. (1944).—A soil map of Australia. *Bull. Coun. Sci. Ind. Res. Aust.* 177.

RIX, C. E. & HUTTON, J. T. (1953).—A soil survey of the Hundred of Kuitpo in the Mt Lofty Ranges of South Australia. *Bull. S. Aust. Land Tax Dept* No. 1.

SPECHT, R. L. & PERRY, R. A. (1948).—The plant ecology of part of the Mount Lofty Ranges. *Trans. R. Soc. S. Aust.* **72**, 91-132.

STACE, H. C. T., HUBBLE, G. D., BREWER, R., NORTHCOTE, K. H., SLEEMAN, J. R., MULCAHY, M. J. & HALLSWORTH, E. G. (1968).— "A Handbook of Australian Soils." (Rellim: Glenside, S. Aust.).

STEPHENS, C. G. (1946).—Pedogenesis following the dissection of lateritic regions in southern Australia. *Bull. Coun. Sci. Ind. Res. Aust.* No. 206.

STEPHENS, C. G. (1961).—The soil landscapes of Australia. *CSIRO Soil Publ.* No. 18.

TAYLOR, J. K. & O'DONNELL, J. (1932).—The soils of the southern portion of the Hundred of Kuitpo. *Trans. R. Soc. S. Aust.* **56**, 3-14.

TAYLOR, J. K. THOMPSON, B. P. & SHEPHERD, R. G. (1974).—The soils and geology of the Adelaide area. *Bull. geol. Surv. S. Aust.* No. 46.

TEPPER, J. G. O. (1880).—The "Bay of Biscay" soil of South Australia and its formation. *Trans. R. Soc. S. Aust.* **3**, 91-98.

TWIDALE, C. R. (1976).—Geomorphological evolution. In C. R. Twidale, M. J. Tyler & B. P. Webb (Eds): "Natural history of the Adelaide region." (Royal Society of South Australia: Adelaide.)

TWIDALE, C. R., DAILY, B. & FIRMAN, J. B. (1967).—Eustatic and climatic history of the Adelaide area, South Australia: Discussion. *J. Geol.* **75**, 237-242.

WARD, W. T. (1966).—Geology, geomorphology and soils of the south-western part of County Adelaide, South Australia. *CSIRO Soil Publ.* No. 23.

WELLS, C. B. (1965).—Patterns in soil geography in and near Adelaide, South Australia. *Trans. R. Soc. S. Aust.* **89**, 117-123.

WRIGHT, M. J. (1973).—Soils of the Mt Lofty Botanic Garden, South Australia. *CSIRO Divn Soils Techn. Pap.* No. 18.

CHAPTER 4

CLIMATE

by Peter Schwerdtfeger*

Historical introduction

South Australia is a land of major climatic contrasts and Adelaide itself occupies an area of which the climate is quite atypical of that of the State as a whole. Although the locations of all of the Australian colonial settlements were indirectly determined by climatic factors, particularly rainfall, in no other case was the choice of the initial site so strongly influenced by these considerations. The location of Adelaide was selected by Colonel William Light within the space of a few weeks during the late spring and early summer of 1836. Very little was known about the annual distribution of meteorological occurrences and it is greatly to Light's credit that he so ably demonstrated his skills in applied bio-climatology. Modern as this discipline appears to be, it is nevertheless clear from Light's own account that he was able to draw conclusions, particularly on precipitation in seasons he had yet to experience, after a thorough inspection of the local plant life. Although he had been enthralled by his first 24 hours at Rapid Bay, some 50 km south of Adelaide, he quickly became aware that the hinterland of the former was unsuitable for the immediate topographical requirements of a growing colony. With the questionable exception of the Encounter Bay region, it is certain that Light discovered the only site in South Australia capable of sustaining a major city dependent on the resources capable of being harnessed in the 19th century.

The first meteorological observations in South Australia were Sir George Kingston's readings of rainfall in Grote Street, Adelaide, which extended from 1839 to 1879. In 1856 Sir Charles Todd was appointed Superintendent of Telegraphs and also Government Astronomer in the State, and began a private series of meteorological observations which were subsequently continued officially. Gentilli (1971) has provided a comprehensive chronological summary of Australian meteorological events and Gibbs (1970) has written a brief monograph on the history of Australian meteorology; both include the developments in South Australia. Although Light is not mentioned in these accounts, presumably because he did not record any quantitative observations, he deserves to be remembered in this context. After Light, Kingston and Todd, 19th Century South Australia lacked scientific, meteorogically innovatory giants such as G. Neumayer, H. C. Russell and Clement Wragge, who initiated major observational and analytical programmes in the eastern states. Wragge's monumental mountain-top observations on Mt Wellington, Tasmania and Mt Kosciusko, N.S.W. are often overlooked (possibly because some of the records have been lost) and it is unfortunate that Mt Lofty was not subjected to a similar study since, as Figure 1 shows, it dominates the Adelaide region.

Rainfall and evaporation

Because of its latitude and exposure to the easterly moving atmospheric pressure system of lows in winter and highs in summer, Mt Lofty (720 m) is the single most important topographical feature causing Adelaide to experience an overall climate which is quite anomalous when compared with those of otherwise similar locations, such as the Yorke Peninsula. Although the southern regions of the State have well defined Mediterranean climates with a strongly polarized seasonal distribution of rainfall and relatively long, dry summers, nowhere is the winter rainfall so high as on the slopes of Mt Lofty and the associated ranges. These thus represent South Australia's major indigenous water resource, while earlier in the State's history they were a valuable source of hardwoods. The mean height of the main ridges of the Mt Lofty Ranges is approximately

* Institute for Atmospheric & Marine Sciences, Flinders University of South Australia, Bedford Park, S. Aust.

Fig. 1. Adelaide and the nearer Mt Lofty Ranges as seen from Gulf St Vincent shortly after dawn, early in summer.

half that of the main peak, and thus only of the same magnitude as the generally prevailing atmospheric condensation level during winter, when it is quite clear that precipitation is likely to be strongly influenced locally. However, orographic uplift results in an amplified wave being formed in the passing airstream, which is probably twice as high as the terrestrial obstacle, so that significant orographic rain benefits the Adelaide Plain and the Mt Lofty Ranges generally for about eight months of the year.

During summer, prolonged dry periods of up to two months are to be expected in the Adelaide region. Any precipitation during January and February is most likely to result from rare thunderstorms, which are more frequent over the plains than in the ranges. The spatial and temporal distributions of rainfall are illustrated by a typical cross-section through Adelaide and the Mt Lofty Ranges from Seaton (near Port Adelaide and the coast) to Tailem Bend on the Murray River. Figure 2 shows the strong correlation between mean annual precipitation and altitude. Figure 3 shows the corresponding rainfall data in the form of percentages for the four seasons, indicating the

striking importance of the Mt Lofty Ranges in inducing orographic rainfall during the winter months; also that this effect diminishes during the warmer months and, as expected, is at a minimum during summer. Figure 3 also shows the relatively greater proportion of summer rain on the plains both east and west of the ranges, which may largely be ascribed to thunderstorm activity. Figures 2 and 3 are based on Bureau of Meteorology records listed in a 1971 publication which also shows the general spatial distribution for the entire Adelaide region. Unfortunately although the Bureau's pluviographic grid is relatively closely spaced, it is nevertheless inadequate to elucidate the striking precipitation gradients which occur throughout the Mt Lofty Ranges, and in particular on Mt Lofty itself. Preliminary conclusions based on three years of observations (1972–74) indicate that the mean annual rainfall in the vicinity of the summit may be as high as 1,600 mm: four times the value recorded near the coast at Port Adelaide! This high total is restricted to an area of only a few km^2, and thus does not significantly influence the total yield of the Mt Lofty Ranges water catchments, which until the mid-1950s sup-

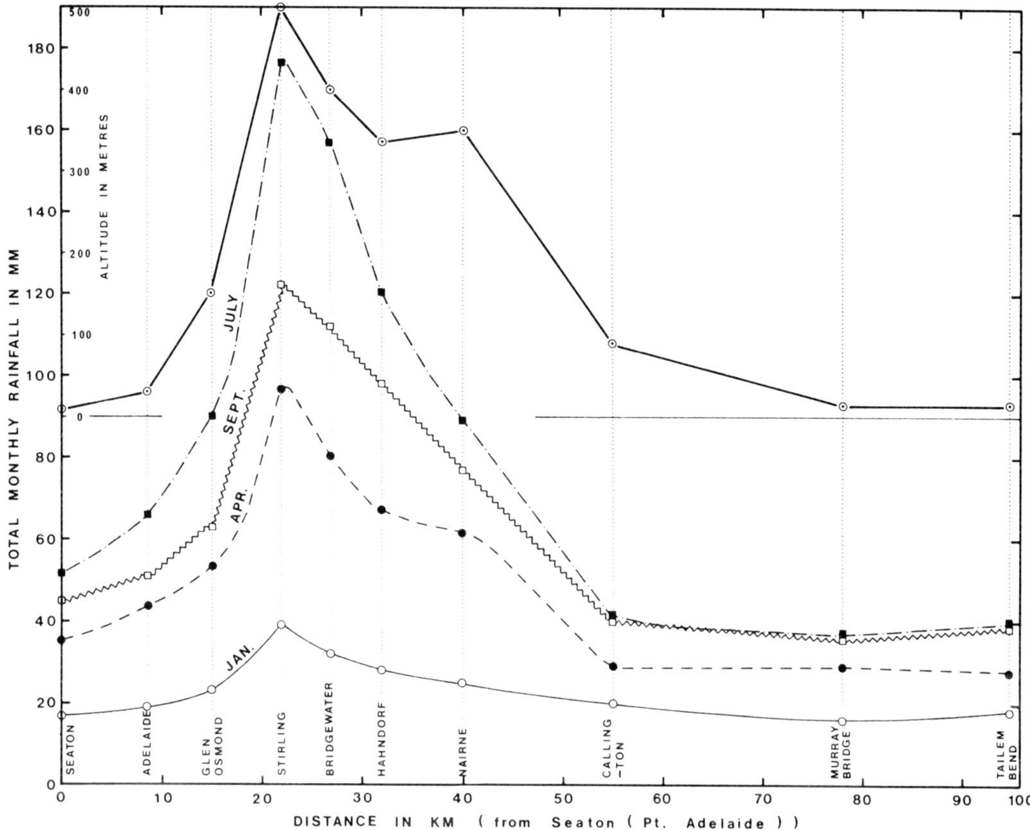

Fig. 2. Mean total rainfall for each of the four seasons (summation yields the mean annual total) for stations on a path between Seaton (near Port Adelaide) and Tailem Bend, including the altitudinal cross-section through the Mt Lofty Ranges.

plied all of Adelaide's water requirements. However, about 70% of this rain falls within a 6-month period, so that some parts of the catchment zones are exposed to precipitation rates averaging 600 mm/month^{-1} during winter. This high rate initiates erosional problems, and seriously affects water quality in the absence of suitable vegetative cover. Unfortunately there is no doubt that the utilization of land in the Mt Lofty Ranges has proceeded since 1839, when timber felling of stringybark eucalypts commenced in Crafers (Hardy 1939), initiating an era of mismanagement which was crowned by a plan endorsed by the South Australian Government in the 1960's to allow urbanization of the high rainfall zone (Town Planning Committee 1962) without adequate consideration of the State's greatest economic asset in the Ranges, namely water. This aspect

has been discussed more intensively elsewhere (Schwerdtfeger 1972a).

A distinct rain shadow effect is observed in the hibernal lee of the Ranges, with a resultant minimum mean precipitation almost directly over the Monarto area, which appears to receive about 300 mm annually. In summer, occasional intensely localised thunderstorms can result in the rapid downpour of a substantial percentage of the annual total. Robin[1] concluded that, on average, once in 10 years a 250 mm cascade in 24 hours must be reckoned with. The Adelaide Plain is less likely to receive such major transient yields. Nevertheless all of the low-lying areas of the Adelaide region, including the near Murray Valley and Yorke Peninsula are not without significant occasional erosional and drainage problems in seasons other than winter.

---

[1] Robin, A. (1975).—Climate of Monarto. Unpubl. address to the Hydrological Society of S. Aust.

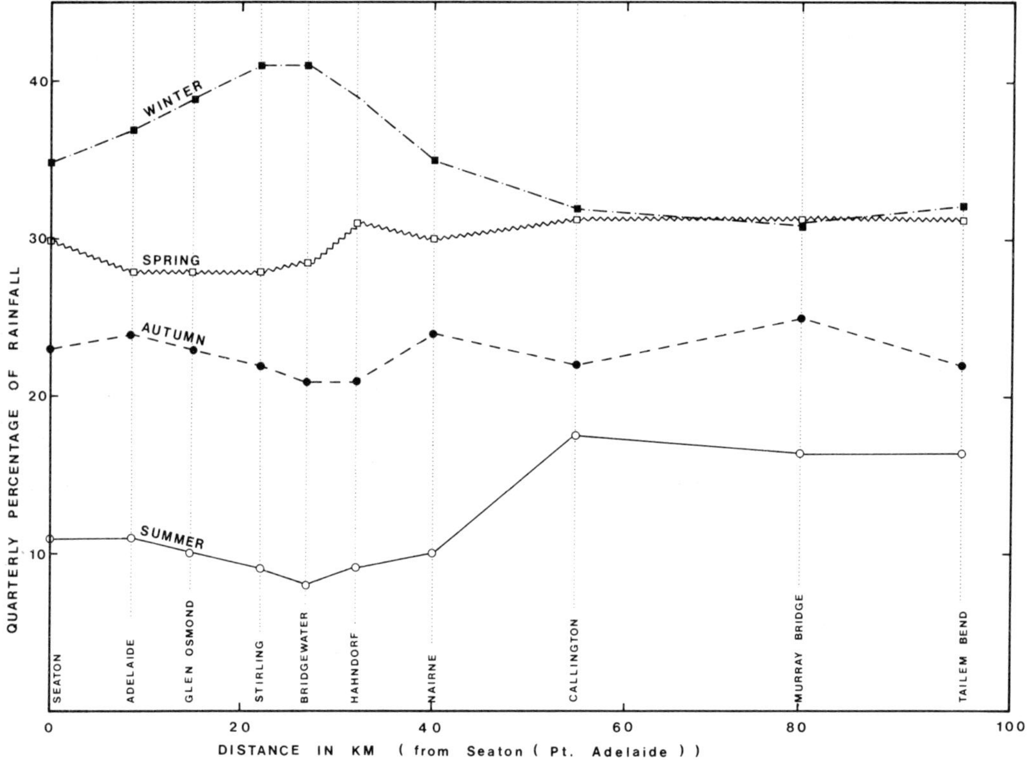

Fig. 3. Fractional mean seasonal distribution of rainfall at stations between Seaton and Tailem Bend.

During the drier spring and summer months the vegetation on the highest ridges of the Mt Lofty Ranges may frequently benefit from a little known precipitation mechanism. This involves the capture of cloud droplets by tall trees on occasions when normal thermodynamic processes are inadequate to cause precipitation. A wide range of both indigenous and exotic plants owe their survival through the summer to this temporal augmentation of effective rainfall.

On no more than one or two occasions each year, between November and February, Adelaide can expect to be reached by tropical air masses whose trajectories over the continent may have commenced anywhere between north-western Australia and the Queensland coast, and yield useful summer rainfall.

Snow rarely embellishes the Adelaide region. Although snow flakes are a probability on at least one occasion per year on Mt Lofty, only rarely do even modest drifts accumulate and remain on the ground for more than one hour. Major snowfalls appear to occur perhaps only once in two decades. On the other hand artificial snow production on a modest scale for recreational purposes would appear to have a good chance of success in carefully selected sheltered valleys in the Ranges.

Although the annual rainfall received by Adelaide and its environs compares favourably for example, with most of non-alpine central Europe, the high rate of evaporation, coupled with the strong seasonal polarisation of precipitation, makes growth for non-irrigated shallow-rooted plants impossible on the Plain for about four months of the year. In the higher levels of the Ranges this period is 2–3 weeks shorter, while on the eastern plains near Monarto it is 3–4 weeks longer. The Bureau of Meteorology has prepared a map of the length of the growing season for most of the region.

Potential evaporation is a function of wind speed, saturation humidity deficit of the air and the energy available for heating, including solar radiation. Thus the general absence of cloud in mid- and late summer and extreme dryness of the air, result in an approximate evaporation rate of 200 m per month in January. This figure has been estimated by the

Fig. 4. Orographic cloud over Adelaide and Mt Lofty, as seen from Yorke Peninsula in early summer.

Bureau of Meteorology (1967) on the basis of very limited information but is useful as a general guide. Although the summer months represent an extreme, the example suffices to illustrate the magnitude of the region's overall water deficit, since the entire annual rainfall in some areas could be lost by evaporation within two months of summer. In July the evaporation rate appears to be about 40 mm/month^{-1} on the plains, but must be taken as being far less than this in the ranges to an extent strongly dependent on location and altitude. Gentilli (1971) has reviewed the literature discussing the significance of rainfall and evaporation in the context of agriculture.

Solar radiation

Few Australian cities have a wealth of data on solar radiation; Adelaide is no exception. The Waite Agricultural Research Institute of the University of Adelaide commenced monitoring global radiation late in the 1960's. Since 1971 measurements of global and diffuse radiation have been conducted by the Flinders Institute for Atmospheric and Marine Sciences (F.I.A.M.S.) at the Adelaide Airport, and at Flinders University of South Australia. Because these two sites are altitudinally separated by about 150 m, significant differences systematically occur in irradiation levels which may be ascribed to air pollutants, particularly in the first 100 m of the atmosphere. These effects have been discussed by Lyons & Forgan (1975).

Present interest in solar radiation covers two main potential applications. The first of these is in elucidating scientific questions on the solar constant and atmospheric attenuation processes, on which Forgan (in press) has prepared further information of relevance to Adelaide. In this connection also arises the question of long term trends, and thus the need for a reliable baseline station free from the effects of both urban and agricultural pollution. Such a station has been established by F.I.A.M.S. in January 1976 at Cape du Couedic on the southwest coast of Kangaroo Island. Because of the recency of this installation, there is at present no information on global trends in atmospheric transparency, based on measurements taken in the Adelaide region. Similarly there is an almost complete lack of knowledge on the spatial distribution of mean irradiational levels in the entire Adelaide region, not to mention the remainder of South Australia. There is a need for these data in the light of current interest in potential applications and methods of solar energy conversion.

From information assembled by Lyons (1974b) the following values may be used as an approximate guide to the mean daily totals of solar radiation fluxes over the near southern suburbs of Adelaide: summer maximum 26.2 $\pm$ 4.2 $\times$ 10^6 J m^{-2}, and winter minimum, 7.2 $\pm$ 3.6 $\times$ 10^6 J m^{-2}. In both instances the $\pm$ indicates the range of fluctuations caused by varying degrees of cloudiness.

The main features of the solar radiation regime are related to cloud cover (apart from shading close to, and in various valleys of, the

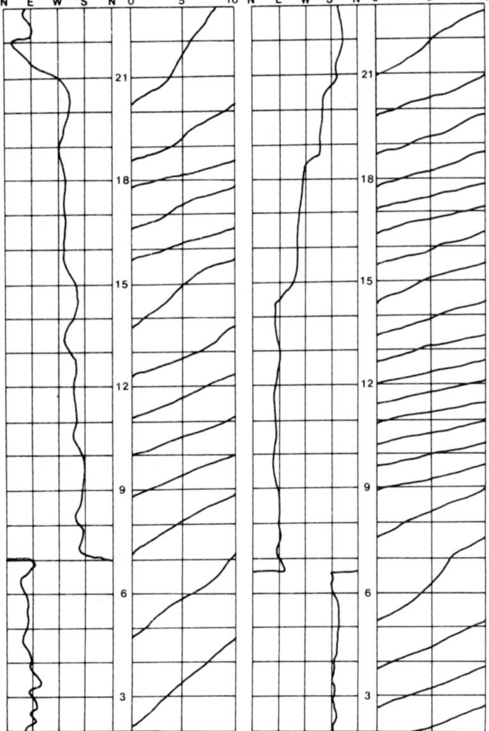

Fig. 5. Simultaneous anemograph records at Somerton Park, Brighton (two left hand columns), and Stansbury (two right hand columns) showing separate sea- and land-breeze system over opposite coasts of Gulf St Vincent. Time of day (hours) is shown by the vertical numbers; integrated wind-path (km), and also direction, scaled horizontally.

Mt Lofty Ranges), once astro-geometrical factors have been accounted for. In winter a substantial cloud cover frequently extends over the entire region, although over the Mt Lofty Ranges the cloud cover is orographically augmented. Because this orographic low cloud extends non-continuously well into spring, when cloud cover on a synoptic scale may have ceased elsewhere in the region, the contrast between radiation totals received at higher levels in the Ranges and on the plains is particularly striking. The cloud associated with Mt Lofty still forms regularly as late as December at heights that may exceed 2,000 m, because of the aerodynamic wave which forms over the Ranges. As shown in Figure 4, this cloud can be viewed from the Yorke Peninsula. It is a most important factor in ameliorating temperature extremes during early summer in the central and near southern urban area. Because the

northern and far southern suburbs do not benefit from this diffuse parasol, higher maximum temperatures are experienced there.

On the eastern side of the Mt Lofty Ranges cloudiness is significantly less than over the Adelaide Plain in winter. In summer a particularly interesting effect, which is of considerable significance to the Monarto area, accompanies the diurnal sea-breeze system as it develops northwards up to the Murray Valley. Here, the clouds capping the sea-breeze cell are seen as a broad belt in the eastern sky, positioned so as to intensify the total afternoon solar irradiation by means of scattering, with a resultant augmentation of ground level temperatures. This mechanism may well underly the thunderstorm potential of the area, which is greater than that of the Ranges and the Adelaide Plain.

Wind speed and direction

Although synoptic scale or geostrophic wind systems have common features all along the coastline of South Australia, the topographically complex location of. Adelaide, between modest mountain and sharply indented gulf, results in peculiarly localised diurnal wind systems, which are small in scale and extremely variable in both time and space. Viewed as a simple location on a synoptic weather chart, Adelaide represents a major aeolian anomaly, particularly in summer. The fact emphasises the need for meteorological observations in Adelaide to be supplemented by a similar monitoring programme on Mt Lofty, where for example the author (1972b) has shown wind directions to correspond closely to the geostrophic.

Climatological observations of wind direction, compiled by the Bureau of Meteorology, have been included by Gentilli (1971), but cover only the daylight times of 0900 and 1500 hours. Unfortunately in this way one of the most important local wind systems (that of the Gully Winds) escapes attention.

On the Adelaide Plain, the dominating hibernal morning wind direction is from the northeast, with somewhat lesser occurrences of northerlies and easterlies. In the light of this fact (established long ago by the Bureau of Meteorology), it is amazing that during the 1960's, State planning regulations allowed housing estates to be created immediately to the south of potentially emissive industrial zones. Consequently during mornings, before the break-up of nocturnal temperature inversions, there is a

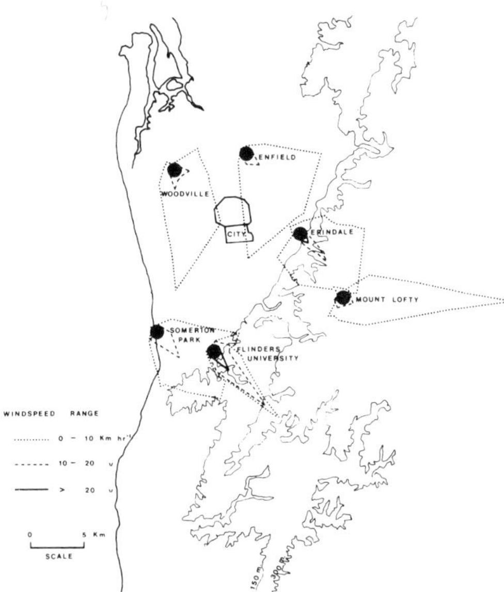

Fig. 6. Typical late summer wind roses for the 2000–0100 hr "temporal quadrant", which highlights the gully winds.

likely channelling of industrial air through these unfortunate residential areas. Figure 1 clearly shows the typical early morning stable stratification of the atmosphere above Adelaide, and suggests that a 2-dimensional air flow analysis over the plain could provide a major tool for regional town-planners. Schwerdtfeger & Lyons (in press) have provided such an analysis, utilising data from the first anemographic grid installed in a major Australian city, and operated for three years from 1972 to 1975.

Unfortunately, wind direction frequencies alone do not distinguish between the directional distribution of aeolian energy, and of moisture-bearing winds, both being of potential architectural significance. The F.I.A.M.S. data library has now digitised the necessary parameters for representative locations on both the Adelaide Plain and Mt Lofty, in order to be able to yield this information in greater detail. In winter high wind velocities tend to coincide in direction with rain-bearing winds, which range from southeast to west. The coastal areas of Adelaide and the higher ridges of the Ranges are particularly exposed to westerly and southwesterly winds.

In summer, the Adelaide Plain is influenced by two local wind systems. The sea breeze is a phenomenon shared by most coastal locations in South Australia. However, the Mt Lofty Ranges act to impede the system's full inland development, as had been shown by Clarke (Gentilli 1971). On the other hand the Ranges also shield the Adelaide coastline from weak easterly winds, which might otherwise result in the reduction or even negation of the sea breeze. The sea breeze is of course able to be augmented, cancelled, or even over-ridden by synoptic systems, but one of the most interesting sea breeze effects peculiar to the region becomes evident only during geostrophically calm periods. On such occasions anemographic readings at Brighton, on the eastern shore of Gulf St Vincent, and Stansbury on the western side, show opposite wind directions over a full 24-hour period (Fig. 5). By tracking balloons and using anemometers respectively released and read on launch traverses of the Gulf, yielding data which subsequently supported numerical models, Physick (1975) has shown that opposing sea breeze cells frequently form over an east-west cross-section of the Gulf. There is a resultant calm region over the water between the two cells (biassed in location toward the Yorke Peninsula whose sea breeze on such occasions, because of the smaller hinterland, is not as strong as that over Adelaide). As may be imagined similar systems are observed to occur over Spencer Gulf (discussed by Schwerdtfeger & Williams 1974), so that Yorke Peninsula actually is able to support two opposing sea breeze systems, with a narrow band of wind calm along the axis of the 50 km wide strip. However, since the hot conditions conducive to optimal sea breeze development occur when high pressure systems lie over southeastern Australia, there is usually a simultaneous weak geostrophic wind. During such days the Adelaide sea breeze may be substantially shielded from this perturbing easterly by the Ranges. On the other hand the sea breezes on the western shores of both Gulf St Vincent and Spencer Gulf become reinforced, while the eastern shore of the latter may then find its weaker system (unprotected by any substantially silhouetting hinterland) cancelled.

The second of the important local summer aerodynamic systems is that of the "gully winds". Simple calculations readily demonstrate that these winds share no similarity with "foehns", the strong winds associated with high alpine mountain chains. During periods of subsidence inversions, which intensify at night when high atmospheric pressure systems lie over southeastern Australia, relatively shallow easterly winds blow toward the Mt Lofty Ranges. Simultaneous thermograph records on Mt Lofty and on the Adelaide Plain confirm that the temperature inversion at night may on such occasions sink well below 700 m. The relatively slowly moving easterly air stream is constrained to move through the lower passes and valleys of the Ranges, where continuity-of-mass requirements result in an acceleration, culminating in exciting wind speeds which can on occasion cause minor damage in those parts of Adelaide's suburbs located near to the terminations of transecting valleys or gullies. The term "gully wind" is thus appropriate.

During the warmer months, when sea breeze activity may on occasion form a relatively well defined "dynamic wall" complementing the "static wall" of the Mt Lofty Ranges, low level northerly winds are funnelled as the air moves southwards over Adelaide. Although further investigations are required, Schwerdtfeger (1972b) has described some of the effects in which low level northerly winds at Christie Downs in the south may reach speeds twice those measured at Enfield north of Adelaide; the distance separating these two stations being only about 30 km. That even the latter site is slightly influenced by funnelling is evidenced by yet lower wind speeds on Mt Lofty.

Although it is impossible to include here all of the features of winds in the Adelaide region, Figure 6 demonstrates the great spatial variations in wind velocities which may be encountered. The method of presenting these data is most important, because averaging over longer periods than the 6-hourly ("diurnal quadrant") interval chosen, would mask most of the phenomena.

Gully winds are a characteristic of the foothills boundary of the Adelaide Plain and similar locations on the western side of the Mt Lofty Ranges. These winds are not experienced on the eastern side. For example, Monarto is not shielded from inland winds which may be cold in winter and unwelcome harbingers of dust in summer. Thus, because of the present nature of land utilisation to the north and northeast of Monarto, the alleviation of potential dust problems in summer will require much more than a local shelter belt of trees.

Temperature and humidity

In any urban region, the natural distribution of temperatures near the ground and in the lower atmosphere is influenced by man's modification of the environment. This modification has two main forms which develop in historical sequence. The first involves the removal of natural vegetation for both agricultural and building purposes, which in the case of expansive, substantial deforestation can have serious consequences. The second stage follows the erection of large buildings. Where environmental perturbations of both of the above types are small scale, even local effects on air temperature may be difficult to observe, except in the case of very low wind speeds or calms. Since Adelaide has a relatively high mean wind speed, environmental changes wrought since the commencenent of temperature records have not substantially influenced mean monthly values. There is no doubt, however, that over shorter periods, such as during specific days and nights, the thermodynamic effects of urbanisation are readily observed on many occasions throughout the year.

Adelaide's natural spatial variations of surface air temperatures, apart from small scale anomalies such as are found in gullies, mainly depend on altitude, proximity to the sea and hence also sea breezes, gully winds and the parasol effect of Mt Lofty's orographic cloud. This is quite apart from synoptic meteorological conditions. Figure 7 shows the approximate distributions of mean maximum temperatures in Adelaide and the near Ranges for the four months which typify each of the seasons. The effects of urbanisation cannot be detected in any of these maps. Nevertheless, it must be emphasised that a map showing extreme (rather than mean) maxima, particularly those observed on calm days, would show strong correlations with urban intensity and sterility. Several observations in the Adelaide region have demonstrated that maximum air temperatures on hot, calm summer days are approximately 5°C greater above substantial waterproofed areas such as are exemplified by major roadworks and parts of central business districts, than in surrounding areas of fully tree-planted gardens. Although such days may occur perhaps only three or four times during each summer month, and hence do not greatly influence the statistics of climatic summaries, the effects are of course of major importance to human comfort, and are far too often dismissed.

The variations in mean minimum temperatures are shown in the four maps constituting Figure 8. In these, the consequences of urbanisation can be seen in all seasons as a 1°C average "heat-island". In the case of Adelaide, this heat-island may largely be ascribed to the increased effective "radiative cross-section" presented by a matrix of taller city buildings

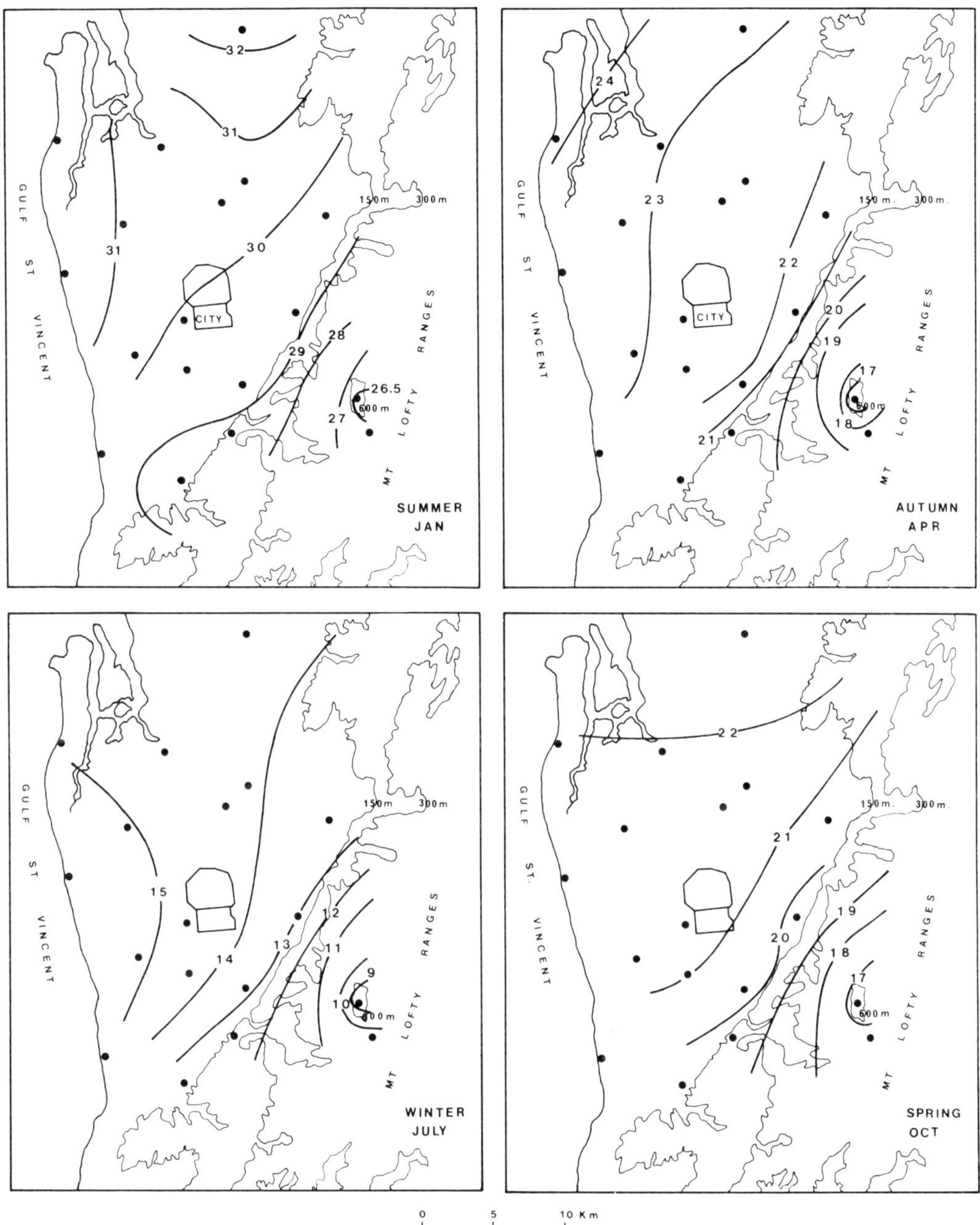

Fig. 7. Approximate contours of seasonal mean maximum temperatures near Adelaide. Meteorological stations of the urban network are indicated by (●).

whose massive heat capacity enables a continued supply of heat to the nocturnal urban atmosphere. Again, in representing mean conditions, the maps do not illustrate the magnitude of this heat-island on occasions of wind calm, measurements of which have been described by Lyons (1974a), indicating that thermal anomalies of up to 5°C may be expected in 10 km, nocturnal north-south traverses which include central Adelaide.

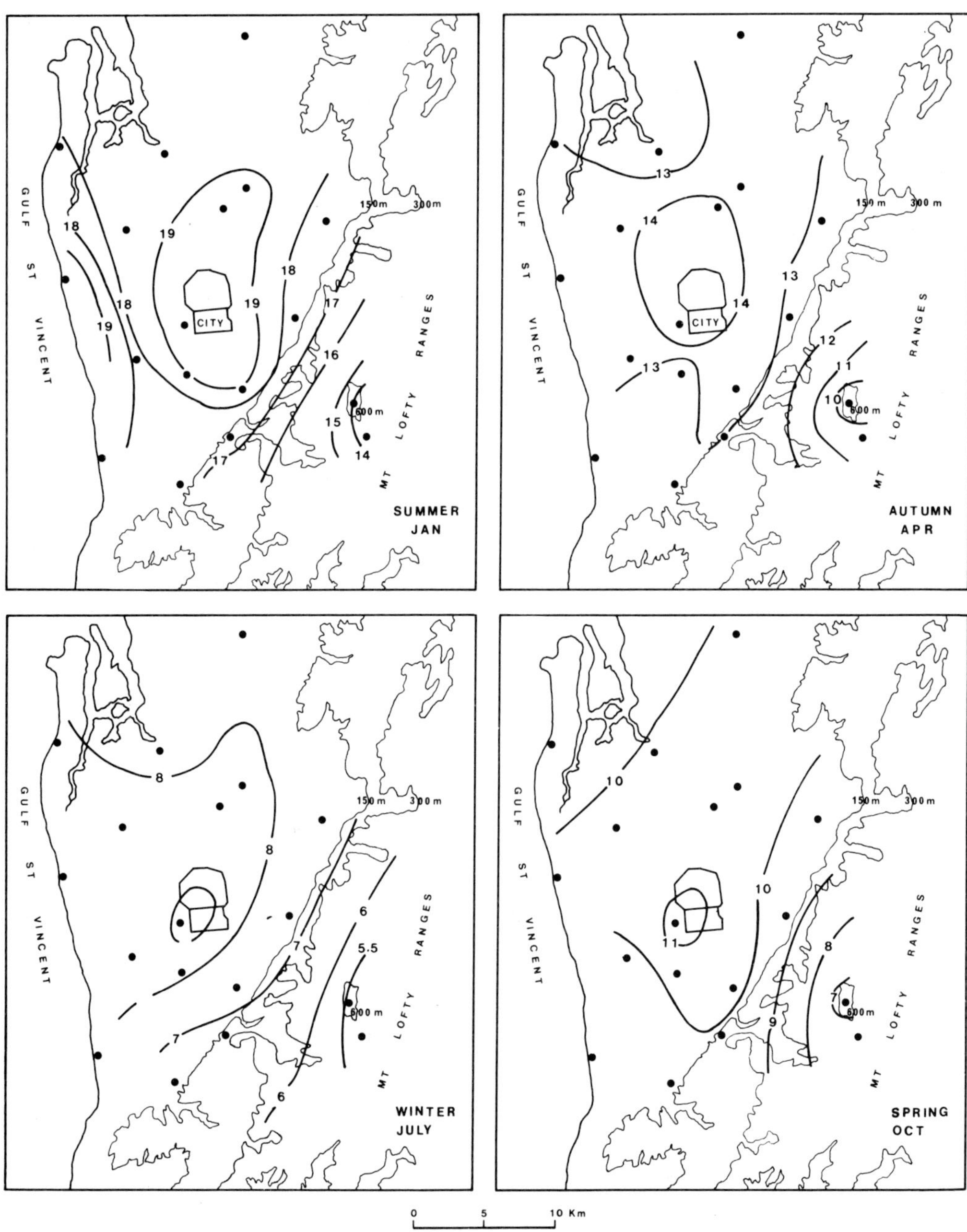

Fig. 8. Approximate contours of seasonal mean minimum temperatures near Adelaide. Meteorological stations of the urban network are indicated by (●).

Just as Adelaide's regional topography results in strong horizontal pluviographic gradients, there is a similarly striking variation in surface temperatures. Although illustrated in Figures 7 and 8, this fact is emphasized by Figure 9, which relates the measured lapse rate to the mean humidity of the atmosphere. The diagram is based on minimum (and hence nocturnal) temperature data to avoid the perturbing effects of insolation, as occurs for example

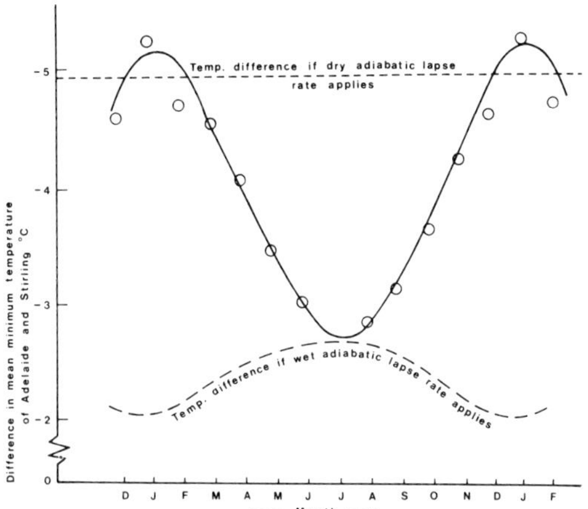

Fig. 9. Differences in monthly mean minimum temperatures observed at Adelaide (altitude 33 m) and Stirling (500 m), approximately 11 km distant, showing relationship between observed lapse rate and theoretical dry and wet adiabatic lapse rates.

on mean maximum temperatures in which local heating of the ground surfaces tends to mask the free atmospheric lapse rate differences. The nature of the seasonal variation in lapse rate shown by Figure 9 indicates that in January, the mean nocturnal atmospheric layer between Adelaide and Stirling is unsaturated, whereas in July the humidity is such that the corresponding layer can be regarded as being almost completely saturated throughout.

Figure 10 shows the diurnal range of humidity throughout the year as observed at the near northern suburb of Enfield. This temporal distribution could be taken to be typical for inland sites on the Adelaide Plain. In the Ranges the amplitude of variations is greater with Mt Lofty being exposed to full saturation for half of the mean winter's day.

This article has not attempted to provide an in-depth discussion of the synoptic meterological systems which determine Adelaide's basic climate. Works by Taylor (1920), Radock[2] and Gentilli (1971), for example, deal with this facet and list a wealth of further references. Rather, in emphasising the importance of perturbatory meso-scale effects of life in the Adelaide region, the aim has been to provide information relevant to the utilisation, maintenance and improvement of the environment. The lack of summer rain, characteristic of mediterranean climates, implies that considerable potential

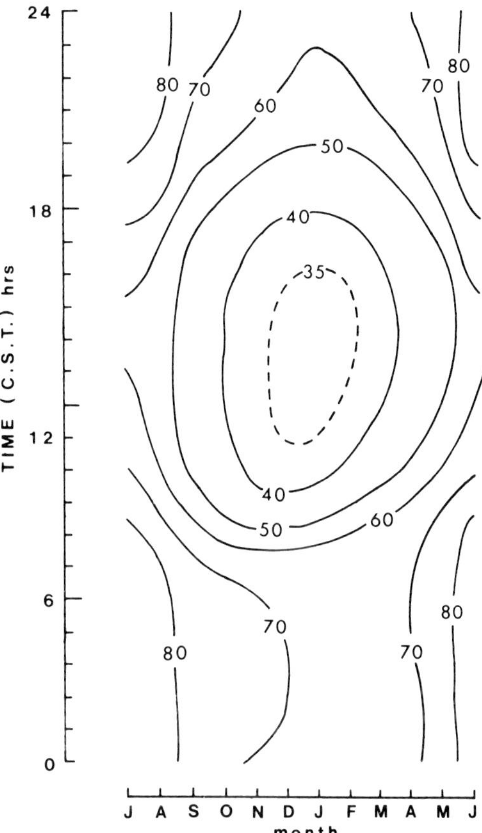

Fig. 10. Mean diurnal variations in relative humidity (%) observed at Enfield, a near northern suburb of Adelaide.

exists in scientifically based creativity in the fields of horticulture and forestry. Their further and closer integration with the entire urban area of Adelaide would lead to a readily sensed and economically valuable amelioration of climatic extremes at ground level, particularly as discerned through winds in all seasons, summer temperatures and the consequences of peaks in precipitation. Adelaide also appears to be well sited to encourage the increased small scale harnessing of both aeolian and solar energy. All of these facts should stimulate an awareness of meteorological paramenters as being among the most important sets of ecological determinants.

Acknowledgments

Most of this paper is based on the results of an investigation supported by Australian

[2] Radok, U. (1952).—A study of meteorological conditions in the free atmosphere over Australia. Ph.D. thesis, University of Melbourne (unpubl.).

Research Grants Committee funds and loans of Bureau of Meteorology equipment to the author at Flinders University during the period 1972–1975. However, the manuscript was written whilst a guest of the Alexander von Humboldt Foundation at Göttingen University, Germany. The author is thus grateful to Mesd. M. McFarlane and J. Lee for having completed basic outlines of references and diagrams from materials located in Adelaide.

References

BUREAU OF METEOROLOGY (1967).—"Average evaporation (maps for each month, Australia)." 3rd Edtn. (Bur. Met.: Melbourne.)

BUREAU OF METEOROLOGY (1971).—"Climate Survey of Adelaide Region 1—South Australia." (Bur. Met.: Melbourne.)

FORGAN, B. W. (in press).—Solar constants and radiometric scales. *Applied Optics.*

GENTILLI, J. (1971).—Climates of Australia and New Zealand *in* "World Survey of Climatology" 13. (Elsevier: Amsterdam.)

GIBBS, W. J. (1970).—The origins of Australian meteorology. *Aust. Met. Mag.* **18**(1), 43-45.

HARDY, M. (1939).—"A history of Crafers." (Crafers Centenary Committee: Adelaide.)

LYONS, T. J. (1974a).—Adelaide's urban climate. *Flinders Instit. Atmos. Mar. Sci. Res. Rep.* No. 12.

LYONS, T. J. (1974b).—A climatic survey of the Adelaide region. *Flinders Instit. Atmos. Mar. Sci. Res. Rep.* No. 14.

LYONS, T. J. & FORGAN, B. W. (1975).—Atmospheric attenuation of solar radiation at Adelaide. *Quart. J. R. Met. Soc.* **101**(430), 1013-1016.

PHYSICK, W. L. (1975).—A numerical model and observations of the sea breeze over a gulf. *Flinders Instit. Atmos. Mar. Sci. Res. Rep.* No. 18.

SCHWERDTFEGER, P. (1972a).—The role of meteorology in urban planning—a case study. *In* "The city as a life system?" *Proc. Ecol. Soc. Aust.* **7**, 143-160.

SCHWERDTFEGER, P. (1972b). — Meteorological measurements relevant in planning urban environments. *Flinders Instit. Atmos. Mar. Sci. Res. Rep.* No. 3.

SCHWERDTFEGER, P. & LYONS, T. J. (in press).—Application of wind field studies to atmospheric pollution control. *Urban Ecol.*

SCHWERDTFEGER, P. & WILLIAMS, S. (1974).—Winds at Whyalla. *Flinders Instit. Atmos. Mar. Sci. Res. Rep.* No. 16.

TOWN PLANNING COMMITTEE (1962).—"Report of the Metropolitan Area of Adelaide." (Govt Printer: Adelaide.)

TAYLOR, T. G. (1920).—"Australian meteorology." (Clarendon Press: Oxford.)

CHAPTER 5

HYDROLOGY: GROUNDWATER

by R. G. SHEPHERD*

History of groundwater development

During the early days of settlement of Adelaide, water supplies were obtained from the rivers Torrens and Sturt, and to a lesser extent from smaller ephemeral streams. As the population grew and the demand for water increased, these sources were often inadequate over a long dry summer. However, the first large storage for water was not constructed until 1860 at Thorndon Park.

Because of the shortage of surface waters, shallow wells were dug adjacent to streams, either into the alluvium of the plains or weathered rock of the hills. Yields were found to be generally small and salinity varied over wide ranges, particularly in the alluvium. By 1880 steam driven percussion drilling plants were in use in the search for groundwater.

On the Adelaide Plains aquifers yielding good quality water were found at depths of 60–120 m and the supplies obtained were relatively large and permanent. Deep drilling in the Adelaide Hills also proved the existence of large supplies of groundwater.

In the Adelaide region groundwater now represents approximately 10% of the total water used. Greatest use is for irrigation from the Tertiary aquifers of the Adelaide Plains; significant volumes are withdrawn from unconsolidated Cainozoic sediments of the Barossa Valley and the Willunga Basin, in addition to the hard rocks of the Mt Lofty Ranges.

In the past water supplies for some towns and for industrial purposes within the Adelaide region have been supplemented by groundwater. Significant quantities are used in some towns in the Adelaide Hills, e.g. Mt Barker. In addition, many schools throughout the area use groundwater for watering ovals and gardens. Most groundwater for industrial purposes and incidentally for watering golf courses is withdrawn from Tertiary sediments beneath the western suburbs of Adelaide.

Hydrogeology

(i) *Lower Proterozoic*

Schist and gneiss, with augen gneisses and metaquartzite are exposed in the area from Meadows to Yankalilla, also from Houghton to the South Para Reservoir, and in small areas south of Uraidla and east of Balhannah. These rocks constitute the crystalline basement and are considered in Chapter 1, where the geology of the region is discussed.

In many places the rocks are deeply weathered or have undergone retrograde metamorphism. As a result they usually have a low permeability, and yields of individual bores are relatively small, usually less than 85 m³/day. Where these rocks occur groundwater is generally used only for stock purposes, but minor irrigation is carried out in parts of the area.

Best yields are normally obtained in the upper part of the aquifer where salinities may be less than 500 mg/l in the higher rainfall areas.

(ii) *Adelaidean*

Because of the wide range of rock types occurring in the Mt Lofty Ranges, the quality and quantity of groundwater available from them varies considerably. Recharge to the groundwater also varies considerably, depending on rainfall, topography, rock texture and structure and soil cover.

Generally the slates, shales, siltstones and tillites of the Umberatana and Wilpena Groups (see Chapter 1) yield relatively small quantities of groundwater except where fracturing or jointing is well developed. Yields generally are less than 430 m³/day, rising to possibly 850 m³/day in the more competent beds. Over most of the Adelaide region salinities are less than 4,000 mg/l, but there are significant areas where salinity is less than 500 mg/l, particularly in the high rainfall areas.

Rocks of the Burra Group occurring in the western part of the Mt Lofty Ranges are low

---

* South Australian Department of Mines.

yielding although in areas where they are suitably fractured yields are good, generally at least 430 m³/day. Sandstones, particularly the Aldgate Sandstone, generally provide low salinity groundwater and yields of up to 1,700 m³/day.

Aquifers within Burra Group rocks are developed for irrigation supplies in Piccadilly Valley, where yields may range from 500 to 700 m³/day.

(iii) *Cambrian*

Rocks of the Kanmantoo Group consist of siltstone, greywacke, shale, limestone and marble. In the Adelaide region they extend from north of Truro through Mt Pleasant in the eastern part of the ranges, and southward through Callington and Strathalbyn to Cape Jervis. In discussing hydrogeology these rocks are considered with the Normanville Group.

Marbles and limestones form the best aquifers as they often contain solution cavities— very large ones were intersected during construction work on the Myponga Reservoir. Strong springs issue from cavities in the marble near Delamere and yields of 1,300 m³/day have been obtained from wells in the area. Salinity ranges from 1,000-2,000 mg/l within the calcareous rocks but is much higher, up to 5,000 mg/l in other rock types.

With rare exceptions groundwater yields of other Cambrian rocks is low, usually less than 350 m³/day. An exception is near Charleston, where a yield of up to 2,000 m³/day was obtained from greywacke, apparently in a strongly fractured zone near the base of the Cambrian.

(iv) *Permian*

Permian sediments which are glacial and fluvioglacial sands, occur mainly on Fleurieu Peninsula, infilling valleys in older rocks from Ashbourne to Cape Jervis.

Permian aquifers consist generally of fine sand of almost uniform grain size. Permeability is low and yields are usually less than 350 m³/day although during initial development the yield may be higher.

Salinity of groundwater in these sediments is usually less than 2,000 mg/l but in the vicinity of streams where recharge occurs salinity may be less than 1,000 mg/l. Groundwater use is restricted mainly to stock supplies but limited irrigation is carried out in some areas, e.g. Mt Compass. In this locality, groundwater occurs at shallow depth and one method of obtaining irrigation supplies is to excavate a large pit with a pump intake installed in the centre. In this way movement of fines towards the pump normally does not occur because of very slow movement of the water, provided the pumping is maintained at a relatively low rate.

(v) *Tertiary Basins*

Adelaide Plains Basin: Tertiary aquifers contain the largest resource of groundwater in the Adelaide region and have been exploited to a considerable degree in the Adelaide Plains Basin. This is a sub-basin within the St Vincent Basin which has a land area of approximately 4,000 km² and extends northward beyond Port Wakefield.

Most development of groundwater has taken place within the Tertiary aquifers occurring west of the Para Fault. In the Adelaide area salinity of the groundwater of these aquifers is generally less than 1,000 mg/l between Port Road and Anzac Highway. Yields of up to 3,000 m³/day have been obtained but average yields would be in the range 1,000–1,600 m³/day.

At times of critical water shortage groundwater has been pumped into the mains from wells in the western suburbs. It has been used to augment surface water supplies on six occasions since 1915 and the volumes pumped are shown in Table 1.

TABLE 1

Adelaide Plains—Groundwater pumped into mains.
1915-1967.

Year	No. of Wells	Quantity Pumped m³
1915	6	5.6 x 10⁵
1934	13	9.23 x 10⁵
1945	11	1.05 x 10⁶
1949	14	2.8 x 10⁵
1950	44	6.28 x 10⁶
1967	40	9.47 x 10⁶

The extensive irrigation area of the northern Adelaide Plains has been the subject of intensive investigations for 12 years. Groundwater use has been exceeding recharge for probably 15 years and currently the pumping/recharge ratio is approximately 3:1. Salinity of Tertiary aquifers is generally less than 800 mg/l over an area of about 160 km². Withdrawal from approximately 1,000 irrigation and industrial wells has been metered since 1970, and now averages 21 m³ x 10⁶/year. Recharge has been estimated, on the basis of stream gauging and bore hydrographs, to be 7.4 m³ x 10⁶/year.

In this situation there is a continuous decline in storage and it is considered that by the year

2000 depletion will be serious and increasing salinity will be a major problem. It is not considered feasible to reduce withdrawal to the rate of recharge, and other sources of water are currently being investigated.

Willunga and Noarlunga basins: Groundwater occurs in Tertiary sands and limestones of the Willunga Basin. The main aquifer is in the North Maslin Sands which are utilised over about 75% of the basin but in the southern coastal part this aquifer contains saline groundwater. There are approximately 800 known wells in the basin, of which about 200 are used for irrigation purposes. Total volume pumped is unknown but is probably less than 4 m^3 x 10^6/year. Recharge takes place along the fault zone forming the eastern boundary of the basin. So far there are no known problems of depletion or salinity increase caused by pumping. However, irrigation is increasing and water balance studies now in progress will indicate the safe yield of the aquifers.

Smaller basins: The Noarlunga Basin is separated from the Willunga Basin to the south by a basement high and the Clarendon Fault. Groundwater occurs within Tertiary sand or limestone and is suitable in places for irrigation purposes.

The aquifer of the Myponga Basin is Miocene limestone from which individual bores may yield up to 1,300 m^3/day and the water is suitable for general irrigation. The limestone overlies Permian sands, which although saturated, yield only small supplies because of their low permeability. Water balance studies in the area have not commenced but at this stage it is believed that no depletion has occurred, because there is no evidence of a decline in storage as shown by the potentiometric level.

The Hindmarsh Tiers Basin is separated from the Myponga Basin by a basement high and the aquifer is a Tertiary limestone, confined beneath a clay bed. Groundwater is used mainly for pasture irrigation and stock watering. Data available indicate that the potentiometric level falls 5–6 m throughout the basin within 2 months of the commencement of pumping. The effect of winter recharge is not yet known but it is believed that at present there are no depletion problems, as recovery of the potentiometric level apparently occurs each year.

Sediments of the Barossa Valley basin consist of clay, silt, sand and gravel ranging in age from early Tertiary (Eocene) to Recent. The aquifer consists of sand with associated lignitic fragments and is generally fine grained but becoming coarser at the base. Salinities vary from 400 to 14,000 mg/l but that portion of the aquifer which has been developed for irrigation has a salinity which is generally in the range 1,000–1,200 mg/l.

Yields of up to 1,700 m^3 per day have been obtained and the groundwater is used mainly for irrigation of vines. No evidence of depletion or salinity increase has been observed. Total groundwater withdrawal is estimated to be approximately 2.5 x $10^5 m^3$/year.

Conclusions

Groundwater is used extensively in the Adelaide region. Currently it represents about 10% of the total volume of water used.

In the northern Adelaide Plains withdrawal is exceeding recharge to the extent of about 14 m^3 x 10^6/year. However, in other basins and in the hard rock aquifers of the Adelaide Hills there is no evidence of depletion. Increased quantities of groundwater could probably be pumped from many aquifers in the Adelaide region, and it is considered that the groundwater component could be safely increased by 50%.

CHAPTER 6

HYDROLOGY OF THE COWANDILLA PLAINS, ADELAIDE, BEFORE 1836

by J. W. Holmes & M. B. Iversen*

Introduction

The flat, swampy land and the back lagoon behind the beach dunes on the coast of Gulf St Vincent, stretching from Glenelg to the Old Port Reach of the Port River (see Fig. 1) used to be known as the Cowandilla Plains, from the aboriginal name for that region. When Colonel William Light began to organise the systematic land survey around Adelaide he referred to this land as low grounds, subject to flooding, and it is so depicted on his 1839 map of "The District of Adelaide, South Australia, as divided into country sections". Much later Fenner (1927) also described this "belt of country, one or two miles wide, running northward from between Glenelg and Morphettville through the reed-beds to Port Adelaide". The area, he said, "is liable to winter floods, is mainly an area of considerable fertility, and must remain sparsely settled mostly used for grazing, for very many years". All of it is situated within 11 km (7 miles) of the City and in the two decades 1955–1975 the whole of the area has been developed for metropolitan purposes.

On the Cowandilla Plains or near environs are now located Adelaide Airport, the Patawalonga and West Beach recreation areas, suburban housing at a density of 10 houses ha^{-1}, and, by most recent utilisation of the lowest swamps, the West Lakes housing region, incorporating, not least in importance, the new Football Park. The hydrology of the Cowandilla Plains has undergone a profound change, never likely to be returned to its natural state. The purpose of this paper is to try to elucidate what may have been its natural hydrological regime before land drainage, reclamation and stream training were undertaken.

Water balance

The Torrens River delivered the largest amount of water onto the area. Three other streams in order of size, Sturt Creek, Brownhill Creek and Glen Osmond Creek debouched independently upon the Cowandilla Plains. Along these four water-courses, together with small tributaries of the Torrens River, First to Fifth creeks, the run-off of a considerable area of the Mt Lofty Ranges was passed through the Adelaide Plains to be ponded in the lagoonal swamps. Openings to the sea at the Patawalonga Estuary and through the Old Port Reach of the Port River discharged the flood flows until in summer and autumn the combined water input could be entirely dissipated by evaporation from the swamp and marsh surfaces. Then sand bars generated by the natural beach regime would, for a time, block off the sea-outlets. One of the purposes of this paper is to suggest a time scale for this regime.

Figure 1 shows the location of the streams whose water balance we attempt to reconstruct, together with Dry Creek and Little Para River, which emptied directly into the reach of Port River east of Torrens Island. The divergence between the direction of Dry Creek and Torrens River is caused by the uplifted Para Block. The convergence towards a common estuary nearer the southern end of the lagoon of the four contributors to the Cowandilla Plains has probably been caused by a gentle but general tilting to the southwest of the downfaulted block of which Gulf St Vincent is but one expression of multiple block-faulting of the Mt Lofty Ranges. Fenner (1927) believed the present Patawalonga opening to the sea to be very young, a suggestion that is consistent with the survival of mangrove tree stumps and estuarine muds on the beach at north Somerton, about 2 km south of the Boat Haven.

The extent of the swamps and marshes of the Cowandilla Plains is shown in Fig. 2 which reproduces ". . . a map of the country between

---

* School of Earth Sciences, Flinders University of South Australia, Bedford Park, S. Aust. 5042.

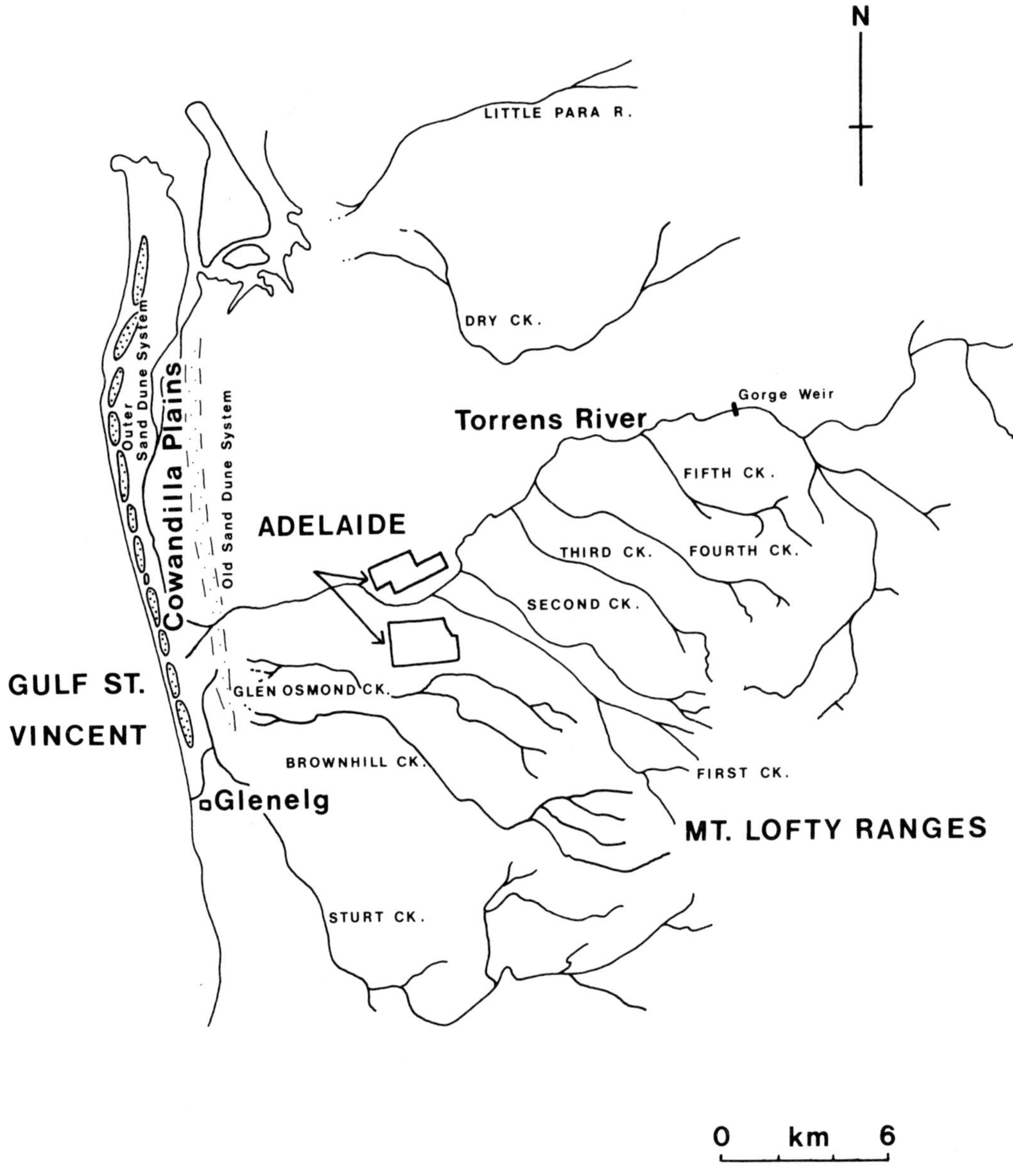

Fig. 1. The Adelaide Plains, showing creeks debouching upon the Cowandilla Plains.

Adelaide and the sea coast" compiled in the Surveyor-General's Office (1882). Urban development and land improvement have eliminated any possibility of modern checks on the indicated boundaries. However, another survey of the region, incorporated into the larger investigation of the soils and geology of Adelaide and suburbs by Aitchison, Sprigg & Cochrane (1954) agrees in most respects with the physiographic detail of the 1882 map. The difference is mainly that the area of estuarine soils which Aitchison and his co-workers mapped is larger than the marshy and swampy areas that were surveyed in the 1880's. The modern extent of the inundated land could of course be less than the extent during some of the last 8,000 years since sealevel approached its present stage. The soils could then preserve some record of past pluvial events but it is not our purpose to pursue this aspect.

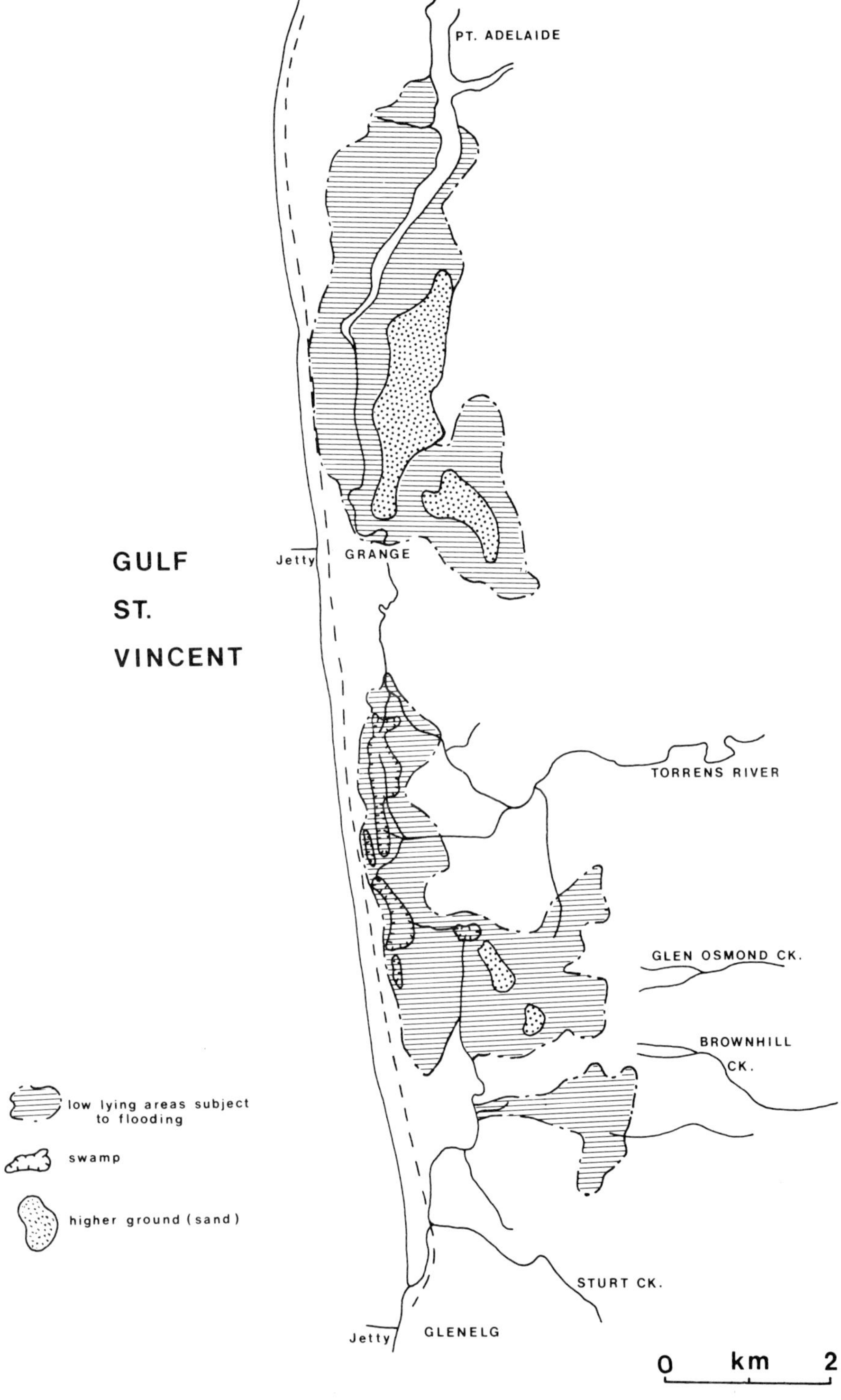

Fig. 2. The country between Adelaide and the sea-coast.

It is important to obtain as reliable an assessment of the area of marsh and swamp as possible, because the evaporation from this area is an important part of the water balance. We have no recourse but to put our trust in the Surveyor-General's Office. The water balance is given by the equation

$$P + I = E + UD + SD + \Delta S, \qquad (1)$$

where the symbols represent the volume per year of water put into or taken out of the Cowandilla Plains by precipitation, P; surface inflow, I; evaporation, E; underground drainage, UD; surface drainage, SD; and ΔS is the increment (positive or negative) of soil moisture change.

The input by rainfall was calculated by the product of the area of the Plains and the mean annual rainfall. The latter was taken to be that of Glenelg, for which the Bureau of Meteorology can supply long-term records. The actual rainfall distributed over the area is probably up to 3% less than Glenelg's rainfall, but other manifest uncertainties make it unnecessary to seek further refinement.

The stream discharge into the area was given by the sum of the discharges of the individual streams. Only one stream, Torrens River at the Gorge Weir has been gauged. Its mean monthly catchment yields are available from the Engineering and Water Supply Department (1969). Gerny (1962) analysed the variation in annual discharge of a number of South Australian streams, including the Torrens River and other hills streams outside our area. Following his suggested empiricisms, the annual discharge of all the streams contributing to the Cowandilla Plains was determined. The ratios of these discharges to that of the Torrens River were obtained during the calculations. The monthly flows of the Torrens River for which data are available were then multiplied by these appropriate ratios to give the mean monthly flows for the ungauged creeks, in the clear understanding that it must be assumed that the annual hydrograph of the Torrens River and those of all the creeks are sufficiently similar for this procedure to be acceptable. Details of this calculation and all others mentioned in this paper are given in Iversen (1973)[1].

The output by evaporation was calculated by the product of the area and the estimated evaporation. It seemed that the best guide to the potential evaporation rate could be the measurements of evaporation from grassland near Murray Bridge reported by Holmes & Watson (1967). The Bureau of Meteorology's data of evaporation of water contained in the Standard Australian Tank suggest close correspondence between Adelaide and Murray Bridge, though the actual magnitude cannot be taken to give evaporation from the land surface. The locale of the Murray Bridge experiment was irrigated pasture on reclaimed swamp (Long Flat Irrigation Area), and it is similar in topographic setting to the Cowandilla Plains.

Actual evaporation was derived as follows. The monthly potential evaporation values for grass given by Holmes & Watson were multiplied by 1.2 for those months during which it was deemed the area was completely flooded. This factor should adjust for the difference in heat budget, principally caused by a smaller albedo of shallow water. When partially flooded the multiplying factor was 1.1. In fact, only for April was the actual evaporation considered to be exactly equal to the potential evaporation rate obtained from the Long Flat experiment. During subsequent months, the area was partially or completely flooded until the sudden onset in December of some soil water deficit. The multiplying factor for actual evaporation was then either 1.0 or reduced linearly to 0.2 depending upon whether the soil moisture deficit was in the range $0 < \Delta S < 40$ mm or between 40 mm and 160 mm. The estimated evaporation had to be arrived at by an iterative method.

A small discharge of groundwater to the sea through the shallow water table aquifer beneath the beach dune was assumed. The total discharge rate, Q, is given by Darcy's Law,

$$Q = k \; b.x. \; \frac{\Delta H}{L}, \qquad (2)$$

in which the parameters adopted were

k, hydraulic conductivity, $= 10^{-5}$ m sec^{-1}, which was measured on disturbed samples;

b, the aquifer thickness, was assumed to be 50 m, from drilling records;

x, the north-to-south extent of the lagoon, was estimated to be 13 km;

ΔH, the hydraulic head difference. was estimated to be 1 m; and

L, the path length for average flow, was estimated to be 250 m.

---

[1] Iversen, M. B. (1973)—"Water and Salt Balance of the Cowandilla Plains." B.Sc. (Hons) Thesis, Flinders University of South Australia. Unpubl.

In determining the soil moisture deficit it was assumed that no deficit remained from the previous summer by the beginning of August. After that month a deficit first existed only when the output due to the actual evaporation and groundwater seepage to the sea exceeded the total input, i.e. when $E + UD > P + I$. Thus the value of the soil moisture deficit would be

$$\Delta S_1 = E_1 + UD_1 - (P_1 + I_1)$$

for the first month of a deficit. For the next month of a deficit, the relation would be

$$\Delta S_2 = \Delta S_1 + (E_2 + UD_2) - (P_2 + I_2).$$

These calculations were similarly repeated until a month was reached (May) for which the total input exceeded the sum of the output due to evaporation, seepage and the residual soil moisture deficit.

Finally, the surface discharge calculations followed similar lines to the determinations of the soil moisture deficits and were made in conjunction with them. Again starting from August, the value of direct stream discharge to the sea was assumed to equal the total input minus the output due to evaporation and seepage, i.e.

$$SD = (P + I) - (E + UD).$$

Then it was assumed that no direct discharge to sea occurred as long as the total input was exceeded by the sum of the output due to evaporation and seepage together with the accumulated soil moisture deficit, i.e.

$$P + I < E + UD + \Delta S, \quad SD = 0.$$

When this condition was first reversed the surface discharge was given by

$$SD = (P_j + I_j) - (E_j + UD_j + \Delta S_{j-1}),$$

where ΔS_{j-1} is the accumulated soil moisture deficit at the end of month $(j-1)$.

Using the methods outlined above the month-by-month water balance of the Cowandilla Plains was calculated to be that shown in Table 1.

Salt balance

The salt budget of the Cowandilla Plains was calculated by estimation of the individual terms shown in equation (3), which, by analogy with equation (1) for the water balance, expresses the expected balance of salt on an annual basis, viz.

$$P.C_p + I.C_1 = UD.C_{UD} + SD.C_{SD} \qquad (3)$$

in which the C's are the indicated contents of salts, that is, dissolved solids minus silica, in mg l^{-1}.

The salt content of the rain was estimated to have a mean value of 9.0 mg l^{-1} consistent with the measurements of Hutton & Leslie (1958) of dissolved salts in rainfall near the southern margins of the Australian continent.

The salt input carried to the area by stream discharge was pieced together from a variety of sources. Only for the Torrens River are records available of mean monthly salt content of the river water, again at a sampling location

TABLE 1
Water budget of the Cowandilla Plains before settlement[1] (in 10^3 Ml month^{-1}).

	INPUTS		OUTPUTS			
	Rainfall	Stream	Evaporation	Surface discharge	Ground-water Seepage	Soil moisture deficit[2]
January	.23	.48	1.42	—	.06	1.61
February	.27	.42	.95	—	.06	1.93
March	.29	.28	.79	—	.06	2.21
April	.51	1.42	.82	—	.06	1.16
May	.87	3.01	.52	2.14	.06	—
June	.90	10.90	.46	11.28	.06	—
July	.81	16.60	.50	16.85	.06	—
August	.77	21.00	.97	20.74	.06	—
September	.66	15.80	1.67	14.73	.06	—
October	.55	7.08	2.04	5.53	.06	—
November	.39	2.21	2.21	.33	.06	—
December	.35	.98	2.11	—	.06	.84
TOTAL	6.60 (460 mm)[1]	80.18 (6020)	14.46 (1005)	71.60 (4965)	.72 (50)	—

[1] Entries in the Table can be converted to a unit area basis by dividing by the estimated area of the Cowandilla Plains. 1.44 x 10^3 km^2; these are given in parentheses for the yearly totals.

[2] Cumulative soil moisture deficits. There cannot be any soil moisture deficit for the whole year, but the term is important on a monthly basis.

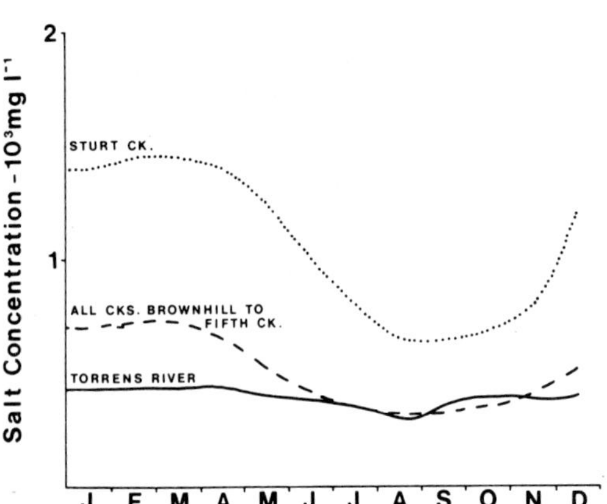

Fig. 3. Seasonal variation of salt content in creek
waters.

at the Gorge Weir. These measurements have
been made regularly since 1925. Imported
water from the River Murray has been pumped
into the Torrens system since the summer of
1953/54, and all records after that date have
been omitted. Various measurements on
samples from Sturt Creek, Brownhill Creek and
First to Fourth creeks were made by the
authors during the years 1967 and 1973. From
these varieties of sources the salt contents of
the streams were estimated to be those shown
in Fig. 3.

The seasonal variation of salt content is
entirely to be expected, as the contribution of
more saline groundwater to base flow of the
streams becomes proportionately larger during
the months of small discharge. There are early
records of a sample of water from the Torrens
River taken in December 1858 that contained
590 p.p.m. of total dissolved solids, and in
January and October 1861 two samples show
720 p.p.m. and 625 p.p.m. respectively. These
analyses would include perhaps 50 p.p.m. of
soluble silica and colloidal organic matter/iron
complexes. Nevertheless, the relatively high
values obtained on samples of the mid-1800s
suggested that, if any significant hydrologic
change has been caused by settlement of the
catchments it is that some leaching of salt has
been promoted.

The annual salt discharge in the groundwater
seepage flow of 0.72×10^3 Ml yr^{-1} was estimated by adding up the monthly flows as follows. For each month when the surface flow
could discharge to the sea the groundwater
intake had the appropriate salt content of the
surface water. For the remaining months,

namely December to April inclusive, the total
salt input by the streams was assumed to be
entirely discharged by groundwater seepage.

The method of estimation requires that the
salt concentration of the groundwater should
be 3,200 mg l^{-1}. This value corresponds accep-
tably with a value of 3,800 mg/l^{-1} determined
by taking a weighted mean value of groundwater salinities for the area, obtained from
records of the South Australian Mines Department. The average sampling depth of the water
so analysed was about 10 m. A few samples had
salinities of 15,000 mg l^{-1} or greater. These
values were omitted because they were believed
to represent either localised examples of stagnant groundwater or underlying saline groundwater associated with the Ghyben-Herzberg
lens, neither of which play any great part in the
salt balance of the region.

The annual salt budget of the Cowandilla
Plains, estimated by the methods detailed
above, is shown in Table 2.

TABLE 2

Salt budget of the Cowandilla Plains before settlement
(in 10^3 tonnes yr^{-1}).

Input		
	by rain	.06
	by streams	34.4
	Total input	34.46
Output		
	by groundwater discharge	2.4
	by surface discharge	32.0
	Total output	34.4

Discussion

Light and his party of surveyors arrived in
South Australia in the brig *Rapid* during
August 1836. The choice of site for the city
was Light's responsibility, and one can perhaps
imagine his anxiety about a secure water supply
during the following months. Port Lincoln,
Rapid Bay and Encounter Bay were even more
poorly endowed with fresh water than the
Adelaide Plains. When Kingston, in walking
the Adelaide Plains, discovered during January
1837 the incised water course of the Torrens
in the vicinity of the present Newmarket Hotel,
it is probable that, by that time in the summer, discharge to the sea either at the Patawalonga or the Port River exits had already
ceased. The weather that summer appears to
have been rather dry following a wetter than
usual spring.

The analysis of salt budget presented in this
paper would suggest that the natural hydrologic regime was vigorous enough to avoid
much accumulation of salt. Existing salt occur-

rences in some localities between Glenelg and Port Adelaide probably owe their provenance more to occasional inundations by sea water than to salt residue from the stream water input.

The modern urban hydrology is very different to that of the natural regime. Spreading of water on the plains has been stopped for many years and all the surface discharge is now taken in earth or concrete-lined drains direct to the sea openings. Garden watering may now replace only a small part of the 6,000 mm of water, formerly ponded or passed through the area each year. The leaching effect of this large natural input has been removed. One consequence could be a slow increase in soil salinity of the region. The new source of salts would be the residue from water applied as garden irrigation. The salt content of the groundwater, which is very shallow, and which derives in part from the drainage from areas higher on the Plains to the east will probably increase in years to come. There will also be local redistribution of salt water from the spread of the sea water in the Patawalonga Lake and West Lakes during wind storms and it may be found desirable to construct common tile-drainage in some areas.

Acknowledgments

We are indebted to officers of the Engineering and Water Supply Department and the South Australian Mines Department for help in assembling unpublished data.

References

AITCHISON, G. D., SPRIGG, R. C. & COCHRANE, G. W. (1954).—The soils and geology of Adelaide and suburbs. *Bull. Geol. Surv., S. Aust.* 32.

FENNER, C. (1927).—Adelaide, South Australia: a study in Human Geography. *Trans. R. Soc. S. Aust.* 51, 193-256.

GERNY, J. S. (1962).—Rainfall and stream flow in South Australia. *J. Inst. Eng. Aust.* 34(3), 45-60.

HOLMES, J. W. & WATSON, C. L. (1967).—The water budget of irrigated pasture land near Murray Bridge, South Australia. *Agric. Meteorol.* 4, 177-188.

HUTTON, J. T. & LESLIE, T. I. (1958).—Accession of non-nitrogenous ions dissolved in rainwater to soils in Victoria. *Aust. J. Agric. Res.* 9, 492-507.

CHAPTER 7

VEGETATION

by R. T. Lange*

Introduction

This account aims to complement the various handbooks which already deal with the land flora and vegetation of the Adelaide region.

Information about local vegetation has accumulated over a long period, during which both concepts and the vegetation itself have changed. Accordingly a resumé is presented tracing from earliest investigations and conditions to the present day, to provide a background against which the reader may appreciate literature on this subject.

Investigations prior to 1924

On his voyage of discovery of 1802 Flinders, three leagues offshore in Gulf St Vincent, described the appearance of Adelaide Hills vegetation and was later quoted by the South Australian Association (1834): *At daylight I recognized Mt Lofty. . . . The coast was . . . mostly low, and composed of sand and rock, with a few small trees scattered over it; but a few miles inland, where the back mountains arise, the country was well clothed with forest timber. . . .* Later Sturt's River Murray expedition of 1829-1830 afforded a view of the Hills from the east. Sturt estimated that five-sevenths of the land south of 34°40′ between the Murray and Gulf St Vincent was *. . . of rich soil upon which no scrub exists,* quoted by Torrens (1835). Knowledge based on more direct experience began to accumulate only about the time of settlement and in the decade 1840–1850.

Immediately following settlement Gouger (1838) wrote *. . . the fact is, I believe, none of our colonists are learned in plants, and I at last got tired of asking the name of any flower I found—it was always an orchis.* Nonetheless significant observations on vegetation were published in this very early period, including some by Gouger himself. For instance, one recognises the Figure 13 scrubs in Wade's description (reported in Gouger 1838): *The only bad land I saw in the province, was in the vicinity of Encounter Bay, and some of the mountains on the road to Adelaide from that place. This was very barren, and covered with dwarf gums.* . . . Similarly recognition of savannah woodland and its values was immediate: *. . . abundance of wood all the way, yet not so thick that agriculture might be pursued without the trouble of clearing* (Light 1839); *. . . the land between the trees, which may be averaged as six or seven per acre, is covered with grass of the richest quality* (Gouger 1838); *. . . an immense crop of grass, but I think it will not stand the summer* (Wade in Gouger 1838). Morphett (1837) gave the Aldinga area this benchmark description *. . . sloping grassland in front, without a single tree for 3 or 4 miles square, of a beautifully bright green in winter and spring, and a golden colour during the hotter months, is surrounded by finely wooded eminences. . . .*

The decade 1840–1850 brought more—e.g. Dutton (1946): *On the hills* (the central ranges) *the most hungry looking soils grow trees of the stateliest dimensions, but of a particular kind only—commonly called stringybark. . . .* Similarly, Angus (1847) published coloured plates of wildflowers and of the grasstree (*Xanthorrhoea*) from Yankalilla, but in particular came scientific contributions. James Backhouse's botanical notes on his visit near Mt Lofty were published in England (1843), descriptions contrasting the grassless and the grassed formations by Behr were published in Germany (1847) and the famous plant taxonomist, von Mueller, was resident from 1847 to 1852. Tate (1879, 1883) and Maiden (1907) reviewed this early progress.

The world-wide study of *flora* had settled into a scientific discipline well before South Australian settlement in 1836, and the task of

---

* Department of Botany, University of Adelaide, North Terrace, Adelaide, S. Aust. 5000.

accounting the flora of the Adelaide region progressed smoothly from settlement on, but no equivalent discipline existed for study of the vegetation. Humboldt (1807) had laid foundations and Grisebach (1838) was attempting to consolidate the association concept but the catalytic publications of Darwin and Wallace lay 20 years ahead and the works of Raunkiaer, Coles, Schimper, Warming and Clements half a century further. This had two important consequences. First, there is little conceptual continuity between 19th and 20th century vegetation studies in the Adelaide region. In a theoretical framework, the former are not the progenitors of the latter which derive from ideas developed overseas, later imported full-fledged. Second, an informal but significant body of vegetation-study developed in 19th century South Australia uninfluenced by academic theory and dictated not by botanical motives but by the practical demands of life in the colony. Both deserve thought before attention is switched to modern studies.

So far as scholarly work is concerned, Tate was the greatest of South Australia's 19th century vegetation scientists. His papers of 1883 and 1879 illustrate vegetation description of that period, as does Schomburgk's summary of 1875. Various local investigators appreciated the association concept in a loose way, and the relatively great importance of soil differences in their determination—e.g., *Mr Tepper directed the attention of the members to the distinct zones of* Eucalyptus. *The relation of the vegetation of the hills to the rocks of which they were composed was referred to in order to show that from the aspect of vegetation conclusions as to the prevailing but hidden rocks could be drawn, just as from the known existence of certain rocks the character of the plants could be inferred.*[1] However, no local author seems to have consolidated such ideas into theory, and it is uncertain whether Tate's reference to certain vegetation . . . *transporting one mentally to the Scotch moors or those of North England* was more or less idle or indicates that he was thinking on the same lines as Raunkiaer. Collectively, local vegetation studies of that period display lack of comprehensive framework. No doubt because of his knowledge of flora, Tate bypassed physical attributes of vegetation as data, used species checklists instead, and achieved great contributions to the definition and comparison of floristic plant regions ("Eremian", "Euronotian", etc.; see Maiden 1907). His contributions were thus to plant geography rather than to community analysis.

It is interesting to note that we owe one of the Adelaide region's earliest "impact-evaluations" not to a botanist but to a geologist. Scoular (1880) provided not only a benchmark description of Munno Para vegetation and its edaphic correlations, but also a keen assessment of vegetation decimation . . . *since the advent of modern husbandry. . . .*

The informal body of knowledge that accumulated independently originated from practical necessity. Initial explorers tended to use the age-old terminology of everyday use in their homelands, thus *turf, meadow, parkland, wood* and so on, misconstrued by intending immigrants to imply equivalence with European types. This led to gross misapprehension of the productive characteristics of South Australian lands, contributing to the need for rapid shift of first settlement from Kangaroo Island to the Adelaide Plains (see Morphett 1837) and later to the misunderstandings of the wheat-farming frontier (see Meinig 1963). On the other hand, informal new terminology linked with opinions about land-use indicator value sprang up as a result of hard experience, thus *kangaroo grass, saltbush, mallee, teatree, polygonum flat* and so on, each implying a kind of plant cover and a kind of agricultural potential. This may not have reached the stature of "science", but it culminated in observations of scientific importance—e.g. Goyder's inferences linking rainfall reliability with native vegetation type, and it achieved a written record in the work of the Land Surveyors. Land surveyed as "Hundreds" was offered to buyers who had not seen it, hence the practice of mapping and annotating vegetation onto survey plans in attempts to convey the land's character. Some of this mapping applied over 20 different botanical terms to quite small areas, and showed vegetation boundaries so accurately that one hundred years later, they fit those seen on modern aerial photographs. Although Prescott (1929) subsequently presented a map abstracting features of this collective effort, the Land Surveyor's studies are neglected today, principally because of their loose approach to plant names. Their records are, however, of great importance in research. In a sense, the spirit of their applied approach is reflected now in

---

[1] Proceedings of the Field. Naturalists Society (1885).—*In Trans. R. Soc. S. Aust.* **9**, 257-300.

publications of the C.S.I.R.O. Soils and Land Use Series.

Modern studies

Sources

Accumulated knowledge dealing with Adelaide region flora (i.e. with distinguishing and identifying the different kinds of plants in the region), is the achievement of collectors and taxonomists too numerous to list. Accumulated knowledge is summarized in Black and Roberson's "The Flora of South Australia" (in four parts), second edition 1943-1957 augmented by Hj. Eichler's supplement (1965). Boomsma's "Native Trees of South Australia" (1972) should be noted. Floristic knowledge also is embodied in reference collections of plants in the State Herbarium of South Australia, where identified specimens may be examined.

In common with parallel world-wide studies, plant-taxonomic endeavour on Adelaide region plants has displayed great consistency and continuity. This reflects partly the nature of the material which long ago led all investigators to but one satisfactory interpretation (the species concept) and partly long-standing agreement among plant taxonomists about procedure (type specimens, priority, binomial nomenclature and so on). In contrast accumulated knowledge dealing with vegetation, i.e. with description and classification of the region's physical clothing or mantle of plants, has a history of varied approach and lack of continuity. This is explained partly by the nature of the material which consists not of individuals to be compared and grouped, but of a physically continuous mantle to be subdivided, and partly (in the Adelaide region) to the failure of imported theory to fit the facts of local vegetation.

The major accounts of Adelaide region vegetations are in Wood's "The Vegetation of South Australia" (1937) and Specht's (1972) rewritten second edition of that book. To these must be added five papers, Adamson & Osborn (1924), Jessup (1946 & 1948; both relevant in part only), Specht & Perry (1948) and Specht, Brownell & Hewitt (1961), plus the theoretical essay by Crocker & Wood (1947).

During the history of these studies, underlying concepts changed and evolved, and to appreciate the literature one must understand how these changes occurred, and when, as the following sections explain. For detail of actual vegetations, which cannot be condensed in any useful way, readers must refer to the publications listed.

Initial imported theories

Three areas of imported vegetation theory are reflected in Adamson and Osborn's pioneer work on Adelaide Hills vegetations. All are based in Darwinian survival of the fittest but lead to different classificatory approaches. First was the relatively "static" Association and Formation theory widely accepted (see Raunkiaer 1934, and Warming 1909) before stress was placed on the dynamics of vegetation.

Any area left without disturbance long enough will as a result of the struggle for existence be vegetated by those plants of the region best qualified to succeed in the given conditions. To the extent that undisturbed landscape presents different sets of conditions in zones or patches, different plant assemblages arise correspondingly, each of "sensibly homogeneous" floristic composition and expressing in physiognomy (= structure and adaptive characteristics of its plant mantle) the underlying conditions. Each such assemblage is a natural *community* and a distinguishable *unit* of the area's collective vegetation, and is called an *Association*. Regions with the same general regime of conditions elicit the same overall structure (forest, shrubland, grassland, etc.) in the associations which vegetate them, grouping such associations into *Formations*.

(It is interesting to note that local 19th century experts argued the reverse, i.e. that treeless areas were low rainfall areas because of the absence of trees; see Meinig 1963.)

Raunkiaer, who also regarded the "sensibly homogeneous" associations (he called them formations) as the units of vegetation, specialized the theory that the adaptational life-styles of their species best expressed the identity of the associations, regardless of species composition *per se*.

Raunkiaer invented a scheme to describe plant life-style. It went further than recognition of arborescent, shrubby, succulent (and so on) life forms, taking into account also which parts of the plants died back in the unfavourable season and in particular how plants carried their perennating organs (buds), whether exposed and elevated, or protected at ground level, or underground, or in seeds alone. He invented terminology for such adaptational types (e.g. chaemaephytes, hemicryptophytes) and determined the proportions in which the species of an association distributed across his fixed "Life-form spectrum" of types. These distributions allowed direct comparison of any associations and their classification on similarity. (Had computer-implemented similarity-analysis been available in Raunkiaer's time, his system would have had wider appeal.)

Third were the very influential ideas of Clements (1916) who similarly recognized as field units the "sensibly homogeneous" tracts of vegetation and invoked the struggle for survival to explain them, but in no static sense. He was so impressed with the evolution of vegetations from one type to another (as may be observed in the reversion of an abandoned field from grassland to shrubland eventually to woodland) that he decided all vegetations must be similarly related in trends of evolution towards stable end points.

In any climatic regime, Clements saw all plant communities as stages in vegetation "organisms" of which the adult stages (the highest types of social organisms capable of evolving under the climatic regime) were termed "climax", and the immature stages "seral" from "succession", the pre-ordained process by which stages succeed one another towards the stable, climax state. Putative successions were deduced, and all classification and nomenclature of vegetation structured to reflect them (see Allred & Clements 1949).

Applications and developments in the Adelaide region

Adamson and Osborn 1924

Osborn was first botany professor at the University of Adelaide and Adamson, a botany professor from South Africa, was a visiting colleague. After half a century their pioneering paper still is used as a main source of information about Adelaide Hills vegetation (e.g. by Maud 1972). Seven divisions were recognized, viz. forests of stringybark, bluegum, manna gum, candlebark, peppermint and rivergum, plus swamp vegetation. Some of these divisions were complexes larger than the Association brought together on floristic similarity—e.g. stringybark forest, in which was included not only structural forest of *E. obliqua, E. baxteri* and *E. goniocalyx* but dwarfed *E. cosmophylla* and *E. baxteri* scrubs as well as *E. fasciculosa* woodland. Each division received detailed description and discussion covering habitat relationships, physiognomic and floristic character (including Raunkiaer analysis), internal variations and modifications. Adamson and Osborn acknowledged all three areas of imported

theory in their summary. Three formations were recognized, viz.: Stringybark, Savannah Woodland and Red Gum (*E. camaldulensis*) formations, and within each a successional interpretation was attempted. For instance in the first, the tall forests were considered climax, *E. cosmophylla* and *E. baxteri* scrubs were assigned stable edaphic (soil-limited) subclimax status and *Casuarina stricta* vegetation in stringybark areas was regarded as seral below *E. fasciculosa* woodland, which in turn was regarded as seral below forest-form stringybark. Raunkiaer's method was used to compare associations but not as a basis for classification *per se*.

Wood 1937

Wood was second botany professor at the University of Adelaide and the first to attempt a "Vegetation of South Australia" which extended treatment also to the coastal and other parts of the Adelaide region. No short account can do justice to Wood's book, which all students of local vegetation must read. The most important point to note here is that Wood adopted a total commitment to Clementsian theory as an overriding preoccupation and to the extent that such theory was misleading in its interpretation of local vegetation, so was Wood's account. So far as factual description of Adelaide region vegetation is concerned, Wood's account greatly augmented Adamson & Osborn's (1924).

Wood 1939

Shortly after issue of his book, Wood (1939) withdrew his commitment to Clementsian theory as the appropriate framework for interpreting overall local vegetation and pursued a fresh theoretical basis which, co-authored by Crocker, was published in 1947.

Crocker & Wood 1947

Acknowledging that Clementsian theory was inappropriate for application to the bulk of South Australian vegetation, Crocker and Wood explained why and proposed a more appropriate theory.

From extensive review and deduction, Crocker & Wood concluded that dramatic historical influences, not Clementsian processes accounted for the occurrence, composition and characteristics of plant communities now present in South Australia. They argued that some species from pre-existing systems, practically wiped out in early Quaternary aridities, survived on a happenstance basis in refuges while the edaphic character of landscape was altered by the aridity. With resumption of wetter conditions, surviving species capable of occupying the modified soil mantle did so under a variety of constraints, e.g. the patterns of rainfall improvement (determining which relict species from which refugia gained access to which new areas), the challenge of the new soil mosaics versus the tolerances of the migrating species, and so on. With the lifting of climatic stress no factor was more important than the soil mosaic in delineating plant distributions.

(Intriguing disjunctions of numerous species, vicarious pairs and the occurrence of much hybridization are consistent with this theory.)

For vegetation classification, Crocker & Wood advocated distinction of Associations as vegetation having constant dominant species in all layers and growing in similar habitats. Recommended internal subdivisions were the Type, where upper stratum dominants alone change locally and the Society, where only lower-stratum dominants change locally. In view of the fact that floristically-allied associations could be found on mosaics of allied soils, Crocker & Wood proposed recognition of the Edaphic Complex to group such associations accordingly. Transition zones were recognized, and overall structural nomenclature (e.g. Shrub Steppe) applied, but succession was no longer a preoccupation.

Jessup 1946, 1948 and Specht & Perry 1948

These papers show a renewed preoccupation with factual observation in which soil studies were pursued in parallel with vegetation analysis and in no less detail. In a sense these accounts over-reach the analysis of vegetation *per se* to deal with the issue of dominant species-soil type relationships. They are most valuable for their presentation of figures, tables, precise maps and data which expose field situations much better than do the generalized accounts of earlier workers. Specht & Perry's account emphasized the significance of ecotones in the Adelaide region and consolidated the edaphic complex concept.

Specht 1972

Successive editions of "The Australian Environment" show development of a system for distinguishing very broad divisions of Australian vegetation, culminating (in the 4th Edn) with Specht's table: "Structural Formations in Australia". Specht used this, together with the Land System concept (a recurring pattern of topography, soils and vegetation) for the framework in his re-write of Wood's book.

The majority of local vegetation was regarded as stable or climax, and was classified directly according to the hierarchical series: "Society", of constant structure and dominants in all strata; "Association", of constant structure and upper stratum dominants but variable lower stratum dominants; "Alliance", of constant structure and related upper stratum dominants but variable lower stratum dominants and "Structural Formation", of essentially similar structure but otherwise variable. Specht embraced all stable South Australian vegetation in 13 such Formations, grouping the unstable, successional coastal forms under Coastal Land Systems. In his tables, discussion, photographs, maps and bibliography, Specht provided our most comprehensive account of Adelaide region vegetation.

Prehistory of Adelaide region vegetation

Accounts of this subject (e.g. Crocker & Wood 1947; Wood 1959; Crocker 1959; Burbidge 1960; Specht 1972) are necessarily deductive and speculative because local palaeobotanical data are so very inadequate. Available to us are either glimpses from many million years ago (e.g. in Eocene time) or from a few thousand years ago, with little in between to connect them. The best course is to consider these data and then examine the essays mentioned above.

Suppose one considers first those features of the region's present flora and vegetation which find explanation in near prehistory. The most self-evident are the disjunctions in species populations—e.g. between Kangaroo Island and the mainland, or of occurrences in the Adelaide Hills of species otherwise restricted to Victoria and other eastern localities. For instance Cleland (1968) verified that the treefern *Dicksonia antarctica,* otherwise absent from South Australia, did occur last century in a few pockets just east of Mt Lofty, and Specht (1972) provided maps showing many similar disjunctions. To explain these, disruptions of former continuous distributions by alterations in the disposition of land, sea, soils and climate are invoked.

Geologists can show that southern Australia is in a warm fluctuation of an ice age. Only 15,000 years ago Gulf St Vincent was dry, the shoreline 70 km or so south of Kangaroo Island and the weather much cooler. Similar fluctuations have occurred repeatedly in

Quaternary time. It is easy to envisage (as does Parsons 1969) how such a history would allow the spread and subsequent interruption of many species populations; leading to the distributions seen today. Quaternary palynological data (e.g. Hall 1967[2], Parsons 1969, Martin 1973, Dodson 1974, Dodson & Wilson 1975) reach back 30,000 years or so to illustrate fluctuations that have occurred in the distribution of local vegetation components, but that is much too short a time to show origins. With regard to Tertiary floras, our best guide to the floristic and vegetational succession comes from palynology (Harris 1971, Stover & Partridge 1973) with occasional glimpses of the macrovegetation itself from such deposits as the Maslin Bay flora (Lange 1970).

At the time of deposition of the Maslin Bay flora (Upper Middle Eocene), southern Australia was at about 50°S and little of what we now regard as Adelaide region geography would have been obvious. The climate appears to have favoured broad-leafed mesomorphic vegetation mixed with conifers, unlike anything in South Australia today. Much of the Maslin Bay Eocene leaf-litter looks like that from forest floors in North Queensland and New Guinea. There are enigmas in these records. For instance, local Tertiary macrofossil floras display a great variety of conifers but not the genera found locally today; similarly the pollen spectra show abundant Myrtaceae but no reliable evidence for this family can be found among the fossil leaves.

Data covering the transition from Eocene floras and vegetations to those of the Adelaide region today are almost entirely palynological. There is very little at all from Miocene and Pliocene time. Also palynology is rather insensitive to key questions like the radiation of *Eucalyptus* of which pollen are very uniform. There are some local macrofossil floras which contain Myrtaceae and maybe ancestral *Eucalyptus* but the ages of these floras are unknown.

Various earlier claims for *Eucalyptus* from local Tertiary macrofossils (e.g. Chapman 1937) are erroneous. It seems best to conclude that the essays already mentioned are speculative, the fossil record insufficiently evaluated, and the prehistory of Adelaide region vegetation likely to remain enigmatic until much more palaeontological work is done.

The native vegetation today

Any traverse of the Adelaide region from Gulf St Vincent east to the Murray Plains intercepts vegetation which is still substantially or partly native in character. The appearance of most is a blend of native and modified features. Figures 1–16 illustrate examples. The idealized traverse starts with (1) gulf coastline, crosses (2) a coastal plain, ascends sharply through (3) western foothills and (4) deeply dissected western slopes to (5) the central Mt Lofty spine fringed with gully-heads of an east-west drainage divide, then descends through (6) broadly dissected highlands to (7) an eastern scarp. Depending on actual traverse route some of the illustrated vegetations might be missed, or alternatives observed, but the theme is: eucalypts to about 25 m high, usually less, most of woodland character (branching low with deep or spreading canopy; not crowded) but some of forest character (with tall boles; crowded) over herb and low shrub layers usually either prominent annual grass (parched straw-yellow in summer) or low (1–2 m high) xeromorphic scrub (grey-green winter and summer) which tends to lack grass, all dry and very fire-prone every summer.

Departures from this theme are introduced by a scatter of specialized habitats. For example, on certain soils the eucalypts are suppressed in stature to scrub forms and in extreme situations are absent altogether; thus some relict sands carry native pine (*Callitris*), skeletal soils about metasediment outcrops carry sheoak (*Casuarina*) and vegetation of

---

[2] Hall, E. A. A. (1967).—"Analysis of a semifossil organic deposit near Palmer, South Australia". M.Sc. Thesis, University of Adelaide (unpubl.).

Fig. 1. Tidal mudflat vegetation of mangroves (*Avicennia*).

Fig. 2. Coastal dune vegetation: a–b, herbaceous sea rocket (*Cakile maritima*), true spinifex (*Spinifex hirsutus*), grey saltbush (*Atriplex cinerea*), sandhill daisy (*Olearia axillaris*), boobyalla (*Myoporum insulare*), coastal wattle (*Acacia sophorae*) and boxthorn (*Lycium*); c, reedbeds (*Phragmites communis*).

Fig. 3. Woodland on relict sand near Normanville: a, honeysuckle (*Banksia marginata*); b, wattle (*Acacia retinodes*).

Fig. 4. Saltflats of samphire (*Salicornia*).

1
4m
2
b
4m
a
c
3
b
4m
a
4
1m

5
b
a
6m
6
b
a
4m
7
6m
8
4m

some of the eastern scarp is almost entirely herbaceous. A short expansion of the overall picture follows, linked to figures 1–16.

Coastal vegetations

The gulf coast presents several distinctive, extreme habitats. First is seaward tidal mud-flat where land plants must cope with anaerobic substrates and regular seawater inundation. Vegetation is mangrove (*Avicennia*) alone, up to 6 m tall (Fig. 1). On the landward side is saltflat typically with succulent samphire (e.g. *Salicornia*) mats 1–2 dm high (Fig. 4). Third is coastal dune where plants cope with beach sand and salt spray in a gradation of situations from water's edge to elevated crest. Vegetation is of corresponding gradational character, from herbaceous sea rocket (*Cakile maritima*) and true spinifex (*Spinifex hirsutus*) near the strand through shrubs like grey saltbush (*Atriplex cinerea*), sandhill daisy (*Olearia axillaris*), boobyalla (*Myoporum insulare*), coastal wattle (*Acacia sophorae*) and boxthorn (*Lycium*) at various positions back from it (Fig. 2, a–b). A fourth extreme habitat (not exclusively coastal) is swampy watercourse, sometimes fringed with paperbark (*Melaleuca halmaturorum*), or occupied (Fig. 2, c) by reedbed vegetation (*Phragmites communis*). None of these habitats support eucalypts or tall vegetation, but where landform carries well-drained loams to near the sea's edge (south of Adelaide) eucalypt vegetation occurs and reflects little of its proximity to the sea except for wind spray pruning (as in Fig. 5, b).

Vegetations of coastal plains

Remnant vegetations are few. Rivergums (*E. camaldulensis*), which may achieve the greatest girths of any native tree in South Australia, persist on some creeklines in native pattern. and scattered individuals from various former vegetations can be found, for instance some native pine (*Callitris*), native cherry (*Exocarpus*), wattle (*Acacia* spp.), rushes (*Juncus*) and grasses, but intact vegetation layers of which Figures 3, 5 and 6 show examples are rare. Figure 3 shows honeysuckle (*Banksia marginata,* a) — wattle (*Acacia retinodes,* b) woodland about 5–7 m tall on relict sands near

Normanville, but the understrata are of introduced plants, viz.: hare's tail grass (*Lagurus ovatus*) and cotton-bush (*Asclepias*). Figure 5 shows the remains of a mallee-box (*E. porosa*) — dryland teatree (*Melaleuca lanceolata*) woodland (a, b resp.) in an agricultural field near Moana, and Figure 6 shows the remains of a sheoak (*Casuarina cristata*) — *E. porosa* woodland near Maslin Beach, with the native grass *Danthonia* still present.

Western foothill vegetation

Typical western foothills situations (e.g. Belair to Eden Hills) carry a sparse grey-barked woodland of peppermint (*E. odorata*) invaded with introduced plants such as olive, bone-seed (*Chrysanthemoides monilifera*), cotton-bush and much alien grass among which various native plants persist. The floors, sides and upper flanks of steep valleys draining through the foothill face may also carry river gum, blue gum (*E. leucoxylon*) and pink gum (*E. fasciculosa*) respectively.

Vegetations of dissected country on either side of the central Mt Lofty Spine

Extensive remnant woodlands and forests of blue gum, pink gum and manna gum (*E. viminalis; E. huberana*) occur in a mozaic according to soils, aspects, rainfalls and drainages. Tall candlebarks (*E. rubida*) have a very limited distribution in the wettest high valleys, river gums line the watercourses and sometimes spread onto adjoining flats, and swamp gum (*E. ovata*) may occur with the river gum. All these species have smooth, light-coloured bark at least on their limbs.

Preservation of their understrata is generally very poor; most of these woodlands have been cleared for farming and the residual native trees combine with introduced pasture to present a groomed, rural, parklike appearance. Alien plants often dominate watercourses and roadside verges, for example willows (*Salix*), blackberry (*Rubus*), gorse (*Ulex*) and broom (*Cytisus*). Summer-parched grass is a prominent feature of this vegetation in particular.

Despite modifications, some or much of the native lesser strata may be distinguished in places, notably in reserves. Where native preser-

---

Fig. 5. Agricultural field near Moana: a, remains of mallee box (*Eucalyptus porosa*); b, dryland teatree (*Melaleuca lanceolata*) woodland.

Fig. 6. Woodland near Maslins Beach: a, *Eucalyptus porosa*; b, *Casuarina cristata*.

Fig. 7. Pink gum as a woodland cover over artificial pasture near Myponga.

Fig. 8. Road verge vegetation near Myponga of golden wattle, blackboy (*Xanthorrhoea*), *Dianella* and native grasses.

9
4m
10
4m
11
10m
12

vation is good, tall undergrowth may be present, for instance of wattle (*A. pycnantha*), honeysuckle (*B. marginata*), blackthorn (*Bursaria spinosa*), native cherry (*Exocarpus aphyllus*) or (on watercourses) blackwood (*Acacia melanoxylon*). Under this may be a layer of lesser shrubs like *Olearia axillaris* or *Spyridium vexiliferum* and under that still smaller shrubs such as guinea flower (*Hibbertia* spp.), scarlet runner (*Kennedia prostrata*) and sedges and tussocks like *Lepidosperma viscidum* and *Dianella revoluta*. There is always a grassy ground layer (e.g. of kangaroo grass (*Themeda australis*), a wide variety of annual herbs and a spring growth of bulb plants, including orchids. Peppermint reoccurs in the east of this dissected country, where sparse sheoak over low grassland is also extensive, and much of the scarp itself is more or less treeless, with irongrass (*Lomandra*) prominent.

As examples, Figure 7 shows pink gum as a woodland cover over artificial pasture near Myponga where road verge vegetation indicates a former substratum including golden wattle (Fig. 8), blackboy (*Xanthorrhoea*), *Dianella* and native grasses. Figure 9 shows a young river gum woodland over rushes (*Juncus*) and the introduced grass, *Pentaschistis thunbergii* on flats east of Charleston while Figure 10 from the same locality shows mixed manna gum—blue gum vegetation approaching forest character, with a wattle-sheoak layer over a sparse *Olearia axillaris* layer over a ground cover of sedge, grass and herbs. Figure 11 shows a woodland-form blue gum tree typical of the dissected high land, with sheoak on a rock outcrop in cleared agricultural pasture, while Figure 12 shows a view typical of this dissected country as a whole, viz.: a parklike scatter of mature trees from the former vegetation (in this case mixed blue and manna-gum near Bull Creek) with most land cleared, and a line of well established aliens (here including willow) on the creekline.

Vegetations of the central divide
The characteristic trees of the Mt Lofty spine and related situations are brown stringybarks (*E. obliqua* and *E. baxteri*) which have dark fibrous bark and often grow crowded in forest form, up to 25 m tall. North of the Torrens Gorge *E. goniocalyx*, of similar character, also occurs. The characteristic substratum is dense sclerophyllous scrub about 1–2 m high, for instance of needle bush (*Hakea rostrata* and *carinata*), *Leptospermum myrsinoides, Platylobium obtusangulum,* bushpea (*Pultenaea daphnoides*), *Isopogon ceratophyllus* and the tussocky *Lepidosperma semiteres*. Annual grass tends to be absent. In places the trees are less tall and dense, or even dwarfed and mixed with low cup gums (*E. cosmophylla*), but the underlying scrub is similar and tends to unify all examples.

Figure 15, southeast of Kuitpo, shows a profile of typical stringybark forest exposed by bulldozing. Figure 16 shows residual *E. obliqua* at Prospect Hill, cleared beneath for pasture. Figure 14 shows stringybark vegetation on Mt Bold watershed where *E. obliqua* is of low stature (a), cup-gum occurs (b) and tussocky *Lepidosperma semiteres* (c) is very prominent. Figure 13 shows dwarfed stringybark (*E. baxteri*) scrub near Hindmarsh Falls, typical of much on southern Fleurieu Peninsula. As a rule the scrubby undergrowths of stringybark vegetations are better preserved than are undergrowths of the pale-barked woodlands because their habitats have been less attractive to farmers and introduced plants alike.

The future
With minor exceptions (Martin 1961; Heddle 1975)[3] computer-implemented philosophies of objective vegetation analysis have been neglected in the Adelaide region. Such approaches contrast with all of the foregoing by attributing importance not necessarily to the physically predominant species but to the central species of statistical association patterns. Data-storage and retrieval have been neglected also; there is no local equivalent of the Victorian work of Churchill and de Corona (1972), for example. It seems likely that these

---

[3] Heddle, E. M. (1975).—"Effects of Man on the vegetation in the National Parks of South Australia." Ph.D. Thesis, University of Adelaide (unpubl.).

Fig. 9. Young rivergum woodland over rushes (*Juncus*) and introduced grass on flats east of Charleston.
Fig. 10. Mixed manna gum/blue gum vegetation with understorey of wattle and sheoak, sparse *Olearia axillaris* and ground cover of sedge, grass and herbs on flats east of Charleston.
Fig. 11. Woodland form blue gum with sheoak on rocky outcrop in cleared agricultural pasture.
Fig. 12. Typical view of dissected country near Bull Creek—mixed blue and manna gum with aliens, including willow, along the creek line.

13
a
4m
14
a
6m
b
c
15
16
25m

aspects will grow in importance and the further pursuit of native associations (based on predominant species) will wane, for the following reasons.

Modifications inevitably will blur and erase the native vegetation patterns of our settlement era and vegetation ecologists will have—indeed, do have—more pressing issues to clarify, particularly emerging trends of new patterns set by the interplay of pre- and post-settlement flora under European man's influence. What are these patterns and are they desirable or undesirable from the viewpoint of wildlife preservation, bushfire risk, appeal as human habitat, and so on?

It is most unlikely that the approaches of the past will be optimal here; instead techniques of data-gathering, storage and objective analysis are called for that embody no preconceptions about relative importance of species (who is to know what will become important?), and that emphasize monitoring of trends. Benchmark studies on emerging "new" vegetations are needed, with the needs of future ecologists to gain full access to the older data records kept well in mind.

If there is to be another contribution to the "natural vegetation" series before that era ends, it would do well to concentrate on benchmark records for future access.

References

ADAMSON, R. S. & OSBORN, T. (1924).—The ecology of the *Eucalyptus* forests of the Mount Lofty Ranges (Adelaide District), South Australia. *Trans. R. Soc. S. Aust.* **48**, 87-145.

ALLRED, B. W. & CLEMENTS, E. S. (1949). "Dynamics of Vegetation". (H. W. Wilson: New York.)

ANGUS, G. F. (1847).—"South Australia Illustrated." (T. McLean: London.)

BACKHOUSE, J. (1843).—"A narrative of a visit to the Australian colonies." (Hamilton, Adams: London.)

BEHR, H. (1847).—On the character of the South Australian flora in general. *Linnaea* **20**(5), 545-558.

BOOMSMA, C. D. (1972).—Native trees of South Australia. Bull. No. 19. (S. Aust. Woods & Forests Dept: Adelaide.)

BURBIDGE, N. T. (1960).—The phytogeography of the Australian region. *Aust. J. Bot.* **8**, 75-212.

CHAPMAN, F. (1937).—Descriptions of Tertiary plant remains from Central Australia and from other Australian localities. *Trans. R. Soc. S. Aust.* **61**, 1-16.

CHURCHILL, D. M. & DE CORONA, A. (1972).—"The distribution of Victorian plants." (Royal Botanical Gardens & National Herbarium & Department of Botany, Monash University: Melbourne.)

CLELAND, J. B. (1968).—*Dicksonia* in the Mount Lofty Ranges. *S. Aust. Nat.* **43**, 27-28.

CLEMENTS, F. E. (1916).—"Plant Succession." (Carnegie Inst.: Washington.)

CROCKER, R. L. (1959).—Past climatic fluctuations and their influence upon Australian vegetation. *Monographiae biol.* **8**, 283-290.

CROCKER, R. L. & WOOD, J. G. (1947).—Some historical influences on the development of the South Australian vegetation communities and their bearing on concepts and classification in ecology. *Trans. R. Soc. S. Aust.* **71**, 91-136.

DODSON, J. R. (1974).—Vegetation history and water level fluctuations at Lake Leake, southeastern South Australia. I. 10,000 B.P. to present. *Aust. J. Bot.* **22**, 719-741.

DODSON, J. R. & WILSON, I. B. (1975).—Past and present vegetation of Marshes Swamp in southeastern South Australia. *Aust. J. Bot.* **23**, 123-150.

DUTTON, F. (1846).—"South Australia and its mines with an historical sketch of the colony, under its several administrations, to the period of Captain Grey's departure." (Boone: London.)

EICHLER, H. (1965).—"Supplement to J. M. Black's Flora of South Australia (2nd Edtn, 1943-1957)." (Govt Printer: Adelaide.)

GOUGER, R. (1838).—"South Australia in 1837; in a series of letters, with a postscript as to 1838." 2nd Edtn. (Harvey & Darton: London.)

GRISEBACH, A. (1838).—Ueber den Einfluss des Climas auf die Begränzung der naturlichen Floren. *Linnaea.* **12**, 159-200.

HARRIS, W. K. (1971).—Tertiary stratigraphic palynology, Otway Basin. *Spec. Bull. Geol. Surv. S. Aust. Vict.*, 67-87.

HUMBOLDT, A. VON (1807).—"Essai sur la geographie des plantes; accompagné d'un tableau physique des regions equinoxiales, Fondé sur des mesures exécutées depuis le dixième degré de latitude boréale jusqu'au dixième degré de latitude australe, pendant les années 1799, 1800, 1801, 1802 et 1803 par Al de Humboldt et A. Boupland." (Levrault, Schoell: Paris.)

JESSUP, R. W. (1946).—The ecology of the area adjacent to Lakes Alexandrina and Albert. *Trans. R. Soc. S. Aust.* **70**, 3-34.

JESSUP, R. W. (1948).—A vegetation and pasture survey of Counties Eyre, Burra and Kimberley, South Australia. *Trans. R. Soc. S. Aust.* **72**, 33-68.

Fig. 13. Dwarfed stringy bark scrub (*Eucalyptus baxteri*) near Hindmarsh Falls.
Fig. 14. Stringy bark vegetation on Mt Bold watershed: a, *Eucalyptus obliqua*; b, cupgum; c, *Lepidosperma semiteres*.
Fig. 15. Profile of typical stringy bark forest exposed by bulldozing southeast of Kuitpo.
Fig. 16. Residual *Eucalyptus obliqua* at Prospect Hill.

LANGE, R. T. (1970).—The Maslin Bay flora, South Australia. 2. The assemblage of fossils. *Neues Jb. Geol. Paläont. Mh.*, 486-490.

LIGHT, W. (1839).—"A brief journal of the proceedings of William Light, late Surveyor-General of the Province of South Australia, with a few remarks on some of the objections that have been made to them." (MacDougall: Adelaide.)

MAIDEN, J. H. (1907).—A century of botanical endeavour in South Australia. *Rep. Aust. Ass. Adv. Sci.* **11**, 159-199.

MARTIN, H. A. (1961).—Sclerophyll communities in the Inglewood District, Mount Lofty Ranges, South Australia. Their distribution in relation to micro-environment. *Trans. R. Soc. S. Aust.* **85**, 91-120.

MARTIN, H. A. (1973).—Palynology and historical ecology of some cave excavations in the Australian Nullabor. *Aust. J. Bot.* **21**, 283-316.

MAUD, R. R. (1972).—Geology, geomorphology and soils of central county Hindmarsh Mount Compass-Milang), South Australia. *C.S.I.R.O. Soil Publ. No. 29.*

MEINIG, D. W. (1963).—"On the margins of the good earth" (John Murray: London.)

MORPHETT, J. (1837).—"South Australia; latest information from this colony, contained in a letter written by Mr Morphett, dated Nov. 25th 1836." (Gliddon: London.)

PARSONS, R. F. (1969).—Distribution and palaeogeography of two mallee species of *Eucalyptus* in southern Australia. *Aust. J. Bot.* **17**, 323-330.

PRESCOTT, J. A. (1929).—The vegetation map of South Australia. *Trans. R. Soc. S. Aust.* **53**, 7-9

RAUNKIAER, C. (1934)."The life form of plants and statistical plant geography." (Oxford Univ. Press: Oxford.)

SCHOMBURGK, R. (1875).—"The flora of South Australia." (Govt Printer: Adelaide.)

SCOULAR, G. (1880).—The geology of the Hundred of Munno Para. Part II. The upland Miocene and fundamental rocks and economics. *Trans. R. Soc. S. Aust.* **33**, 106-128.

SOUTH AUSTRALIAN ASSOCIATION (1834).—"South Australia: outline of the plan of a proposed colony to be founded on the south coast of Australia; with an account of the soil, climate, rivers, etc." (Ridgway: London.)

SPECHT, R. L. (1972).—The vegetation of South Australia." 2nd Edtn. (Govt Printer: Adelaide.)

SPECHT, R. L. & PERRY, R. A. (1948).—The plant ecology of part of the Mount Lofty Ranges. 1. *Trans. R. Soc. S. Aust.* **72**, 91-132.

SPECHT, R. L., BROWNELL, P. F. & HEWITT, P. N. (1961).—The plant ecology of the Mount Lofty Ranges, South Australia. 2. The distribution of *Eucalyptus elaeophora. Trans. R. Soc. S. Aust.* **85**, 155-176.

STOVER, L. E. & PARTRIDGE, A. D. (1973).—Tertiary and late Cretaceous spores and pollen from the Gippsland Basin, southeastern Australia. *Proc. R. Soc. Vic.* **85**, 237-286.

TATE, R. (1879)—The natural history of the country around the head of the Great Australian Bight. *Trans. Proc. phil. Soc. Adel.* **1**, 94-128.

TATE, R. (1883).—The botany of Kangaroo Island. *Trans. Proc. R. Soc. S. Aust.* **6**, 116-171.

TORRENS, R. (1835).—"Colonization of South Australia." (Longman: London.)

WARMING, E. (1909).—"Oecology of plants." (Clarendon Press: Oxford.)

WOOD, J. G. (1937).—"The vegetation of South Australia." (Govt Printer: Adelaide.)

WOOD, J. G. (1939).—Ecological concepts and nomenclature. *Trans. R. Soc. S. Aust.* **63**, 215-223.

WOOD, J. G. (1959).—The phytogeography of Australia (in relation to radiation of *Eucalyptus, Acacia,* etc.). *Monographiae biol.* **8**, 291-302.

CHAPTER 8

THE ABORIGINAL INHABITANTS AND THEIR ENVIRONMENT

by R. W. ELLIS*

The setting

The first European navigators to visit the shores of the Australian continent discovered an environment which, though not exhibiting dramatic evidence of the fact, had already been modified by human activity. Indeed, the concept of a "natural" environment is in almost all circumstances a problematical one.

Archaeological studies indicate continuous human occuption of the Australian continent for at least 40,000 years; sufficient time for even the limited tool-kit of environmental modifiers utilised by Aboriginal man to have had some discernable effect. Certainly the area surrounding the present site of metropolitan Adelaide was the scene of deliberate Aboriginal environmental manipulation, almost entirely dependent upon the use of fire.

Fire and its impact

Flinders (1814) reported fires, presumably lit by Aboriginals, on several occasions while following the south-eastern coast of Australia; the earliest colonists settling near Adelaide discovered to their concern that during the dry summer months, the Kaurna tribespeople who occupied the area, deliberately lit huge bush fires, to encourage the new, tender plant growth, which subsequently attracted the game upon which they depended and as a means of collecting animals overcome by the fire or smoke. The extent of these fires was considerable, as is indicated by the following description of such a fire in February, 1837:

"Before next day's sunrise a great change took place in the landscape before us. The watchers on deck beheld a fire on one of the hills, which seemed to spread from hill to hill with amazing speed. All on board were now awake and on deck looking at this grand, and yet to those, who knew not its cause, fearful conflagration, as it seemed as if the whole land was a mass of flame. In the morning a great change had taken place; the whole range was black as midnight, except where trees were burning, and shortly after we landed, the mystery was explained. At the end of summer as this was, the natives had set fire to the long grass to enable them more easily to obtain the animals and vermin on which a great part of their living depends." (Finlayson 1903.)

Significantly, it was only two years after this event that the European colonists took legal action to limit this traditional practice—perhaps the first action in the colony which might be described as "environmental litigation". An Aboriginal man, known to the Europeans as Willamy, was arrested on a charge of "wilfully and maliciously setting fire to the grass". (S.A. Gazette & Colonial Register, 9th March, 1839.)

While such apparently anarchical practices on the part of the Aboriginal inhabitants were quickly suppressed by the European authorities, it was not long before the benefits of controlled burning were recognised by the immigrant settlers. Just three years after Willamy's prosecution one anonymous observer was moved to comment:

"These fires, which generally occur in January or February, burn off not only all the dry grass and weeds which the cattle will not eat, but also the leaves and fallen branches of trees; and not infrequently entire trees of large growth are brought down, and partly consumed. . . . Whatever alarm or danger may be produced on such occasions, one beneficial result is certain, the annual top dressing of the soil of fertilizing ashes. Many a fair tree, however, would pass unscathed by fire, but for the experienced and interested eye of the settler. He finds some heavy trees which he wants to take down, and carefully examines their lower stems, and some blackened spots where the fires have taken effect . . . as sure as he sees these tokens, he knows he can burn down such trees by applying fire to these places. . . . I am aware opinions have been expressed that the fires injure the grass, but this is contrary to my experience. Every year our grass is partly burnt, and every year it has become more closely planted." (The South Australian Magazine (1842), 401-2.)

---

* South Australian Museum, North Terrace, Adelaide, S. Aust. 5000.

How great was the effect of thousands of years of such periodic firing upon the vegetational and faunal complexes of the area, prior to European settlement, must remain conjectural. Furthermore, bushfires ignited by natural agencies may well have resulted in even greater immediate impact had this annual firing of the bush not occurred. It seems reasonable, however, to suggest that the open grasslands described by the early colonists in the Adelaide area were a direct product of such "fire-stick farming": the large expanses of shoulder-high grass covering these plains, which so impressed the first European settlers, may have been of anthropogenic origin. The contrast between the low-lying *Eucalyptus* shrubland of the then uninhabited Kangaroo Island and the *Eucalyptus* savannah-woodland of the continually occupied mainland appears also to add weight to this suggestion.

It is difficult to picture the scene described by Stephens (1889):

"The site of the present Norwood was then a magnificent gum forest with an undergrowth of Kangaroo Grass, too high in places for a man to see over. In fact people had lost their way in going from Adelaide to Kensington in those days."

Nor need the result be related directly to the effect of fire upon vegetation. An alteration in soil structure, which might subsequently have influenced vegetational complexes may also have arisen from the use of fire; this in part, has led other students to describe the practice of firing of bushland as "proto-agriculture". Similarly, increased erosion, particularly gullying (see Chapter 2) may also have resulted from the practice of regular firing and consequent vegetational depletion.

Other environmental modifications

Several other results of the influence of Aboriginal man upon the environment may also be suggested as a result of inferences from archaeological evidence. The introduction of the dingo, believed to have occurred about 8,000 years ago, may also have directly influenced the original faunal assemblages of the Australian mainland. It has, for example, been suggested that its introduction, following the separation of Tasmania from the mainland, may have hastened the disappearance of the Thylacine and, in more recent times, *Sarcophilus,* both of which were extinct on the mainland at the time of European settlement, but present in Tasmania where the dingo was absent. Similar claims have been made with regard to a human role in the extinction of some of the giant marsupials, an event which in at least several cases demonstrably postdates the arrival of Aboriginal man (Milham & Thompson 1976). Certainly, evidence strongly suggests the extinction of the elephant seal in Tasmania was a direct result of human over-exploitation in prehistoric times, while changes in shell-fish populations, inexplicable in terms of climatic or other natural environmental change, may have been induced by the activity of coastal societies with patterns of food gathering similar to those described for the Kaurna of the Adelaide district (Swadling 1976).

Unlike the later European migrants, Aboriginal man's limited and specialised technology, with the exception of his use of fire, has made little directly attributable alteration to the Australian environment. Micro relief features, such as Aboriginal arrangements of stones, fish traps, midden heaps, areas cleared of vegetation and hollowed trees utilized as shelters, may still be identified in some areas of the Adelaide Plains, but it requires, in most cases, an informed observer to trace the existence of these modified features.

The Kaurna

The group occupying the Adelaide region most commonly referred to themselves collectively by the name *Kaurna*. It should perhaps be noted that various alternative names were employed both by themselves and their neighbours. These are listed in detail by Tindale (1974).

Several attempts have been made by field workers to define the boundaries for the Kaurna territory. Perhaps the most widely accepted is that of Tindale (1974) who describes a territory exhibiting a close correlation between cultural boundaries and geomorphological and vegetational features. To the west, the territory was bounded by the shores of Gulf St Vincent, running from Cape Jervis in the south to Port Wakefield in the north. The boundary continued in the northwest, along the eastern side of a group of hills now known as the Hummocks, to perhaps as far north as Snowtown and Crystal Brook. To the east and southeast of the Adelaide Plains the territory was bounded by the high ridges and dense stringy bark forests of the Mt Lofty Ranges. Further north, the eastern boundary was defined by the low-lying hills of the mid-north region which intersect the mallee-covered plains.

The use of natural phenomena to delineate territorial boundaries is a feature common to many Australian groups. It derives logically from a mode of existence which is based upon an environmental awareness and the intelligent utilization of natural sources of food, water and raw materials for continued human survival and the fabrication of artefacts. It also coincides with a mythology and intellectual life which places considerable emphasis upon natural features. These mythologies provide dramatic accounts of the creation of topographical features recognised in the landscape and this explanation of origin in turn imbued them with even greater significance in the belief and oral traditions of the society.

Organisation

The group referred to as the Kaurna, while sharing a common language, social organisation and belief system, was for most purposes simply an amalgamation of more basic collective units. In 1842 the entire group probably numbered no more than 700 individuals. Indeed, even prior to the 1836 settlement it is believed that the effects of European occupation elsewhere in Australia had resulted in a rapid decline of numbers in this and other areas surrounding the River Murray, along which a smallpox epidemic is believed to have spread from eastern Australia (Stirling 1911).

The landowning units were, it seems, these basic local groups probably numbering no more than 30 individuals, who collectively constituted the group referred to as Kaurna. While the basic local units appear to have been patrilineal descent groups, kinship and social interaction was based upon the operation of two, named, exogomous, matrilineal moieties, *Kararu* and *Matheri*. The Kaurna, unlike the Murray societies, practised circumcision, and their closest social ties appear to have been with their northern neighbours who shared similar customs and systems of social organisation. The Mt Lofty Ranges, which separated the Kaurna from the Murray Plains, appear traditionally to have operated as an important social and cultural barrier.

In order to clarify the system of land ownership operating amongst the Kaurna, it is perhaps useful, following Stanner (1965), to draw a distinction between what he calls an "estate" and a "range". The "estate", according to Stanner, was the traditionally recognised locus ("home", "country", "ground"), of some kind of patrilineal descent group forming the core or nucleus of the territorial group. This area seems usually to have been more or less geographically continuous. The "range" normally included the "estate", but in some circumstances the two could be practically disassociated. The "range" appears to have been determined by considerations of survival, and might extend by a common understanding into the territories of neighbours during times of need. The two together, Stanner suggests, constitute a "domain"—an ecological life space. Reference to the historical descriptions of the Kaurna suggest that Stanner's distinctions are appropriate. Two terms which are relevant to our concerns are recorded in vocabularies collected from the Adelaide groups. The first, *Yerta*, perhaps equivalent to Stanner's "range", is occasionally encountered as meaning a definite territory or as a general term for land. The second, *Pangkarra*, has a more limited meaning, roughly equivalent to the "estate". This is given as, "a district or tract of country belonging to an individual, which he inherits from his father. (*Ngarraitya paru aityo pangkarilla;* 'there is abundance of game in my country'.) As each *pankarra* (sic) has its peculiar name, many of the owners take that as their proper name, with the addition of the term *burka;* for instance *Mulleakiburka, Karkulyaburka, Tindoburka,* etc. (sic). Another mode of giving names to themselves is to affix the same term, or *itpinna*, to the surname of one of their children; as *Kadlitpiana* (Captain Jack), *Wauwitpinna, Wirraitpiunna,* etc." (Teichelmann & Shurmann 1840).

Both the "estate" and the "range" (*pangkarra* and *yerta*), appear to have been named and delineated units of land and, in the case at least of the *pangkarra*, to have been derived from ecological considerations. Reference to vocabularies suggest, for example, that *wirra itpiunna* is derived from *wirra*—"forest", "scrub", while elsewhere Teichelmann & Shurmann (1840) give *Mullawirraburka* (an individual known to the Europeans as "King John"), literally "dry—forest (mallee), oldman".

The transference of *pangkarra* names, derived from environmental considerations, into the social domain, with the addition of the term "burka"—"old (mature) man" or *itpinna,* does, however, emphasise one feature of Kaurna systematics—the apparent operation of cross domain systems of classification.

Crossing and linking the estates and ranges of the basic groups was an extensive mythological system and cosmology which provided a systematic ordering of topographic features and vegetational zones into different categories or classes, often linked by a common mythological origin or creative being. These systems of classification, derived from the operation of Kaurna cosmologies, interacted with the ecologically determined landscape classifications.

Several of the names of localities appear to have been derived from the creative heroes or events associated with them: *Yurebilla* ("two ears"—Mt Lofty; asociated with a creative hero and since corrupted as a European place name into Uraidla), *Tarndana* ("South Adelaide"—associated with *Tarnda* "red kangaroo" who introduced cicatrisation "and was later transformed into a kangaroo") (Gell 1903).

An investigation of the various resources available provides us with an extensive inventory of named areas and features of the Adelaide Plains and surrounding areas (e.g. Teichelmann & Schurmann 1840; Wyatt 1878; Black 1920). Unfortunately it is often difficult to distinguish from these documentary sources whether the names relate to particular features, vegetational complexes or territorial units, so that their usefulness to sustained enquiry is limited. It is clear that in a number of cases culturally significant areas or territories were named, and probably delineated, on the basis of ecological considerations, while in others the recognition of landscape units was based upon mythological associations.

Aboriginal utilization of the environment

The Kaurna, despite their relatively narrow tribal territory, enjoyed considerable variation of food resources and utilizable raw materials. The flora and fauna ecosystems of the region varied from the marine systems of the coastline, the coastal sand dunes, the marshes and lagoons, and the riverine estuaries and littoral zones, to the plans and undulating uplands of the north, and the ranges to the east.

Besides the regional variation, seasonal variation in food supplies also occurred and influenced the periodic distribution of groups.

One early commentator, in a discussion of the habits of the tribal groups of the lower north of South Australia, notes that "The bokra (a small kangaroo rat) afforded their chief supply of animal food at all times, but more especially during the summer months . . ." (Browne 1897). Similarly, other commentators have recorded that the gum of various trees formed an important part of their diet during the summer months. Tindale (1974) also records that during the winter months the Kaurna hunted possums and collected grubs in the wet sclerophyll scrub covering the Mt Lofty Ranges.

Journal entries available for study would seem to indicate a preference by the Aboriginal people for the coastal areas in summer, and for the timbered areas during the winter months. This is not to suggest a complete depopulation of the coastal camps, as it is possible that the large catches of fish during the spring and summer months enabled the Aboriginals to congregate in large groups, but with the coming of winter and a shortage of food, shelter and firewood, these groups may have been forced to disperse and spread themselves more thinly along the coast, thus giving the appearance of a decrease in population at any one centre.

Early descriptions of the economic activities of the Kaurna indicate the common division of labour based on sex; men hunting and providing meat (*paru*), and women providing the vegetable foods (*mai*), which formed the major part of the diet of the group, though there are occasional references to men occupied in collecting vegetable foods which might otherwise be the concern of the women. Similarly, two Aboriginal women engaged by Colonel Light to follow the *Rapid* up the Gulf during his search for a suitable site for settlement, provided fresh meat for the vessel from a kangaroo they had captured.

Vegetable foods formed the greater part of the diet because of their reliability and the diversity of available sources. There is no evidence of the existence of special avoidances with regard to these foods.

Seasonal variation in the availability of vegetable foods did occur and knowledge of this was apparently exploited by the Kaurna, who appear to have regulated their movement and ritual life on the basis of anticipated periods of surplus which would enable large and relatively fixed gatherings of people.

Although meat comprised a relatively small part of the diet, it was accorded considerable importance among Aboriginal groups. Strict rules of conduct determined its preparation and the types of animals and parts of a carcass which could be consumed by people in various status categories.

Seasonality was also reflected in the material culture of the group. During the summer months, when habitation was close to the coast, the shelters constructed by the Kaurna were often little more than windbreaks, a few branches laid in semi-circular form, against which the head was placed while sleeping, while a small fire was located in the centre of the circle. If inclement weather threatened, the same semi-circular form was maintained while the sides were elevated against a framework of interlocking branches and covered with bark, grass, seaweed and earth, meeting at the top to form an arch.

With the approach of autumn, or *Wadlwor-ngatti*: "the time of building huts against fallen trees", more substantial shelters were built. During this time there was a general movement inland towards the foothills within the tribal boundary and large trees were incorporated into the larger waterproofed semi-circular hut called *Wattowadli*. In some cases, large, burned and hollowed trees may have served as shelters.

During the winter months large skin or seaweed cloaks were worn to protect the Kaurna from the rain and cold. Skin rugs were made from possum, rabbit bandicoot, wallaby or kangaroo skins. It is believed they preferred the more pliable rugs made from the skins of smaller animals such as the possum.

Naturally occurring materials were also widely used in the manufacture of domestic items and weapons. The most important of these materials were probably the hard native timbers. Archaeological evidence suggests that over a period of time wood gradually replaced stone as the most important material in the modern Aboriginal tool-kit.

Stone was employed at the time of European contact for the manufacture of a crude flaked knife and in the form of a hafted flake fastened with gum to a wooden handle. Unworked stones with natural cutting edges were used for chopping, while women employed stones, usually waterworn pebbles, in a pestle and mortar action to grind seed.

The use of natural fibres in basketry was highly developed in this region of South Australia. From one of the rushes growing along watercourses the Kaura made a round, coiled mat which was used as a kind of ground covering and also as a type of cloak. Several different kinds of baskets were also manufactured, using the same coiling technique. Nets were also constructed from vegetable fibre and were used for the purpose of trapping game and for fishing.

Kaurna perceptions of the environment

Every civilization tends to overestimate the objective orientation of its thought. When we make the mistake of thinking that people of other cultures are governed solely by organic or economic needs, we forget that they level the same reproach at us, and that to them, their own desires for knowledge seem more balanced than our own. Aboriginal society was interested in the natural environment not only because of its use value. Indeed, many of the named plants and animals were of no economic value to them. Their thought, like ours, was founded on the demand for order. Things were not *known* because they were *useful,* they were *useful* because they were *known.* It is occasionally forgotten that the activity of classification and ordering of the environment is not exclusively the prerogative of the natural scientist—rather it is an essentially human concern. Not only was this, as were other areas of Australia, a known and occupied area of land, with delineated and named areas, it was also the locus of an intellectual activity aimed at creating a cultural order out of the apparent chaos of nature.

The plant and animal assemblages of Australia were, it seems from modern investigation of Aboriginal cosmologies, organised into systematic taxa, which while predicated upon different principles from our own, were none-the-less the product of intellectual and systematic enquiry. Similarly, the refinements of ritual and mythology, for which Australian groups are widely known, and which to the outsider may appear to be the disordered and illogical activity of a primitive people, should be seen as a concern to assign every single creature, object or feature to a place within a class—in much the same way as scientific taxonomic studies in our own society.

Unfortunately, however, the process of enquiry aimed at deriving the general principles of a folk taxonomy is a particularly difficult one, requiring reference to pertinent and extensive studies in semantics, linguistics, kinship systems and biosystematics. It is perhaps unnecessary to emphasize that the original inhabitants of the area discussed in this collection are today no longer represented by people with direct links to the original community. Their dispossession from the land rapidly resulted in their collapse as a distinct social group.

Thus it is no longer possible to collect information first-hand which might assist us in the enquiry into traditional taxonomies or environmental classifications. Those resources which are available are entirely archival and, while of great interest, are often of limited value to enquiries in this area.

Kaurna taxonomic systems

Languages are symbolic and pragmatic systems, which, within any particular social framework, contain cross-references and reverberations of meaning derived from the experiences of that society and its members. For this reason, while individual elements of language systems derived from two different societies may in some sense be "translation equivalents", they may play non-equivalent roles in their respective semantic systems.

Vocabularies of Aboriginal languages which seek to provide simple "translation equivalents" in English are therefore thwart with danger. Even apparently simple concepts such as human body parts may not be exact equivalents in the two conceptual systems.

Most people are today aware that terms used in Aboriginal kinship systems and translated into our degenerated kinship terms may leave wide areas of disagreement untouched. Terms translated as "mother" or "father" may, in Australian systems be extended to people who are clearly not the individual speaker's progenitors, but people standing for example in a relationship of "brother" or "sister" (again non-equivalent translation terms) to the actual progenitors.

Amongst vocabulary entries recorded for the Kaurna are environmental terms subsequently translated as "sea-shore", "mallee-scrub", "gum-forest" or "pine-forest". Many of these have been transposed into what, to us, is a separate social domain, in the naming of persons directly associated with ecologically defined territories. This fact illustrates an important sense in which Aboriginal societal members may be classified as "environmentalists". Kaurna methods of landscape classification, at least implicitly, indicate that total environmental and human interrelationship were pre-eminently their concern. The mediation of this relationship appears to have been achieved through the operation of extensive myth systems within which man and nature were interwoven in narrative and symbol.

Kaurna animal and plant classifications, implicit in the assigning of names to these classes, suggest a further way in which culturally unique attributes may operate, revealing interactions of otherwise apparently separate semantic domains.

Reference to early vocabularies collected from the Kaurna will, for example, provide four quite different names for the red kangaroo, *Macropus rufus*. Upon further enquiry we discover that they result from a subdivision of the species class into male/female, old/young. Separate names also exist for male or female emu, young or old, while a dingo might be described as a "wild" dog (and thus *paru*— "meat") or as a "camp" dog. No similar classification is discovered in European taxonomies which would enable a "translation equivalent", other than that already suggested, to be offered. In fact, the naming process in these instances is the result of a cross-domain interaction of a culturally specific set of attributes which, in order to be made explicit, require enquiry into the Kaurna social system.

The Kaurna possessed the common Australian division of labour based upon sex: men hunting and providing meat, *paru* (which included the larvae of various invertebrates), and women providing the vegetable foods, *mai,* which formed the major part of the diet of the group.

A second important social distinction which existed amongst the Kaurna was one based upon age. Amongst male members of the group this was more correctly a distinction based upon progression through several stages of initiation, although in traditional society the two were practically indistinguishable. For females, maturity was reflected in their bearing of children—a reflection of their innate creativity. Five stages of development marked a male youth's progress into manhood. The final stage amongst the Kaurna was signalled by a man becoming *wilyaru,* identified by a series of scars upon his body. More significantly, perhaps, each of these stages in a male youth's development represented different progressions in the acquisition of traditional knowledge (and attendant privileges and responsibilities)—the *wilyaru* lore and mythologies being known only to the fully initiated male members of the group.

Male activities therefore were imbued with a special significance, and not surprisingly *paru,* "meat", was itself the subject of particular prohibitions, resulting in the distinction between animals not the subject of prohibitions (*paru*) and *paru koonyoonda* (those which were). Vegetable foods, existing as they did within a "female" domain, were not the concern of such prohibitions. The prohibitions themselves were articulated in terms of social status and sex. Briefly, the restrictions of diet which are recorded for the Kaurna are: only adult men were permitted to eat the *barti* or grub collected from the body of a tree, *wortabarti,* whereas all were permitted to eat the grubs collected from the roots of plants, *koope.* The dingo, *waroo kadli,* and the forearm of a kangaroo might be eaten only by the fully initiated males. Newly circumcised boys might not eat fish, and young unmarried men were not permitted to eat the female kangaroo. Girls and women, until the birth of the second child, were forbidden to eat possums and emu.

While such prohibitions may seem arbitrary and illogical to people unfamiliar with Aboriginal society, they are explicable by reference to the symbolic associations of animals or their body parts. Reference to neighbouring groups with similar prohibitions helps throw light upon these attributions and their operations. Thus circumcision is associated with rain making— the foreskin being employed in ritual activities associated with this activity. Fish are associated with large quantities of water and therefore if a newly circumcised boy were to consume fish, unwanted rain might be threatened at a time when groups were gathering together for ceremonial activity. (In northern, non-coastal areas, duck replaces fish as the prohibited meat.) Similarly, the possum possessed a particularly strong grip and was difficult to dislodge from a refuge; therefore for a young woman having only her first or second child, consumption of possum flesh would place the woman in danger of difficulties in delivery.

This brief digression enables us to suggest socially significant differentiations which have been transposed into the animal taxonomic system. Thus, in a society where prohibitions exist on the consumption of animals, based upon the sex of that animal and where considerations of the age of that animal may also be significant, it is no longer difficult to appreciate the operation of four sets of names for the one species. While these distinctions are irrelevant to our own classification, to a society such as the Kaurna with prohibitions operating upon axes of sex and age, they are essential.

Plant classification

The example we have employed in order to demonstrate the operation of attributes in a system differing markedly from our own is perhaps an extreme illustration of the difficulty one confronts in transposing domains from one semantic system to another. It is nevertheless a useful one to introduce a more detailed examination of the naming of plants by the Kaurna, in so far as in this realm also, similar distinctions appear to operate.

Students have commented that amongst other Aboriginal societies different names are occasionally encountered for plants which botanists would say were regional variations of the same species. It is difficult to discover if such distinctions existed for the Kaurna based upon such regional variation, although reference to vocabularies again indicates that different specific names were given to the different parts of a single plant. These terms furthermore existed in addition to the general forms of reference roughly equivalent to our terms "blossom", "fruit", "stem", or "root", and are not compounds formed from a single name (such as "orange" and "orange tree"). Thus *Xanthorrea* (*quadrangulata* ?) had at least three specific, named elements, *Kurru*—"body", *yutuka*—"flower", *paipola*—"young, sweet leaf heart" (edible), and a fourth, *pinyatta*— "honey", which may have been a general term for sweet substances, as it appears also to have been applied to European sugar. Separate specific names may also have existed for the red gum, *Eucalyptus camaldulensis,* describing various stages in its growth.

In contrast to this, one term or name may be found to apply to apparently unrelated elements of discrete semantic systems. Thus the name *Karko* is recorded in Kaurna vocabularies as meaning "she-oak" (*Casuarina stricta*), as well as the dish or container fabricated from its timbers. Such a connection is not difficult for us to understand, and there is little likelihood that in normal speech distinct references would be confused by the Kaurna. However, the same name was also apparently used for a particular type of red ochre, and here the connections are unclear and not explained.

The operation of family/genus or genus/species distinctions familiar to us, do not appear to be widely represented in Kaurna plant or animal terminologies. Hints may be discovered of "family tree" type schema in some areas, but they are often blurred by "cross-domain references". For example, the term *barti* was used as a general reference to "grubs", whereas this classification was subdivided into *Koope*—grubs discovered in the "roots of plants", *wortabarti,* grubs discovered in the "bodies of trees", *taingilla*—grubs "discovered in soil", *gadla barti*—"native bee" (literally "wood grub") (Teichelmann & Schurmann 1840). The necessity for the distinction between edible grubs discovered in the body of trees, and those in root systems has been discussed in relation to food avoidances, but the apparent family association of the native bee is not readily apparent.

A similar, but "cross domain" transformed family/genus operation occurs in the association—*parna*—"star", *parnatu*—"autumn, time of *parna* star" and *parnappi*—"mushroom" (literally "star dust")—probably referring to the common puff-ball.

A further refinement of the "cross domain" metaphor also appears to operate between particular human body parts and parts of plants. The connections inferred in Kaurna plant terminology have been deliberately sought as a result of investigations amongst other South Australian groups.[1]

Conclusion

It is perhaps too much to be hoped that today, almost 150 years since the first European settlement on the plains of Adelaide, the basic structure of Kaura environmental classification or plant and animal taxonomies might be derived. However, one thing at least has been achieved: a realisation that Aboriginal man sought to encompass and manipulate his environment, both as active agent and through an intellectual possession. This feature distinguishes all human societies from other systems in nature, for the latter always reproduce their own internal relationship in some characteristic composition or shape, and passively integrate other adjacent structures. As an active agent upon his physical environment, Aboriginal man changed his external relations with nature and his mutual relations in production. In turn, these changes altered his other social relations, his activities, his thought and himself.

References

BLACK, J. M. (1920).—Vocabularies of four South Australian languages. *Trans. R. Soc. S. Aust.* **44,** 76-93.

BROWNE, H. J. (1897).—Anthropological notes relating to the Aborigines of the Lower North of South Australia. *Trans. R. Soc. S. Aust.* **21,** 72-73.

FINLAYSON (1903).—Reminiscences by Pastor Finlayson. *Proc. R. Geog. Soc. S. Aust.* **6,** 39-55.

FLINDERS, M. (1814).—"Voyage to Terra Australia." London. (Facsimile, State Library of S. Australia: Adelaide.)

GELL, J. P. (1903).—South Australian Aborigines: the vocabulary of the Adelaide tribe. *Proc. R. Geog. Soc. S. Aust.* **6,** 92-100.

MILHAM, P. & THOMPSON, P. (1976).—Relative antiquity of human occupation and extinct fauna at Madura Cave, southeastern Western Australia. *Mankind* **10**(3), 175-180.

STANNER, W. E. H. (1965).—Aboriginal territorial organisation: estate, range, domain and regime. *Oceania* **36**(1), 1-26.

STEPHENS, E. (1890).—The Aborigines of Australia—being personal recollections of those tribes which once inhabited the Adelaide Plains of South Australia. *J. R. Soc. N.S.W.* **23,** 476-503.

STIRLING, E. (1911).—Preliminary report on the discovery of native remains at Swanport, River Murray: with an inquiry into the alleged occurrence of a pandemic among the Australian Aboriginals. *Trans. R. Soc. S. Aust.* **35,** 4-46.

SWADLING, P. (1976).—Changes induced by human exploitation in prehistoric shellfish populations. *Mankind* **10**(3), 156-162.

"SYLVANUS" (1842).—Periodical fires of South Australia. *S. Aust. Mag.* **1**(11), 401-404.

TEICHELMANN, C. G. & SCHURMANN, C. W. (1840).—"Outlines of a grammar, vocabulary and phraseology of the Aboriginal language of South Australia." (Native Location, Adelaide.)

TINDALE, N. B. (1974).—"Australian Aboriginal Tribes." (Australian National University Press: Canberra.)

WYATT, W. (1878).—Some account of the manners and superstitions of the Adelaide and Encounter Bay tribes. *In* Woods, J. P., "The Native Tribes of South Australia." (Wigg: Adelaide.)

---

[1] Hercus, Austin & Ellis in manuscript.

CHAPTER 9

TERRESTRIAL FAUNA AND AQUATIC VERTEBRATES

by M. J. Tyler,* G. F. Gross,† C. E. Rix‡ & R. W. Inns*

Introduction

There are a number of generalizations that can be made about the composition of the animal life in and around any Australian city. Historically the process of construction and cultivation, of clearing, deforestation and quarrying so essential for city growth and suburban development must destroy the habitats of a variety of animals, and so eliminate or seriously deplete the population of many species. This inevitable outcome is frequently accompanied by the totally reprehensible actions of hunting and killing some animals to the point of extinction. At the same time exotic animals such as the fox have been introduced, effectively contributing to the same result.

Initially the creatures at the greatest risk are those with limited dispersal ability, scattered distribution or which need specialized niches. Those which men can eat are also at risk and this attribute resulted in creatures such as the Emu, *Dromaius novaehollandiae,* and the malleefowl, *Leipoa ocellata,* being placed under great pressure from the advent of the first colonists.

Following 140 years of European contact the fauna of the Adelaide region has still not attained any degree of stability. House erection on the Hills Face Zone, highway construction with its associated high noise levels, increasing problems of pollution, and the introduction of additional exotic animals continue to modify habitats or introduce additional competition for limited resources.

The nature of the fauna remaining in the Adelaide region can be attributed to several factors. A native animal survives because there has been minimal disturbance of its habitat, because it lacks qualities desired by human beings, or it has the ability to live in close proximity to man.

The Adelaide region provides a diverse variety of habitats ranging from mangroves, salt pans and marshes to forests with comparatively high rainfall. The fauna of the region is comparably diverse, but it is principally an integral part of the Bassian zoogeographical province centred upon southeastern Australia, and it has close ties with the geographic area east and south of the Great Dividing Range. For many animals the Adelaide region approximates the northern limits and/or western limits of their distribution in South Australia.

In the case of the terrestrial creatures, and in particular 10,000 or so invertebrate animals, it is generally possible to recognise distinct faunae associated with each of the various biomes. In contrast the freshwater environments, within their relatively small geographical area, do not provide such diversity of niches.

The following notes are intended to provide a broad introduction to the major components of the fauna.

Aerial and terrestrial invertebrates

(a) *Mangroves and samphire swamps*

This fauna is not well known. Saline and brackish pools are frequent in both formations, and breed substantial populations of the saltmarsh mosquito *Aedes vigilax* and also of *Aedes camptorhynchus,* the latter able to breed in fresh water also and more widely distributed. Both very rarely breed in larger bodies of saline water not sheltered by vegetation and subject to wave action such as the I.C.I. salt pans.

Small blackish and greyish flies belonging to several families of acalyptrate diptera including Ephydridae and Lauxaniidae, are common around samphire plants and there is fauna of several species of ant and occasionally saldid bugs on the saline sands and clays between the

* Department of Zoology, University of Adelaide, North Terrace, Adelaide, S.A. 5000.
† South Australian Museum, North Terrace, Adelaide, S. Aust. 5000.
‡ 42 Waymouth Avenue, Glandore, S. Aust. 5037.

samphire bushes and pools. Species of Lauxaniidae and Tethinidae (both families of acalyptrate flies) have been recorded from mangroves in other parts of Australia but it is not known whether any are here.

(b) *Sandhills and sandy beaches*

The beaches are not a favoured habitat for creatures who do not spend some of their time in the sea. Patches of stranded sea grass occasionally breed populations of kelp flies (Coelopidae) and also members of the Tethinidae, another family of acalyptrate flies. One species of earthworm, *Pantodrilus matsushimensis,* occasionally occurs in or under the patches. In the rest of the Adelaide region, especially on the plains or in cultivated areas of the hills the commonest earthworms are *Eisenia rosea* and *Alblobophora caliginosa.*

The sandhills appear to represent an extension of the fauna of the arid regions into the Adelaide area and their fauna is substantially different from those of any other parts of the region. Some species present are largely restricted to coastal sand dunes such as the small bug *Anaxilaus vesiculosus* and other bugs like *Dictyotus caenosus* and *Cermatulus nasalis* are more abundant there than elsewhere.

One exotic species of large black beetle, *Blaps polychresta,* has become locally abundant in parts of the sandhills and on the airport. This beetle originates from the eastern Mediterranean region and is frequently associated with stored wheat, but this association is most likely with the associated rat population as the beetles feed on the manure of rats, mice and rabbits (Matthews 1975). Infestation of *Blaps* started around the grainports on the western side of Yorke Peninsula and eventually spread to the Adelaide coastal region. It has been reported to excavate the mortar between courses of brick or stone to make a cavity for pupation.

(c) *Adelaide Plains*

Amongst the unaffected native species are the brown trapdoor spider, *Aganippe subtristis;* wolf spiders, *Lycosa* spp.; the caterpillar hunting wasp, *Podalonia suspiciosa,* and the spider hunting wasp, *Sceliphron laetum; Hermetia pallidipes* and *Neoexaireta spinigera,* both stratiomyid flies which now breed in compost heaps but which would formerly have fed as larvae in the rotting grass layer, the painted lady butterfly, *Cynthia kershawi,* and the Australian admiral butterfly, *Vanessa itea;* several species of skipper butterfly and such white ants as *Coptotermes acinaciformis* which fed on old and rotting eucalyptus, sheoak, etc., but now also feeds on fence posts and the woodwork of sheds and dwellings.

In the vacant lots and parklands still occur some of the old grassland suite such as the bugs *Nysius vinitor, Oxycarenus* (two species), the grass seed mimicking *Stenophyella macreta* and *Dieuches notatus* (all Lygaeidae) together with *Erythroneura ix* (Cicadellidae), *Lepyronia moerens* (Cercopidae) and *Oliarus* sp. (Cixiidae), the two beetles *Corticaria adelaideae* (Lathridiidae) and *Phalaerus fimitarius* (Phalacridae), the native fruit fly *Tephritis pelia* and the green head ant *Chalpoponera metallica,* the latter also common around lawns.

Amongst the native species which have clearly benefited we might mention the following. The coreid *Mictis profana* or St Andrews Cross bug, an inhabitant of xeric environments (in Australia at least) where it lived on acacias, but which now may be seen in large numbers on the introduced *Cassia floribunda,* and also sometimes on citrus. The cottony cushion scale coccid *Icerya purchasi,* a native originally feeding on wattles but which attacks citrus, the arctiid moth *Phalaenoides glycine,* whose larvae defoliate vines and as a consequence its predator the pentatomid bug *Oechalia schellengergii.* Another arctiid moth *Ardices glatygni,* also which has benefited is more likely to be known to you by its caterpillar, the black and brown hairy woolly bear.

Introduced insects of note include Red Scale of citrus, a coccid (*Aonidiella aurantii*), various species of thrips and aphids, *Anthrenus* beetles which feed on dried animal furs and hair including woollen carpets, three beetles of the family Curculionidae, namely the white fringed weevil *Graphognathus leucoloma,* the curculionid beetle *Otiorrhynchus cribricollis,* Fuller's rose weevil *Pantomorus cervinus* and the sitona weevil *Sitona humeralis,* the cabbage white butterfly *Pieris rapae* and the monarch or wanderer butterfly *Danaus plexippus,* the former feeding on cabbage and other brassicas, the second on milk weed and swan tree, the house fly *Musca domestica* which has displaced the bush fly over all except the fringes of the whole region we are considering, and several other flies of medical importance including *Stomoxys calcitrans* (the stable fly) and *Fannia canicularis* (the lesser house fly).

Introduced molluscs are common and are a severe problem all over this area. In the drier grasslands of the parklands and vacant blocks the two snails *Cochlicella acuta* and *Theba pisana,* both of which aestivate in the drier months on fence posts and the stems of the taller weeds such as *Malva,* are widespread. Where there is some sort of ground cover such as sheets of iron or bricks, or where shrubs or higher annuals are grown and regularly watered, the garden snail *Helix aspersa* can be very abundant. In gardens the two slugs *Lehmannia flava* which is large and yellowish and the smaller and black *Milax gigates* are both abundant. For a fuller account of the insects of homes and gardens see Gross & Matthews, in press.

(d) *Savannahs and Savannah woodlands*

In general this formation has a similar fauna to that of the drier grasslands of the Adelaide Plains but with some important differences. The introduced snails do not occur other than rarely in the formation and the garden snail and the two slugs mentioned earlier occur only in gardens. The introduced monarch butterfly on the other hand occurs in greater numbers due to the greater abundance of one of its food plants, the broad leafed cotton bush (*Aclepias rotundifolia*) in the formation.

Amongst the endemic species, a few appear which are also found in the formation which follows but which do not occur in the lowland grasslands. Amongst those we may mention the hopper ant *Myrmecia pilosula,* the satyrid butterfly *Geitoneura klugii* and a small suite of arthropods associated with eucalypts, of which the scorpion *Lychas marmoreus* found under the bark of those species of gums with white trunk and peeling bark, and *Rhicnopeltella sarah,* a eulophid wasp forming galls on the leaves, rate a mention.

(e) *Dry sclerophyll forest*

The most comon trapdoor spider is now *Missulena occatoria,* whose male is bright red on its anterior half and purple posteriorly, whereas the female is uniformly black. Millipedes of several species are much more common and in the wetter patches of leaf litter amphipods may sometimes be found. Prominent "new faces" in the bugs (Hemiptera) include the acanthosomatid *Duadicus pallidus,* the lygaeid *Ontiscus obscurus,* the predaceous reduviid *Leana australis lictor,* the mirid *Pseudopantilius australis,* the cicadellids *Cephaleus minutus* and *Ectopiocephalus australis,* the ricaniid *Salona panorpaepennis* and the butterfly bugs (Flatilae) *Euphanta ruficeps* and *Siphanta c.f. granulicollis.* Two quite large

beetles, the lagriid *Lagria grandis* and the weevil *Cherrus ruficornis* are not seen in this region outside of this formation nor are the lauxaniid fly *Sapromyza hirtiventris* or the fruit fly *Trypanea glauca.* Amongst the Hymenoptera we now see the two sawflies *Clarissa variabilis* and *Neourys affinia,* the inch ant *Myrmecia pyriformis* which can sting severely, and a species of sugar ant, *Camponotus aeneopilosus.* The thynnid wasps peculiar to the formation include *Rhaphigaster pugionatus* and *Elidothynnus melleus.* The blue or green ant, *Diamma bicolor,* whose female frequently exceeds 20 mm length, is brilliant green or bluish with brown legs, and able to sting, is now only commonly found in this formation. It is a member of the Thynnidae, not of the family of the ordinary ants.

In the Lepidoptera the little greenish moth *Pollanisus viridipulverentula* (Zygaenidae) which feeds on *Leptospermum* species and the common brown butterfly *Heteronympha m. merope* are also either restricted to this formation or only uncommonly found outside of it.

Native species of land snail are more abundant in this formation than in those preceding. They may occasionally be found in damp patches under logs and stones. Most frequently seen are *Exilibadistes sutilosa* and *Strangesta gawleri.*

Owing to the large numbers of land invertebrates there are few comprehensive texts available and those which are, cover only certain groups. Notable amongst them are the works by Cotton (1959) on Mollusca, Gross (1975, 1976) on portion of the Hemiptera, Common & Waterhouse (1972) on butterflies, and Womersley (1939) on primitive insects. Each of these works cover much wider areas than the Adelaide region but include all species known to occur here at the time of their publication.

Aerial, terrestrial and aquatic vertebrates

(a) *Birds*

The bird component of the fauna of the Adelaide region has been severely modified in recent decades. For example, the reedbeds which extended from the Patawalonga to the river and inland as far as Camden and Fulham have been totally destroyed. Up until the late 1920's enormous flocks of the following species assembled there: *Cygnus atratus* (black swan), *Tadorna tadornoides* (mountain duck), *Anas superciliosa* (black duck), *A. gibberifrons* (grey teal), *A. castanea* (chestnut teal), *A.*

rhynchotis (shoveler), *Malacorhynchus membranaceus* (pink-eared duck) and *Aythya australis* (white-eyed duck). By modern standards of duck populations the flocks were almost unbelievable. When disturbed they filled the air with a conglomeration which extended from horizon to horizon. *Burhinus magnirostris* (stone curlew), *Porphyrio melanotus* (eastern swamp hen) and *Gallinula tenebrosa* (moorhen) frequently fed with the domestic poultry at East Terrace, Henley Beach, only about 50 m from the western margin of the Reedbeds at that point. *Rallus philippensis* (banded land rail) was often seen and in the years when *Gallinula ventralis* (black-tail native-hen) moved south it was in such vast numbers that the whole landscape appeared to be moving.

With wide congregations of waterfowl there were occasionally a few strangers. Small parties of *Anseranas semipalmata* (pied goose) and *Cereopsis novaehollandiae* (Cape Barren goose) were seen at irregular intervals. In addition the conditions seemed to attract flocking nomads from the northern parts of the continent. Vast flocks of *Artamus personatus* (masked woodswallow) and *A. superciliosus* (white-browed woodswallow) appeared and nested at irregular but not infrequent intervals. Less frequently small numbers of *Epthianura tricolor* (crimson chat) and *E. aurifrons* (orange chat) visited the area; sometimes together, but not usually together, and the last being the more frequent.

With the passing of the reedbeds complex all of these visitations ceased and it is only over the past 15 years or so with the extension and "maturation" of the I.C.I. saltfields, the construction of the Bolivar Sewage Works, and the rehabilitation of the "lake" at Buckland Park has there been any sign of a return to the region of many species which had not been recorded since the 1930s. On present indications it would seem that this complex, if maintained in conjunction with extensive tidal flats which adjoin the seaward side of the mangroves (which in turn form the boundary between the complex and the sea), will become the most important bird conservation area in the region and indeed in South Australia.

Two species which are now widespread and occur in considerable numbers but which prior to the first decade of this century had not been recorded from the regions are *Ocyphaps lophotes* (crested pigeon) and *Cacatua roseicapilla* (galah). Both of these profited from the clearing of what was formerly fairly densely tree-covered country both within the

region and, what is probably even more important, in the lower and middle north. This allowed them to extend southwards from the upper and far northern areas which had hitherto been their range within the State. The better rainfall status of their new territory combined with the extensive plantings of cropping and pasture plants which furnish large quantities of edible seeds, has provided them with a much more stable environment. Their numbers have greatly increased in recent years and appear to be still increasing.

Another species which is an even more recent arrival in the region is the cattle egret referred to earlier. To date sightings have been limited to single birds on two or three occasions at Meadows and at Buckland Park.

A species which has profited as much if not more than any other is the silver gull. Rubbish disposal methods, the picnicking habits of the general public and the extensive tilling of the soils for cropping and pasture development have created new sources of food supply in such quantities that the gulls have greatly increased their breeding rate. They have also largely changed their life-style. While it seems likely that all gulls return to the seashore at times, it is not a daily occurrence with many of them and certainly a vast quantity of their food is obtained from terrestrial sources.

The list of intercontinental migrants is much greater today than even 20 years ago, but it is probable that the great increase in the intensity and standard of bird observation in this period has been the major factor. It seems likely that such species as *Tringa brevipes* (grey-tailed tattler), *T. stagnatilis* (marsh sandpiper), *T. terek* (terek sandpiper), *Calidris subminuta* (long-toed stint), *Limicola falcinellus* (broad-billed sandpiper) and *Philomachus pugnax* (ruff) visited this area in small numbers at irregular intervals long before they were first recorded here.

Most of the resident passerines which inhabit the sclerophyll forest, the savannah woodland and particularly the mallee have been reduced in numbers. The depletion of vegetation in the last mentioned is virtually complete with only a few very small pockets of the original trees remaining, and even in these the shrubby understorey has been eliminated by grazing stock. The savannah woodlands have suffered equally and for many years bird populations were substantially reduced. However, the exotic shrubs and fruit trees, combined with the recent extensive plantings of native trees and shrubs in suburban streets and gardens, have provided sources of food which are probably more diverse and more evenly distributed throughout the year than was the case under the original vegetation. As a result *Rhipidura leucophrys* (willie wagtail), *Meliphaga penicillata* (white-plumed honeyeater), *Phylidonyris novaehollandiae* (yellow honeyeater), *Anthochaera chrysoptera* (little wattlebird), *A. carunculata* (red wattlebird) and *Grallina cyanoleuca* (magpie lark) are much more numerous in the metropolitan area than they were when the greater part of the area was used for farming purposes and probably as plentiful as under the original conditions.

The small undergrowth and ground-loving species such as *Malurus cyaneus* (blue wren) and *Acanthiza chrysorrhoa* (yellow-rumped thornbill) have succumbed to the cats in all but the remnants of natural vegetation in the extending northward from Port Adelaide to Port Wakefield.

The last mentioned strip of country supports two taxa which are much more dependent upon a specific habitat than almost any of the other terrestrial species: *Sericornis frontalis osculans* (spotted scrubwren) and *Acanthiza iredalei rosinae* (samphire thornbill). Both occur in the samphire flats bordering the inner margin of the mangroves and, while both do move into the edge of the nitraria and lignum which adjoin the eastern sides of the samphire flats, it is not to any great extent. Destruction of the samphire habitat would undoubtedly see the extinction of both of these birds.

The extinctions that have occurred within the region since 1836, in addition to those mentioned earlier, have been *Burhinus magnirostris* (southern stone-curlew), a victim of the fox, *Pezoporus wallicus* (swamp parrot), one of the very early casualties which was probably in very small numbers as only two or three records are known, probably *Neophema chrysogaster* (orange-bellied parrot) and possibly *Lathamus discolor* (swift parrot). Other species have not been recorded for some time, but they have always been irregular in appearance and in such small numbers that they are probably best regarded as accidental to the area rather than birds of the region.

(b) *Mammals*

At least nineteen species of native mammals have disappeared from South Australia since settlement began (Aitken 1970). Of the two monotremes, only the spiny ant-eater, or echidna, *Tachyglossus aculeatus,* is still widespread, although not particularly common, while the platypus, *Ornithorhynchus anatinus,* which was once found in the Torrens and Onkaparinga Rivers has not been reported for many years.

Within the Mt Lofty Ranges two species of small carnivorous marsupials of the family Dasyuridae occur. One of these, the dunnart, *Sminthopsis murina,* has only been seen a few times and is probably restricted to the Fleurieu Peninsula; the other, the yellow-footed marsupial mouse, *Antechinus flavipes,* is common through most of the Mt Lofty Ranges. Yellow-footed marsupial mice have a rather dramatic life history, as all of the males die towards the end of each year, just after the breeding season. It seems likely that their demise is associated with a number of complex hormonal, metabolic and behavioural changes that occur during the breeding season, although its function remains unknown (Inns, in press). Another species of carnivorous marsupial, the native cat, *Dasyurus viverrinus,* was once quite common on the Adelaide Plains but due mainly to destruction of its habitat, it is now absent from South Australia.

In the early days of settlement the rabbit-eared bandicoot, *Macrotis lagotis,* was also quite common at Pinkie Flat along the Torrens, and on the Adelaide Plains. These beautiful animals are now restricted to a small area in central Australia. The short-nosed bandicoot, *Isoodon obesulus,* however, still survives in some pockets of uncleared natural scrub that can be found throughout the Adelaide Hills and Fleurieu Peninsula, although it is seldom seen because of its secretive nature and nocturnal habits.

Although three species of pigmy possum are known to occur within South Australia only one, *Cercartetus concinnus,* lives in the Mt Lofty Ranges. Pigmy possums are insect and nectar feeders and have the ability to become torpid when exposed to low temperatures during winter. Torpor is a means of conserving energy and involves a lowering of body temperature and heart rate, similar to the physiological changes that occur in hibernating animals, but not as prolonged.

The ring-tailed possum, *Pseudocheirus peregrinus* (Fig. 1), is quite widespread throughout southeastern Australia, and can be found in areas of dense natural scrub throughout the Mt Lofty Ranges. Small numbers of these attractive animals live in some of the parklands

Fig. 1. Ringtail possum *Pseudocheirus peregrinus.*

around Adelaide. A study of ring-tailed possums was made in Victoria by Thomson & Owen (1964), who found them to be comparatively social animals, building communal nests out of leaves and twigs. The composition of groups sharing nests usually consists of an adult male, one or two adult females which may both have pouch young, and immature offspring from the previous year.

Hunting has not seriously affected many of the native mammals but two species in particular were quite heavily persecuted due to their much desired fur. The brush-tailed possum, *Trichosurus vulpecula,* was removed from the protected list from June to September in 1920; in this interval 100,000 were killed (Wood Jones 1924). Despite this the possum has remained one of the most common and urbanized of any of the native mammals.

The koala, *Phascolarctos cinereus,* was also hunted for its fur so that it was necessary to reintroduce it to South Australia. Small colonies are becoming established in parts of the Adelaide Hills.

The only species of macropod, or kangaroo, that can still be seen in the Adelaide region is the black-faced grey kangaroo, *Macropus fuliginosus melanops,* although three other species once inhabited this area. The grey kangaroo breeds throughout most of the year although births tend to be concentrated in the summer

months (Poole 1975). The young remain in the pouch for approximately 45 weeks.

The Kangaroo Island wallaby, *Macropus eugenii,* was formerly one of the most numerous and widespread of the mammals in the southern parts of this state but no longer exists in the Adelaide region.

The clearing of scrub on the Adelaide Plains also seems to have resulted in the extinction of two species of rat kangaroos, although the introduction of the European rabbit, *Oryctolagus cuniculus,* may have hastened this extinction. The two species were the brush-tailed rat kangaroo, *Bettongia penicillata,* and Lesuer's rat kangaroo, *B. lesueri.*

Of the placental mammals native to the Adelaide region there are three species of rats and seven species of bats.

The southern bush rat, *Rattus fuscipes greyii,* is found throughout the Mt Lofty Ranges in areas where there is a thick ground cover in the form of vegetation or fallen logs. The swamp rat, *R. lutreolus,* has a more limited distribution, being restricted to the wetter coastal areas of the southern ranges and the south-east of the State. These two species are very similar in appearance but may be distinguished by the colour of the hairs on the feet. Those of *R. fuscipes* are usually silvery-white while *R. lutreolus* have dark brown feet. Both species breed during the spring and early summer and have between four and six young in a litter (Taylor & Horner 1973).

The third native rodent is the water rat, common along most watercourses in the State. However, due to the alteration of many creek systems in the settled areas its range has been reduced considerably in recent years. Although the water rat is aquatic in habit it makes extensive pathways and burrows into the banks of the streams where it lives. These rats are entirely carnivorous and eat fish, freshwater crustaceans and shell-fish.

There are seven species of bats that can be found in the Adelaide region:

white-striped mastiff bat	*Tadarida australis*
western little mastiff bat	*T. planiceps*
Gould's wattled bat	*Chalinolobus gouldii*
chocolate wattled bat	*C. morio*
little brown bat	*Eptesicus pumilus*
bent-winged bat	*Miniopterus schreibersii*
lesser longer-eared bat	*Nyctophilus geoffroyi*

Aitken (1975) gives a good description of these species and a key to their identification.

A number of introduced species of mammals also occur in South Australia. Some have become serious pests while others have adversely affected populations of some native species. The rabbit, *Oryctolagus cuniculus,* was introduced into eastern Australia in the early days of colonization. It began to spread in the 1860s and, after rapidly covering the lower half of the continent, became a serious agricultural pest despite various eradication methods. Besides destroying natural vegetation and pastures they competed with the rat kangaroos, then present on the Adelaide Plains, for food and burrow sites. They were probably partly responsible for the disappearance of these small mammals and also the rabbit-eared bandicoot. The hare, *Lepus europaeus,* although present in South Australia has not increased in vast numbers. Neither have two species of deer, the fallow deer, *Dama dama,* and the red deer, *Cervus elaphus,* which were deliberately released but have been controlled, at least partly, due to the efforts of hunters. Another introduced herbivore, the goat, *Capra hircus,* has become a feral animal in some of the wilder parts of the Mt Lofty Ranges but is not yet a serious pest, as in other parts of the State.

Two species of introduced rodent ,the house mouse, *Mus musculus,* and the black rat, *Rattus rattus,* are widespread and in some areas are probably in direct competition with native species although their influence is hard to assess. The brown rat, *R. norvegicus,* is present but is not very common.

It is probably the carnivores, the domestic cat, *Felis cattus,* and the fox, *Vulpes vulpes,* that have had the most direct influence on the numbers of small mammals. In a study of the food habits of feral cats in Victoria Coman & Brunner (1972) found that in undisturbed forest situations there was a heavy dependence on the small native mammals such as marsupial mice, native rats and possums. Foxes were also found to prey upon these animals in Sherbrooke Forest Park in the Dandenong Ranges of Victoria (Brunner, Lloyd & Coman 1975) although rabbits did form the major part of their diet. It is likely that a similar situation occurs in the Adelaide region.

Although settlement has produced a decline in the number and abundance of native mammals there are still many extremely interesting species inhabiting the Adelaide Hills region.

Hence it is essential that further clearing of natural bushland be curbed and that a study be made on the effective control of the introduced animals. Only by such measures can we ensure the continued survival of our remnant mammal fauna.

(c) *Freshwater Fishes*

In the absence of any major rivers within the Adelaide region there are relatively few native freshwater fishes. Nevertheless some occur in all of the permanent creeks, and the opportunity exists to observe them close to the city.

The most recent State handbook by Scott, Glover & Southcott (1974) lists four local species of minnows and native trout, all of which are species of *Galaxias.* One species, *G. schomburgkii* (named in honour of Richard Schomburgk, the first Director of the Botanic Garden, Adelaide), has not been reported since its discovery in 1868. It seems likely that in reality there are three species with a total of four names. Other small fishes of the Adelaide region include the gudgeons, such as the big-headed gudgeon, *Philypnodon grandiceps.* The gudgeons are widely distributed in South Australia, being found in the creeks, rivers and reservoirs.

The River Torrens has a variety of fishes including the native blue-spot goby, *Lizagobius galwayi,* which has a bright blue spot in its first dorsal fin. The Torrens is the home of a number of introduced species including the redfin, *Perca fluviatilis,* the tench, *Tinca tinca,* the golden carp, *Carassius auratus,* the mosquito fish, *Gambusia affinis,* and the rainbow trout, *Salmo gairdneri.* The rainbow trout and other species of trout have been introduced into many permanent waters in the Adelaide region.

Little is known of the use of freshwater streams as breeding sites by jawless marine fishes called lampreys. These are elongate, eel-like creasures up to about 50 cm long, characterised by a large, anterior, suctorial disc and a single, median nostril. In other States the young stages or Ammocoetes larvae are commonly found in great numbers in freshwaters but the extent of their seasonal occurrence in waters of the Adelaide region is not known. Prior to construction of the city weir the short-headed lamprey, *Mordacia mordax,* entered the city section of the Torrens. Other native species no longer taken there are the congolli, *Pseudaphritis urvilli,* and the native trout, *Galaxias maculatus.* In a register at the South Australian Museum H. M. Hale notes (on 16.i.1929) that the congolli "has practically disappeared from

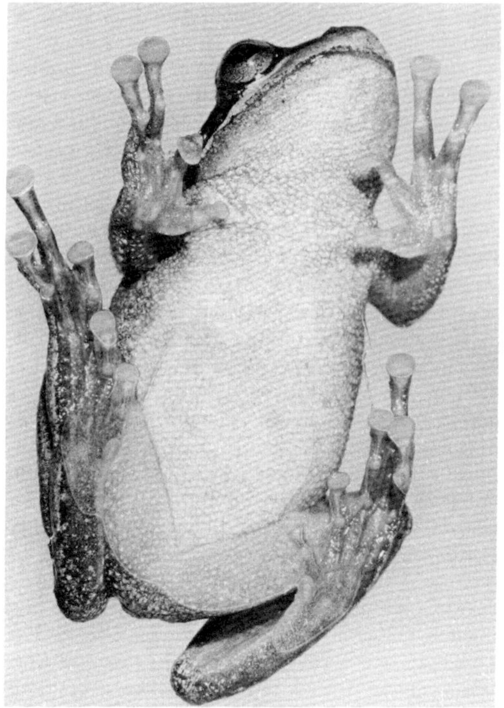

Fig. 2. Tree frog *Litoria ewingi* climbing pane of glass.

the Torrens at Adelaide, although twenty years ago it was plentiful enough there".

(d) *Frogs*

Only eight species of frogs occur in the Adelaide region. All are described and illustrated in a field guide by Tyler (1966). Of the total of eight species, one (Peron's tree frog *Litoria peroni*) has been taken only adjacent to the Hindmarsh River at Victor Harbour, and must have originated from the River Murray where the species is fairly common. The remaining seven species are more widely dispersed in the Adelaide region and most can be observed quite readily.

The *Limnodynastes* species, the marbled frog, *L. tasmaniensis,* and the bullfrog, *L. dumerili,* are well known, because of the conspicuous spawn clump that they lay on the surface of still or slowly moving water. The spawn is in the form of a nest of foam resembling undegraded detergent. *Limnodynastes dumerili* has large, protuberant, oval glands on the upper surface of its tibiae. These glands secrete a creamy, poisonous fluid which is presumed to have a protective function, aiding the animals' defence from enemies. The smallest species are the brown froglet, *Ranidella signi-*

fera, and the short-legged toadlet, *Pseudophryne bibroni,* neither of which exceed 30 mm in snout to vent length. The latter species is common in the Mt Lofty Ranges, and is unique amongst local anurans in laying its eggs out of water, in depressions covered with leaves or beneath damp debris. It is only after the onset of heavy winter rains that the tadpoles hatch from the egg membranes, and are washed into pools.

The best known of the tree frogs is *Litoria ewingi,* which is a brown or grey species with yellow and often black markings on the back of its thighs. This species is closely associatd with man and quite abundant in the metropolitan area. In suburban gardens it lives amongst dense vegetation and favours well-watered areas. It calls throughout the year.

Until recently the sole record of *Litoria raniformis* (formerly called the golden or bell frog, *Hyla aurea*) was an introduced population at Salisbury. However, about seven years ago someone released some of these frogs near the upper reaches of the Onkaparinga River. There is now a large population which is in the process of spreading throughout the Adelaide Hills and, via the Onkaparinga River, to the Noarlunga Plains.

Finally there is the water holding frog *Neobatrachus pictus* for which Adelaide is the type locality. This rotund frog avoids dry summer conditions by digging below the surface of the soil, and then forming an impermeable cocoon of dead skin around itself. The cocoon impedes the rate of water loss to the surrounding soil.

(e) *Reptiles*

A variety of lizards, snakes and tortoises occurs in the Adelaide region. The first State monograph of this reptile fauna written by Waite (1929) is long out of print and has now become a collector's item. Houston (1973) has provided a brief synopsis of the State reptile fauna, whilst there is now a detailed field guide for the entire continent by Cogger (1975).

Amongst the lizards there are three particularly large species. *Tiliqua rugosa* (the sleepy lizard, shingle back or stumpy-tail) occurs on the Adelaide Plains and is characterised by its gross scales. Commercial exploitation of this species initiated the steps that led to reptiles being protected in South Australia. In many Adelaide suburbs the blue tongue lizard *T. scincoides* is a common inhabitant of gardens. Like *T. rugosa* it has a large, blue tongue and the habit of inflating its lungs with air when approached. The third large lizard is a dragon of the family Agamidae: *Amphibolurus barbatus,* better known as the dew lizard, bearded dragon or frilled neck. This species exhibits the fatal habit of stopping and standing in the path of on-coming traffic.

There are numerous small skink lizards of the family Scincidae, many of which have an elongate body form, reduced limbs and digits, and progress in an undulating snake-like manner. They possess fracture planes in caudal vertebrae, enabling them to carry out autotomy: a process of self-mutilation by which their tails can be dismembered when the animals are in danger of capture. By this technique the amputated tail is left behind wriggling violently, whilst the lizard escapes to safety.

There are a number of legless lizards in the Adelaide region. Although they resemble tiny snakes, these creatures are in reality lizards whose limbs become vestigial. Kluge (1974) revised these creatures comprising the family Pygopodidae, and recorded a new species from the Adelaide suburb of Tea Tree Gully.

Local geckos range from the spectacular to the cryptically marked. In the undisturbed areas the thick-tailed gecko *Underwoodisaurus milii* is quite common. Disturbed from its hiding places beneath large rocks or fallen timber, it raises itself up on the tips of its tiny limbs and barks at any intruder. In contrast *Phyllodactylus marmoratus,* the marbled gecko, is likely to be found hidden beneath the loose bark of eucalypts. A dull grey coloured animal, it uses the tiny hooks to climb rapidly, and so escape from intruders.

The snake most commonly encountered is the venomous brown snake, *Demansia textilis,* which occurs in many Adelaide suburbs. This creature is hunted by domestic cats whose extreme susceptibility to the venom provides a practical test of the skills of the local members of the veterinary profession.

Acknowledgments

We are indebted to Mr C. J. M. Glover, Mr P. Aitken and Mr S. Parker of the South Australian Museum for advice and comments on the manuscript.

References

AITKEN, P. F. (1970).—Mammals. *In* "South South Australian Year Book", (Commonwealth Bureau of Census & Statistics: Adelaide).

AITKEN, P. F. (1975).—Two new bat records from South Australia with a field key and checklist to the bats of the state. *S. Aust. Nat.* **50,** 9-15.

BRUNNER, H., LLOYD, J. W. & COMAN, B. J. (1975).—Fox scat analysis in a forest park in south-eastern Australia. *Aust. J. Widl. Res.* **2,** 147-154.

COGGER H. G. (1975).—"Reptiles and amphibians of Australia". (Reed: Sydney.)

COMAN, B. J. & BRUNNER, H. (1972).—Food habits of the feral house cat in Victoria. *J. Widl. Manage.* **36**(3), 848-853.

COMMON, I. F. B. & WATERHOUSE, D. F. (1972). —"Butterflies of Australia". (Angus & Robertson: Sydney.)

COTTON, B. C. (1959).—"South Australian Mollusca, Archaeogastropoda". (Govt Printer: Adelaide.)

GROSS, G. F. (1975).—"Plant feeding and other bugs (Hemiptera) of South Australia— Heteroptera—Part 1". (Govt Printer: Adelaide.)

GROSS, G. F. (1976).—"Plant feeding and other bugs (Hemiptera) of South Australia— Heteroptera—Part 2". (Govt Printer: Adelaide.)

GROSS, G. F. & MATHEWS, E., (in press).—Insects of South Australian homes and gardens. *In* "South Australian Year Book, 1976". (Govt Printer: Adelaide.)

HOUSTON, T. F. (1973).—Reptiles of South Australia. A brief synopsis. *In* "South Australian Year Book, 1973". (Govt Printer: Adelaide.)

INNS, R. W. (in press).—Some seasonal changes in *Antechinus flavipes* (Marsupialia: Dasyuridae). *Aust. J. Zool.*

KLUGE, A. (1974).—A taxonomic revision of the lizard family Pygopodidae. *Misc. Publ. Mus. Zool. Univ. Michigan* (147), 1-22.

MATHEWS, E. (1975).—The Mediterranean beetle *Blaps polychiesta* Forskal in South Australia (Tenebrionidae). *S. Aust. Nat.* **49**(3), 35-39.

POOLE, W. E. (1975).—Reproduction in the two species of grey kangaroo, *Macropus giganteus* Shaw and *M. fuliginosus* (Desmarest) II Gestation, parturition and pouch life. *Aust. J. Zool.* **23:** 333-353.

SCOTT, T. D., GLOVER, C. J. M. & SOUTHCOTT, R. V. (1974).—"The marine and freshwater fishes of South Australia". 2nd edtn. (Govt Printer: Adelaide).

TAYLOR, J. M. & HORNER, B. E. (1973).—Reproductive characteristics of wild native Australian *Rattus* (Rodentia: Muridae). *Aust. J. Zool.* **21,** 437-475.

THOMPSON, J. A. & OWEN, W. H. (1964).—A field study of the Australian ringtail possum *Pseudocheirus peregrinus* (Marsupialia: Phalangeridae). *Ecol. Monogr.* **34,** 27-52.

TYLER, M. J. (1966).—"Frogs of South Australia". (South Australian Museum: Adelaide.)

WAITE, E. R. (1929).—"Reptiles and Amphibians of South Australia". (Govt Printer: Adelaide.)

WOOD JONES, F. (1924).—"The Mammals of South Australia Part II. The bandicoots and the herbivorous marsupials". (Govt Printer: Adelaide.)

WOMERSLEY, H. (1939).—"Primitive Insects of South Australia". (Govt Printer: Adelaide.)

CHAPTER 10

FRESHWATER INVERTEBRATES

by K. F. Walker*, J. E. Bishop*, R. J. Shiel* & W. D. Williams*

In the Adelaide region there are several kinds of inland aquatic environment, all containing a diversity of invertebrate animals. Most of these animals will be unfamiliar to casual observers, for some knowledge of sampling methods is necessary, particularly to collect the smaller, more cryptic forms. A dip-net like those sold by aquarium dealers is an effective tool for beginners. With a larger, more robust net the "kick-sampling" method may be used: the animals are dislodged by shuffling and kicking the substrate and amongst water plants, and the net is held so as to trap the disturbed material. The net contents are best emptied into a shallow dish and sorted using a rubber-bulb pipette or tweezers. Another effective way of sampling shallow water habitats is to remove rocks and submerged wood temporarily from the water and inspect these for animals. A mesh size of about one-quarter of a millimetre generally is satisfactory, but for the microscopic plankton a net of the finest mesh is needed. If preserved samples are wanted, 70% ethanol is probably the best all-round preservative.

From the viewpoint of limnologists (ecologists interested in inland waters), the Adelaide area has not been well studied, and there is much to be discovered by enthusiastic naturalists. However, the range of animal types is so great that it would not be possible to do justice to all groups in these few pages. To illustrate the diversity of the fauna, Table 1 presents a list of invertebrates recently recorded from various stations in the River Torrens.

Perhaps the best way to encourage studies and offer most assistance is to consider the fauna of the Adelaide region in the context of a simplified taxonomic key. The key will assist in determining the general group to which any free-living (non-parasitic) invertebrate belongs. As each group is mentioned, some supplementary information is given: the common name (if any), common or conspicuous local species, a basic reference for further study, and an indication of the kind of environment in which the group has representatives according to the following symbols:

A ponds

B lakes

C streams

D saline lakes (salt concentration greater than one-tenth seawater)

E temporary waters (e.g. rainwater pools)

Outline sketches of some representative animals are also included. A book that will generally be helpful for all groups is *Australian Freshwater Life* (Williams 1968). Where no reference is given for any group dealt with in the following key, consult Williams' book to begin with.

To use the key, start at the first couplet and choose the alternative that best applies to the unknown animal. Then consult the couplet number shown at the end of the line. Repeat this procedure until a final conclusion is reached. The number in brackets shows, for easy reference, the preceding couplet.

Key to the aquatic stages of major groups of free-living invertebrates in Australian inland waters

1 (—) Minute to microscopic animals consisting of a single "cell" with simple internal structure; usually solitary, but may be colonial; attached or free-swimming ....
.... phylum PROTOZOA
 A large group for which little is known locally. Many species, however, are

---

* Department of Zoology, University of Adelaide, North Tce, Adelaide, S. Aust. 5000.

cosmopolitan, and much of the information available for other countries is relevant (Jahn 1949; ABCDE).

—   Animals with many cells 2

2 (1) Small to moderately large forms of indefinite shape; always attached, encrusting submerged rocks, plants, etc.; often "spongy" to touch ....
.... phylum PORIFERA
Freshwater sponges (Racek 1969; ABC).

—   Not as above 3

3 (2) Attached and free-swimming forms. If attached, then solitary or colonial, minute to small, individuals having a short stem and terminal tentacles. If free-swimming, then solitary, small, transparent, with an umbrella-shaped disc ....
.... phylum COELENTERATA
Inland representatives of this group are of three general types: the solitary, attached *Hydra* and its allies (ABCE), the colonial attached forms such as *Cordylophora* (ABCE), and the free-swimming, solitary, freshwater jellyfish *Craspedacusta sowerbyi* (Thomas 1951; AB).

—   Not with either of these combinations of characters 4

4 (3) Small to large, usually attached, colonial forms; either encrusting and mat-like, or branching and twig-like, or gelatinous and large; individuals with numerous retractile tentacles around mouth and main body mass enclosed ....
.... phylum POLYZOA
Pipe moss (ABC).

—   Without this combination of characters 5

5 (4) Unsegmented, without jointed appendages, and without a shell 6

—   Either segmented, or with jointed appendages, or with a shell, or with any of these features combined 13

6 (5) Long, worm-like, not flattened in cross-section 7

—   Not as above 9

7 (6) Small forms (usually less than 3 cm long), often brightly coloured, and with a protrusible proboscis armed with a small hooked spine or "stylet" ....
.... phylum NEMERTEA
Rarely encountered, but typically found amongst submerged vegetation. No published records from South Australia (ABC).

—   Without a proboscis, and not brightly coloured 8

8 (7) Small "worms" (usually less than 1 cm long), moving with whip-like motion; more or less round in cross-section; posterior end tapering to a fine point ....
.... class NEMATODA
Round worms. Usually associated with bottom sediments in a wide variety of habitats (ABCDE).

—   Thread-like forms, from 10 cm to about a metre in length but always very thin, usually dark brown to blackish, posterior end not tapering conspicuously
.... class NEMATOMORPHA
Horse-hair worms. Often found as a tangled mass (ABCE).

9 (6) Small to moderately large, elongate flattened forms, moving by a gliding motion .... class TURBELLARIA
Flatworms. Often found on submerged rocks in flowing water or shallow standing water. *Dugesia* is among those often encountered locally (ABCDE).

—   Small to microscopic animals not as above 10

10 (9) Small (up to 1 cm) organisms associated with larger crustaceans; with 2-6 anterior tentacles .... class TEMNOCEPHALIDEA
Relatives of flatworms. Often found in the branchial chamber or on the body surface of freshwater crayfish. Not parasites, as they might appear (ABC).

—   Not as above 11

11 (10) Microscopic organisms with a short, stout cylindrical body and 4 pairs of stumpy unjointed legs .... phylum TARDIGRADA

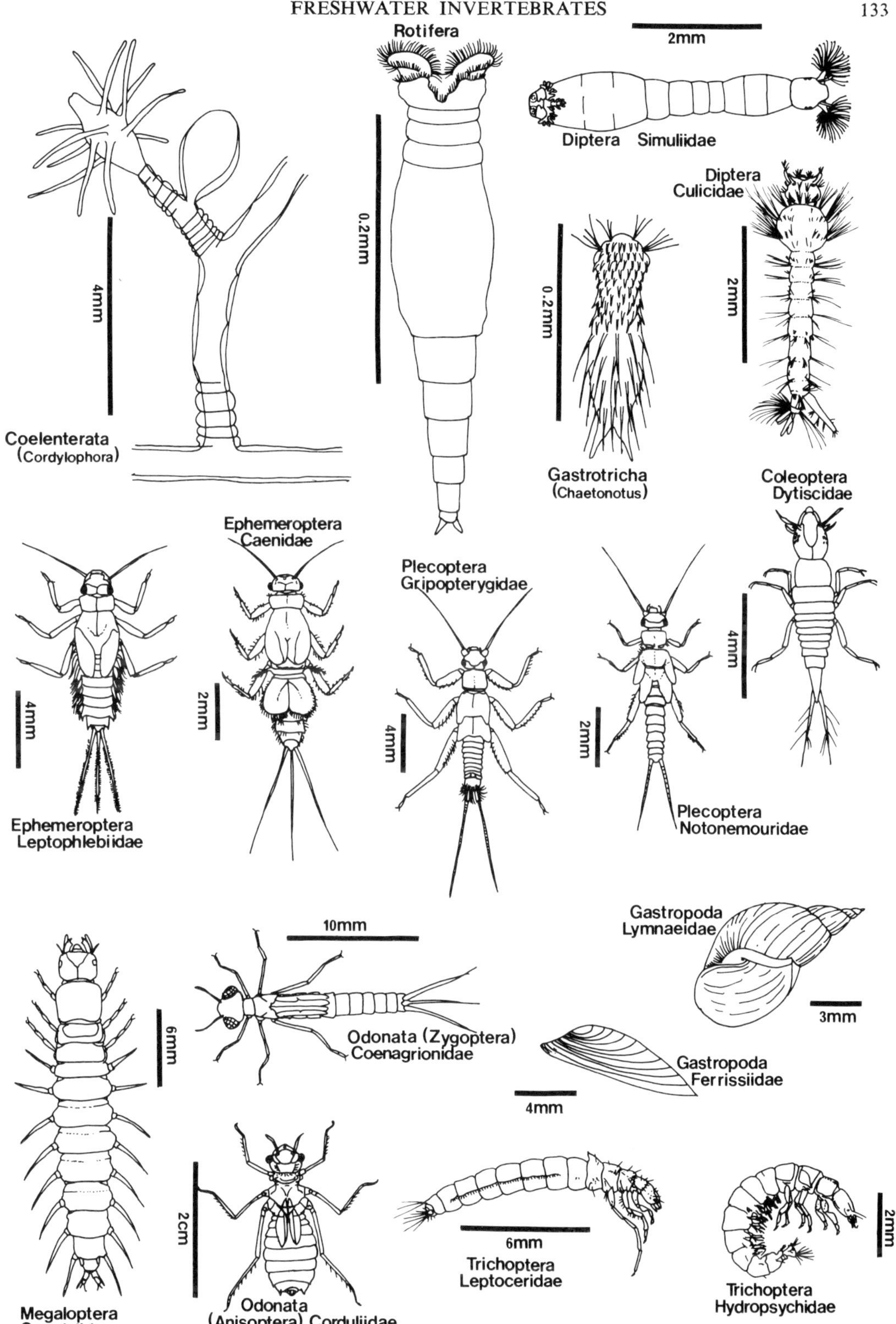

Fig. 1. Some representative freshwater invertebrates: minor groups and some insects.

"Bear animalicules." Usually found crawling amongst the surface film associated with mosses and similar plants (ABCD).

—   Organisms without 4 pairs of stumpy unjointed legs 12

12 (11) Free or sessile, solitary or colonial organisms with a body usually moderately elongate and cylindrical; three main body regions often discernible as a distinct head, trunk and foot; foot when present frequently with 2 terminal "toes", whilst head bears circularly arranged cilia; body sometimes superficially segmented .... class ROTIFERA
Rotifers or "wheel animalicules" (Edmondson 1959; ABCDE).

—   Free and solitary, short to moderately long organisms with a well-defined head usually separated from body by a distinct neck; general body surface covered by a cuticle often with spines, scales, or plates; often with 2 backward projections posteriorly .... class GASTROTRICHA
Usually found associated with Protozoa and other micro-organisms forming communities on the bottom of small, still water bodies (ABE).

13 (5) Small to large animals with a hard shell that is spherical, coiled, cap-like, or bivalved; animal itself of soft consistency, unsegmented, and *without jointed appendages* (phylum MOLLUSCA) 14

—   Not with the above combination of characters 15

14 (13) Shell coiled, spherical, or cap-like, and of one piece ...... class GASTROPODA
Snails and limpets. The small limpet *Pettancylus* is often found on rocks in flowing water. Common snails include the flat-coiled *Gyraulus,* and the spired *Bulinus, Physa, Physastra,* and *Potamopyrgus.* Another snail, *Lymnaea tomentosa,* is implicated in the transmission of sheep liver fluke (Cotton 1953; Lynch 1965; ABCD).

—   Shell bivalved (two pieces) .... class BIVALVIA
Clams and mussels. Small clams such as *Sphaerium* and *Corbiculina* are common in local streams. Mussels include the widespread *Velesunio ambiguus* (McMichael & Hiscock 1958; ABC).

15 (13) Minute to large, segmented (sometimes indistinctly so) animals without jointed appendages (phylum ANNELIDA) 16

—   Minute to large, segmented (sometimes indistinctly so) animals, frequently with a hard protective skin or cuticle; typically with jointed appendages (phylum ARTHROPODA) 17

16 (15) With bristles on at least some segments, and without suckers ....
.... classes OLIGOCHAETA and POLYCHAETA
True worms. Oligochaetes ("few bristles") such as *Tubifex* and *Branchiura* typically are found in mud-bottom communities. Polychaetes ("many bristles") are much less common in fresh waters (Brinkhurst & Jamieson 1971; ABCD).

—   Without bristles, but with suckers at one or both ends of the body ....
.... class HIRUDINEA
Leeches. The brown-striped *Limnobdella australis* is common in creeks and dams (ABC).

17 (15) In adult form, with 4 pairs of jointed legs, body usually globular and indistinctly segmented .... class ARACHNIDA
Water mites and spiders. Water mites, belonging to the Hydracarina, are small spherical organisms often found in shallow still water. Among the spiders, *Tetragnatha* is typically found amongst fringing growths of sedges, and the nursery-web spider *Dolomedes* lives around the bases of emergent plants. The latter species can inflict a nasty bite (Main 1976; ABC).

—   Not with the above combination of characters 18

18 (17) With 3 pairs of jointed legs, legless, or with few to several pairs of stout unjointed legs; body typically divided into a head bearing a single pair of antennae, a thorax with 3 pairs of jointed legs (and one or two pairs of wings in adults), and an abdomen (class INSECTA) 19
(A useful reference for Insecta is C.S.I.R.O. 1971.)

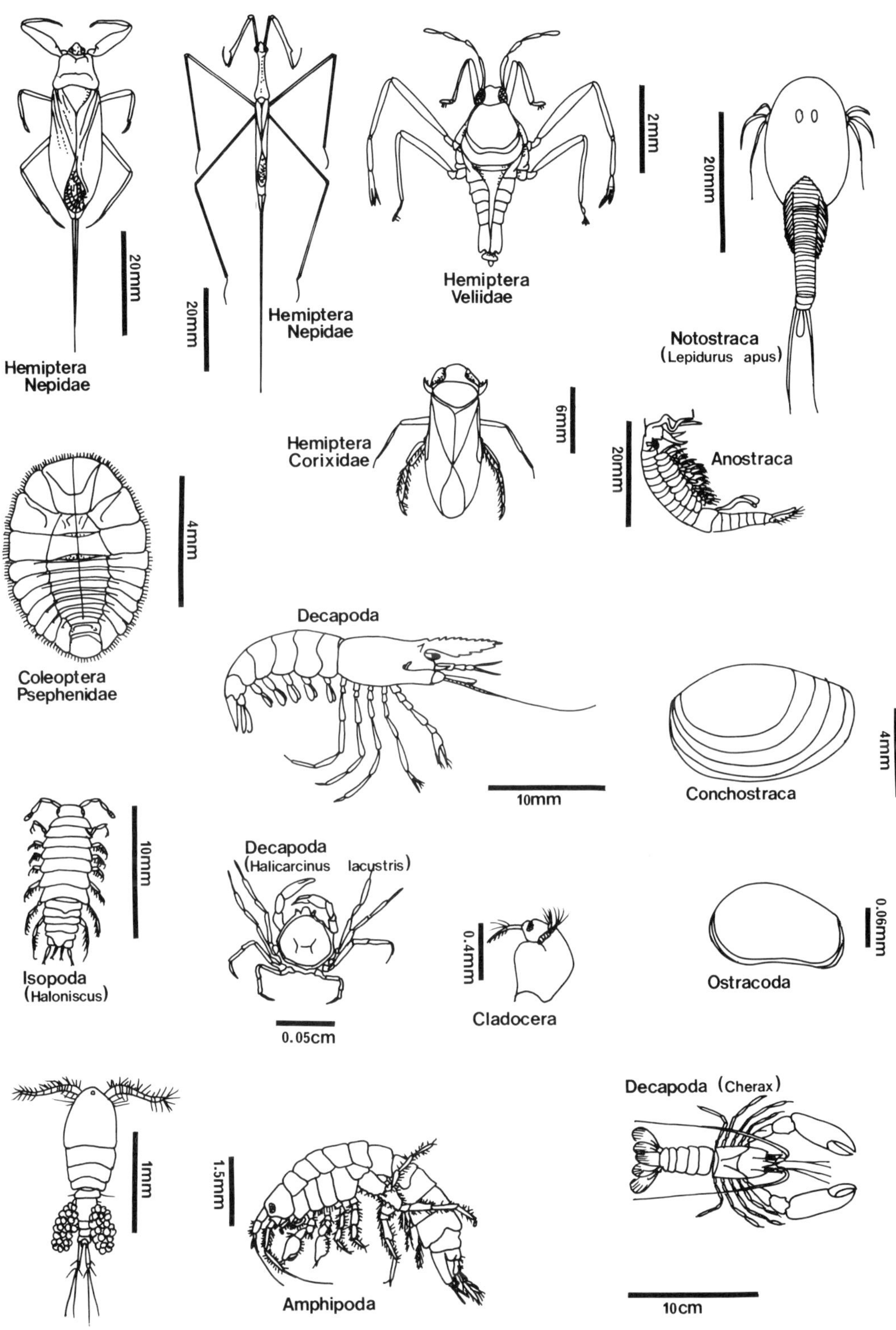

Fig. 2. Some representative freshwater invertebrates: some insects and crustaceans.

—   Usually more than 3 pairs of legs; never legless; typically, body with regions as above but head with 2 pairs of antennae and thorax never with wings (class CRUSTACEA) 33

        (A useful reference for Crustacea is Bayly, Bishop & Hiscock 1967.)

19 (18)   *Aquatic* adult forms, with wings 20

—   Aquatic immature stages (nymphs, larvae, or pupae) without wings, or wingless adults 21

20 (19)   Forewings hard, chitinous, and folded along back so that typically there is a straight and prominent dorsal longitudinal suture; usually with biting mouthparts; antennae usually obvious .... order COLEOPTERA

        Adult beetles. A very large group including the gyrinids (whirligig beetles), dytiscids (diving beetles), and hydrophilids ("water lovers"). The large, shiny black *Hydrophilus latipalpus* is the largest member of the Hydrophilidae in Australia. Among the dytiscids, the creamy brown *Eretes australis* is commonly found in shallow temporary waters (C.S.I.R.O. 1971, ABCDE).

—   Forewings not folded in the above manner; mouthparts clearly suctorial or head triangular; antennae often small .... order HEMIPTERA (in part)

        Adult bugs. Another large group, among which the most conspicuous species are the "water scorpion" *Laccotrephes tristis*, the needle-bug *Ranatra dispar*, "water boatmen" of the genera *Sigara* and *Agraptocorixa*, and the freshwater "backswimmers" of the genus *Anisops*. Bugs of the belostomatid genus *Diplonychus* are interesting in that the female cements her eggs on to the back of the male (C.S.I.R.O. 1971; Lansbury 1969, 1970, 1972; Knowles 1974; ABCE).

21 (19)   Small insects (less than ½ cm long) dwelling on the surface film; usually with a long forked springing appendage at posterior end of abdomen................... order COLLEMBOLA

        Springtails (ABCE).

—   Not as above 22

22 (21)   Mouthparts beak-like, suctorial, and piercing, or head triangular in front view; antennae often small; abdomen not terminating in more than one (or two closely applied) projection(s) .... order HEMIPTERA (in part)

        Nymphal and adult bugs (see couplet 20).

—   Without the above combination of characters 23

23 (22)   Legs well-developed; with 2 or 3 long terminal abdominal processes either flattened or whip-like, or with abdomen ending in 3 triangular processes; developing wings visible as wing pads in more mature nymphs 24

—   Without the above combination of features 26

24 (23)   Antennae long, usually longer than forelegs; abdomen *typically* ending in 2 long cerci or tails; most commonly found in running waters .... order PLECOPTERA

        Nymphal stoneflies. A small group normally associated with clean, clear streams where the nymphs occur on stones and among leaf litter in the current. Occasionally found in dams. Not widespread in S.A., but quite common in Mt Lofty and S. Flinders Ranges. *Dinotoperla*, with a rosette of gills at the end of the abdomen, and *Austrocerca*, without gills, are common genera. Adults grey to black, slow-flying, carry their wings folded flat over the back (Hynes & Hynes 1975; BC).

—   Without the above combination of features 25

25 (24)   Abdomen ending in 3 (rarely 2) flattened tails or 3 triangular processes; mouthparts include a hinged jaw apparatus (the "mask" or labium) folded beneath head .... order ODONATA

        Nymphal damsel flies, with delicate terminal gills, and dragonfly nymphs, with less conspicuous triangular abdominal extremities, are found in a wide variety of habitats, amongst vegetation, under stones, in the mud, of streams, lakes and dams. Larvae are predatory on other invertebrates, tadpoles and fish. *Hemicordulia, Hemianax* and *Austrolestes* are common forms (Watson 1962; ABC).

—   Abdomen ending in 3 long cerci or tails; mouthparts without a hinged mask .... order EPHEMEROPTERA

Nymphal mayflies. *Atalophlebia,* flattened nymphs with lateral gills on the abdomen; *Baetis* with streamlined, hump-backed nymphs and *Tasmanocoenis,* "hairy" nymphs with the dorsal gills protected by a cover, are the common forms. *Baetis* in hard-bottomed streams, the others also in farm dams, reservoirs, lakes on a variety of substrates. No mayflies are found in salt or brackish waters (C.S.I.R.O. 1971; ABC).

26 (23)   With 3 pairs of jointed legs on thorax 27

—   Without jointed legs on thorax, but sometimes with unjointed legs on abdomen or thorax .... order DIPTERA

Larval and pupal two-winged flies. A very large group including the mosquitoes, gnats, long-legged crane flies, sand flies, biting and other midges, black-fly, soldier and robber flies. The larvae are of various forms found attached in flowing water (black-fly), free-swimming in any standing water (mosquitoes) and living in soft muddy bottoms in any body of water (gnats and midges). Some forms found in saline waters. Various groups involved in the transmission of parasitic and viral diseases (malaria, encephalitis, etc.) (C.S.I.R.O. 1971, ABCDE).

27 (26)   Inactive forms (pupae) 28

—   Active forms 29

28 (27)   Legs separately encased and not fused to body; animal often in case of plant or mineral matter .... order TRICHOPTERA

Pupal caddis flies (see couplet 32).

—   Legs fused to body; whole animal enclosed in a cocoon .... order LEPIDOPTERA

Pupal moths (see couplet 31).

29 (27)   Each abdominal segment bearing laterally a pair of unjointed or jointed processes, often hairy; mouthparts strongly developed for biting; abdomen posteriorly with either a long terminal process or a single pair of hooked prolegs (take care not to confuse some larval Coleoptera with this group) .... order MEGALOPTERA

Larval alder and dobson flies. *Austrosialis,* with a long "tail" and *Archichauliodes* and *Austrochauliodes,* both with terminal hooked prolegs, are the Australian representatives. Larvae occur principally in streams, occasionally in still water (ABC).

—   Without the above combination of features 30

30 (29)   Small insects with slender piercing mouthparts about half as long as the body .... order NEUROPTERA

Lacewing larvae. A small group usually found associated with freshwater sponges. *Sisyra* restricted to clean, silt-free streams and ponds. A few semi-aquatic free-living *Kempynus* occur (ABC).

—   Mouthparts not as above 31

31 (30)   Abdominal segments 3–6 and terminal one each with a pair of stumpy "prolegs"; often living inside case made of plant matter ...... order LEPIDOPTERA

Larval moths. *Cataclysta* and *Nymphula* are representative of the moths with aquatic larvae. Usually found in vegetated areas of fresh and slightly saline streams and still-water where case-building materials available (ABC).

—   Without abdominal prolegs, or with them but arranged differently 32

32 (31)   A pair of hooked prolegs on last abdominal segment; often living in case of plant or inorganic material .... order TRICHOPTERA

Larval caddis flies. A large Order with many free-living and case-building members. Common in all fresh waters and a few occur in mildly salt streams and lakes. Leaf and stick case builders are most often found among emergent vegetation; stone-ballasted case builders and free-living net spinners in hard-bottomed streams (C.S.I.R.O. 1971; ABC).

—   Not as above; larvae of a wide variety of forms .... order COLEOPTERA
         Larval beetles. Variety of forms from the flattened larvae of Psephenidae ("water
         penny") to the prominently jawed larvae of the Dytiscidae. Others grub-like
         with no hardened parts. Found in all types of aquatic habitat including salt and
         brackish streams and lakes (ABCDE).

33 (18) Animals with leaf-like appendages, at least 19 segments, no carapace, and
         stalked eyes; very delicate forms of moderate size swimming back downwards
         .... order ANOSTRACA
         Fairy shrimps. The commonest inhabitants of the salt pans north of Adelaide.
         Some species also occur in temporary fresh waters (DE).

—   Not with the above combination of features 34

34 (33) Animals with leaf-like appendages and numerous segments, but with a large
         dorsal shield (carapace) and sessile eyes .... order NOTOSTRACA
         Tadpole shrimps. Typically found in shallow temporary fresh waters. *Lepidurus*
         is the common genus found in temperate S.A. (E).

—   Not with the above combination of features 35

35   Animals with a bivalved carapace 36

—   Animals not enclosed (partly or wholly) in a bivalved carapace 38

36 (35) Bivalved carapace enclosing trunk and its appendages, but not head ....
         .... suborder CLADOCERA
         "Water-fleas." Numerous genera and species occur in the Adelaide region .The
         group occurs in almost all standing water bodies of relative permanency but is
         also occasionally found in temporary waters. Most representatives are planktonic
         (free-floating) (ABE).

—   Bivalved carapace usually enclosing entire animal 37

37 (36) Animals with never more than 2 pairs of trunk appendages, small size ....
         .... subclass OSTRACODA
         "Seed-shrimps." Small animals that are normally benthic in habit (bottom-
         living). Superficially they are remarkably similar to minute bivalved molluscs
         (ABD).

—   Animals with several to numerous trunk appendages, small to moderately large
         size .... suborder CONCHOSTRACA
         "Clam-shrimps." Almost entirely confined to temporary fresh waters. Of rare
         occurrence in the metropolitan area (E).

38 (35) Microscopic to small, pear-shaped animals, without a carapace, but with the
         first and sometimes second thoracic segment fused to the head; abdomen
         4-segmented, each segment, except the last, lacking appendages; often with
         well-developed antennae .... subclass COPEPODA
         Microscopic to minute, normally free-floating forms found in standing waters.
         Most waters that are relatively permanent have representatives (ABD and occa-
         sionally E).

—   Not with the above combination of features 39

39 (38) Small crustaceans temporarily ectoparasitic on fish; with a flat, broad, bilobed
         carapace covering most of the body, only 4 thoracic appendages, an unseg-
         mented and limbless abdomen, and a suctorial mouth ....
         .... subclass BRANCHIURA
         "Fish-lice." There appear to be no S.A. records, though representatives have been
         recorded from other Australian States (AB).

—   Not with the above combination of characters 40

40 (39) With a carapace covering the back and sides of the thorax ....
         .... order DECAPODA
         Freshwater shrimps, yabbies and crabs. The shrimps are represented by *Paratya
         australiensis,* which is very common in the River Torrens and often fresh coastal
         streams near Adelaide. Yabbies (freshwater crayfish) are known from both still
         and running waters in the area (Riek 1969). Crabs (*Halicarcinus lacustris*) occur
         in the coastal streams (Walker 1969; ABC).

—   Without a carapace covering the back and sides of the thorax 41

41 (40) Small to moderately large animals with little difference in form between thoracic and abdominal segments, with exopodites on walking legs .... division SYNCARIDA

> No representatives of this group have been recorded from S.A. thus far, though it would be surprising if representatives of the syncarid bathynellaceans were found not to occur near Adelaide. These are minute interstitial forms. The occurrence of other syncarid representatives should also not be ruled out.

—   Not with the above combination of features 42

42 (41) Gills on some thoracic appendages, pleopods (abdominal appendages) slender and never plate-like, body usually laterally compressed .. order AMPHIPODA

> The only genus so far found in S.A. is *Austrochiltonia*. Other genera may possibly occur in the Mt Lofty Ranges. *Austrochiltonia* is widespread and occurs in all manner of permanent, fresh lowland waters (ABC).

—   No gills or thoracic appendages; some pleopods (abdominal appendages) plate-like and modified for respiration .... order ISOPODA

> There are two representatives of this order known from the Adelaide region: *Haloniscus*, a slater-like form found in the salt pans north of the city; and a form not unlike *Heterias*, a much smaller isopod apparently confined to small streams (CD).

TABLE 1
Invertebrate fauna collected recently in the River Torrens

				Torrens Gorge	Highbury-Athelstone	Windsor Gardens-Marden
Coelenterata	Hydroida		*Chlorohydra ?viridissima*		+	
Nematoda				+		
Mollusca	Gastropoda	Planorbidae	*Glacidorbis* sp.	+	+	+
			other		+	+
		Lymnaeidae	*Lymnaea* sp.	+		
		Hydrobiidae			+	+
	Bivalvia	Sphaeriidae			+	
		Corbiculidae	*Corbiculina angasi*			+
Annelida	Oligochaeta	Naididae	*Naidium* sp. no. 1	+		+
			Naidium sp. no. 2	+		
			Chaetogaster sp.	+	+	
			others	+	+	+
		Aelosomatidae	*Stylaria* sp.	+		
		Tubificidae				+
Arachnida	Aranae	Pisauridae	*Dolomedes* sp.		+	
	Acarina	Porohalacaridae			+	
		Hydracarina			+	+
Crustacea	Cladocera	Daphniidae	*Daphnia carinata*			+
			Ceriodaphnia quadrangula		+	
			Simosa exspinosa	+	+	
		Macrothricidae	*Ilyocryptus spinifer*			+
		Chydoridae	*Chydorus sphaericus*	+	+	+
			Alona sp.	+	+	+
			Alonella sp.		+	
			Pleuroxus sp.		+	
			Biapertura sp.		+	
			Pseudochydorus globosus		+	
			others		+	
	Ostracoda		*Stenooyrpis* sp.	+		
			others	+	+	
	Copepoda	Calanoida	*Calcimoecia* sp.	+	+	+
		Harpacticoida	*Attheyella* sp.	+		

TABLE 1—(*continued*)

			Torrens Gorge	Highbury-Athelstone	Windsor Gardens-Marden
	Cyclopoida	*Acanthocyclops vernalis*	+	+	
		Paracyclops sp.	+		
		Macrocyclops albidus	+	+	+
		Ectocyclops sp.	+		
		Eucyclops speratus	+	+	+
		Tropocyclops prasinus	+		
	Isopoda	*Heterias* sp.			+
	Amphipoda	*Austrochiltonia* sp.	+		+
	Decapoda	Atyidae	*Paratya australiensis*		+
		Parastacidae	*Cherax* sp.		+
Insecta	Collembola			+	
	Ephemeroptera	Baetidae	+	+	
	Odonata	Anisoptera	*Hemicordulia* sp.	+	+	+
		Zygoptera	+		+
	Plecoptera	Nemouridae	+		
	Hemiptera	Gerridae		+	
		Veliidae	+	+	+
		Corixidae	*Micronecta* sp.	+		
	Coleoptera	Dytiscidae	+	+	+
		Hydrophilidae			
	Diptera	Tipulidae			+
		Culicidae			+
		Simuliidae		+	
		Chironomidae	+	+	+
		Stratiomyidae			+
		Syrphidae			+
	Trichoptera	Calamoceratidae	+		
		Leptoceridae	+		
		Hydropsychidae	+		
		Hydroptilidae	+		+
	Lepidoptera	Nymphalidae	+		

In our survey of the Torrens in May 1976, different invertebrate communities were found at each of three sampling areas. Species typical of clear, cool and flowing waters were found in the Torrens Gorge, where a stony bottom and lack of emergent vegetation precludes many of the species found downstream below the Athelstone market gardens. There a sandy bottom, reedbeds, and nutrient inflow from the gardens provide an abundance of microhabitats. At Marden, further downstream, tubificid worms and syrphid (hover-fly) larvae dominate. These groups are commonly found in areas of organic pollution, resulting here from runoff into the river from extensive urban areas, with accompanying deterioration of water quality and decrease in species diversity.

Rather than being merely series of names, lists of invertebrates recorded from freshwater habitats may suggest community composition at the time of sampling, feeding relationships and habitat conditions. Within any body of water, variations in species diversity and abundance reflect changing environmental conditions. It is therefore important that sampling be carried out over sufficient time to cover possible seasonal and man-made variations. These may be induced by changes in water levels, light intensity, day-length, temperature, vegetation, nutrient enrichment and pollutants.

References

BAYLY, I, A. E., BISHOP, J. A. & HISCOCK, I. D. (Eds.) (1967).—"An illustrated key to the genera of Crustacea of Australian inland waters." Special publication Aust. Soc. Limnology, No. 1.

BRINKHURST, R. O. & JAMIESON, B. G. M. (1971). —"The aquatic Oligochaeta of the world." (Oliver & Boyd: Edinburgh.)

COTTON, B. C. (1953).—"South Australian shells." 3rd ed. (South Australian Museum: Adelaide.)

C.S.I.R.O. (1971).—"The insects of Australia." (Melbourne University Press: Melbourne.)

EDMONDSON, W. T. (1959).—Rotifers. *In* W. T. Edmonson (Ed.), "Freshwater Biology". (Wiley: New York.)

HYNES, H. B. N. & HYNES, M. E. (1975).—The life histories of many of the stoneflies (Plecoptera) of south-eastern mainland Australia. *Aust. J. Mar. Freshw. Res.* **26**, 113-153.

JAHN, T. (1949).—"How to know the Protozoa." (Wm C. Brown Publ.: Dubuque, Iowa.)

KNOWLES, J. N. (1974).—A revision of the Australian species of *Agraptocorixa* Kirkaldy and *Diaprepocoris* Kirkaldy (Heteroptera: Corixidae). *Aust. J. Mar. Freshwat. Res.* **25**, 173-191.

LANSBURY, I. (1969).—The genus *Anisops* in Australia (Hemiptera-Heteroptera, Notonectidae). *J. Nat. Hist.* **3**, 433-458.

LANSBURY, I. (1970).—Revision of the Australian *Sigara* (Hemiptera-Heteroptera, Corixidae). *J. Nat. Hist.* **4**, 39-54.

LANSBURY, I. (1972).—A review of the Oriental species of *Ranatra* Fabricius (Hemiptera-Heteroptera: Nepidae). *Trans. R. Ent. Soc., London,* **124**, 287-341.

LYNCH, J. J. (1965).—The ecology of *Lymnaea tomentosa* (Pfeiffer 1855) in South Australia. *Aust. J. Zool.* **13**, 461-473.

MAIN, B. Y. (1976).—"Spiders." (Collins: Sydney.)

McMICHAEL, D. F. & HISCOCK, I. D. (1958).—A monograph of the freshwater mussels (Mollusca: Pelecypoda) of the Australian region. *Aust. J. Mar. Freshw. Res.* **9**, 372-508.

RACEK, A. A. (1969).—The freshwater sponges of Australia (Porifera: Spongillidae). *Aust. J. Mar. Freshw. Res.* **20**, 267-310.

RIEK, E. F. (1969).—The Australian freshwater crayfishes (Crustacea: Decapoda: Parastacidae), with descriptions of new species. *Aust. J. Zool.* **17**, 855-918.

THOMAS, I. M. (1951).—*Craspedacusta sowerbyi* in South Australia, with some notes on its habits. *Trans. R. Soc. S. Aust.* **74**, 59-65.

WALKER, K. F. (1969).—The ecology and distribution of *Halicarcinus lacustris* (Brachyura: Hymenosomatidae) in Australian inland waters. *Aust. J. Mar. Freshw. Res.* **20**, 163-173.

WATSON, J. A. L. (1962).—"The dragonflies (Odonata) of south-western Australia." Handbook No. 7. (Western Australian Naturalists Club: Perth.)

WILLIAMS, W. D. (1968).—"Australian freshwater life. The invertebrates of Australian inland waters." (Sun Books: Melbourne.)

CHAPTER 11

PHYSICAL OCEANOGRAPHY OF GULF ST VINCENT AND INVESTIGATOR STRAIT

by JOHN A. T. BYE

Introduction

The coastal waters of South Australia may be defined as occurring over the continental shelf on which the depth is less than 200 m. Beyond this contour the depth increases rapidly on the continental slope, reaching over 5,000 m about 500 km from the coastline in the South Australian Basin.

To the west of Cape Carnot at the tip of Eyre Peninsula, the region forms part of the Great Australian Bight; to the east, however, as far as the Victorian border, no name has been given to the complex zone which includes Spencer Gulf, Gulf St Vincent, Investigator Strait and Encounter Bay. This zone, which comprises 10^5 km^2 of water, is of fundamental importance to South Australia, and forms a geographic entity, part of which is the subject of this paper. To emphasise this entity, it will be called a sea, which will be referred to as the South Australian Sea (Fig. 1). The area is similar to that of the Irish Sea, with which there are strong resemblances. There are considerable data on all regions of the sea except the approach waters beyond Kangaroo Island, where the important interactions with the deep ocean occur. Within the sea, Gulf St Vincent has a surface area of 6.8×10^3 km^2 with a mean depth of about 21 m, and the corresponding figures for Investigator Strait are 6.1×10^3 km^2 and 34 m (Fig. 1).

Recently, considerable controversy has been generated concerning the extent of the coastal waters over which the State of South Australia has the power to apply fisheries legislation, so that the question of definition entertained above may not be just of academic interest.

The first detailed hydrographic charts of Investigator Strait and Gulf St Vincent were of course made by Matthew Flinders (Flinders 1814) in 1802. He entered Investigator Strait from the west near Cape Spencer and, after making several traverses of the Strait, completed a counterclockwise sweep of Gulf St Vincent, leaving through "the private entrance" (Backstairs Passage). In all, Flinders travelled about 1,000 km in a time of 18 days and recognised the line joining Troubridge Point and Cape Jervis as marking the approximate entrance to the Gulf.

Tidal observations were made off Kangaroo Head, near Port Wakefield, and in Pelican Lagoon. The tide at these locations at the time of Flinders' visit have been estimated, using contemporary tidal analysis procedures (Table 1), and the results further enhance his reputation as an observer. The second survey of coastal waters was made by Captain Lipson in HMCS *Yatala* in 1850, during the intervening years many hazardous shoals having been charted. The major survey, however, was made between 1862 and 1880 by Captain Hutchinson and Lieutenants Goalen and Guy in HMS *Beatrice*. This Admiralty survey, which included a detailed examination of the approaches to Port Adelaide in 1875, forms the basis of present day hydrographic charts of the region (Ingleton 1944). Further surveys have been conducted from time to time with Naval vessels including HMAS *Barcoo*, HMAS *Moresby*, HMAS *Gascoyne*, and HMAS *Diamantina*.

Following the charting of the sea, the next objective was the accurate prediction of tidal levels, especially at Port Adelaide. This had been achieved by about 1890. Following this birth of physical oceanography, there seem to have been few planned investigations until the last twenty years.

Recent work has been mainly concentrated in the Universities, the formation in 1966 of the Horace Lamb Centre at Flinders University

---

* School of Earth Sciences, Flinders University of South Australia, Bedford Park, S. Aust. 5042.

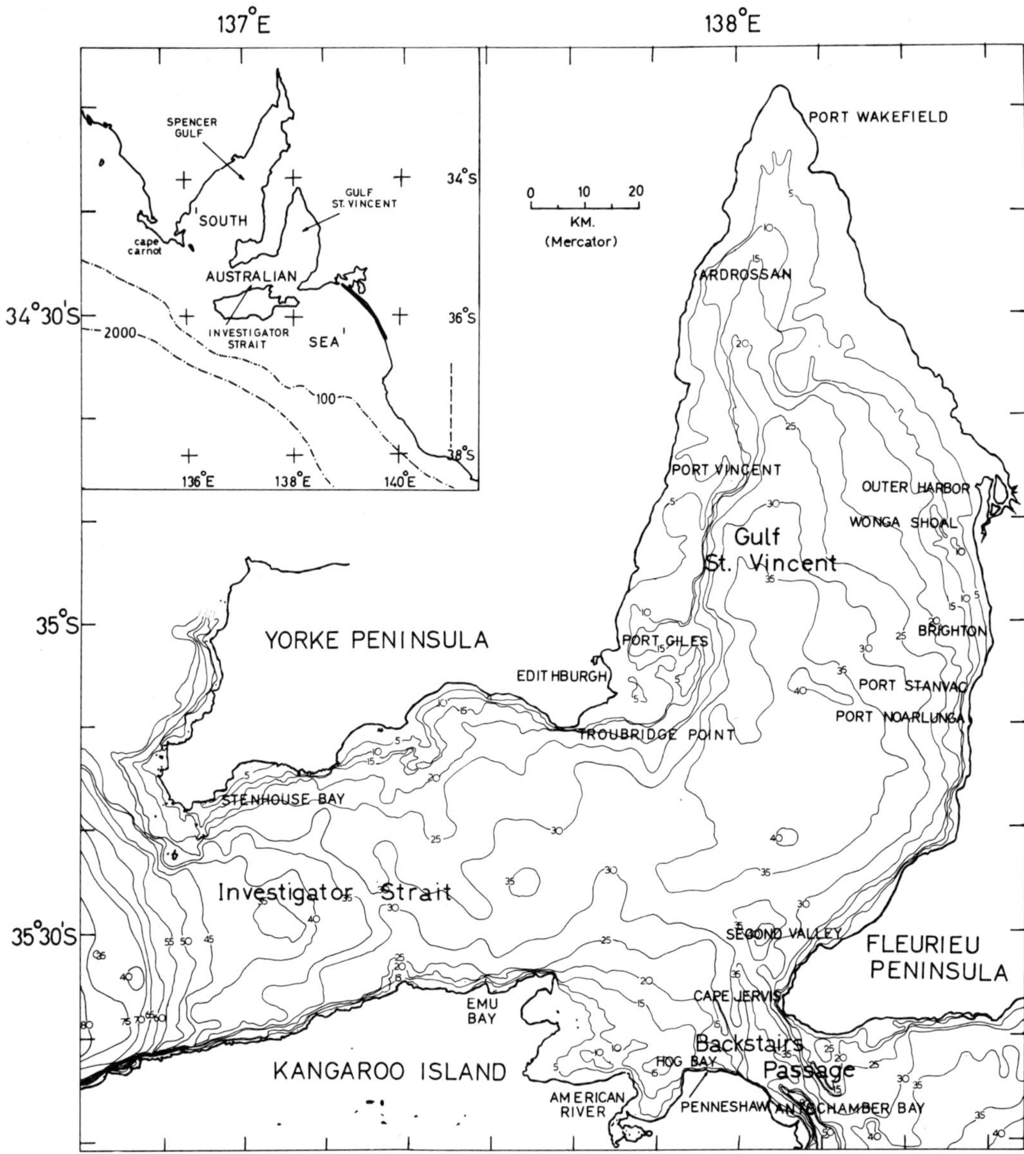

Fig. 1. South Australian Sea. Inset: bathymetry of Gulf St Vincent and Investigator Strait, showing ports mentioned in text. Contours in metres.

being especially important for the development of physical oceanography. Recently independent centres have also been established, stimulated by governmental, industrial and environmental needs for more information on the sea. The new wave of investigations has begun to cover all aspects of physical oceanography, and provide some perspective of the coastal waters as an integral part of the State's resources.

Many of the important studies are only possible with expedition vessels, of which MV *Saori*, the Fisheries patrol vessel MV *Warrendi*, and the Flinders University motor launch MV *Scipio* should be mentioned. The fundamental

TABLE 1
Tide levels at tide stations occupied by Matthew Flinders in 1802

Near Port Wakefield, 30th March 1802: Low water observed to occur at 1000. (Prediction of low water based on tidal constants for Ardrossan: 1002.)

Pelican Lagoon, 5th April 1802: High water observed to occur at 0600. (Prediction of high water based on tidal constants for American River: 0512.)

need for vessels for oceanographic research in the South Australian Sea remains.

Tides

The water level variation along the coast of South Australia is mainly due to the astronomical tide, although important meteorological tides also occur.

The two main species of astronomical tide behave very differently in the ocean between Australia and Antarctica. The diurnal tide travels towards the west. The semi-diurnal tide, on the other hand, travels towards the east, but is almost normally incident on the coast (Irish & Snodgrass 1972). The tidal constants of the two most important semi-diurnal constituents (M_2 and S_2), and the two most important diurnal constituents (K_1 and O_1) are shown in Table 2.

The difference in behaviour between the two tidal species is consistent with the strong dissipation of tidal energy occurring for the semi-diurnal tide in the South Australian Sea, which has resonances near the semi-diurnal period.

Mathematical computation (Tronson 1975) indicates that the fundamental resonance of Spencer Gulf is 15.9 hours, and of Gulf St Vincent with Investigator Strait is 9.4 hours.

In Spencer Gulf both species of tide behave as progressive waves such that high water (or any other phase of the tide) occurs successively later towards the head of the gulf. Thus high water occurs at Port Augusta about 6 hours later than at Port Lincoln.

The tides of Gulf St Vincent, however, are quite different, because of the dual connection with the open sea of Investigator Strait. For the semi-diurnal tide there is almost simultaneous high water at either entrance, and an apparent standing oscillation is set up between the two ingressing tidal waves within the Gulf. This causes the time of high water everywhere within Gulf St Vincent to be almost identical. At a higher resolution the co-tidal lines run approximately north-south within the gulf, with a phase increase of about 10° (20 mins) from west to east (Fig. 2(a)). As in Spencer Gulf the tidal range increases toward its head (Table 2). The diurnal tide behaves in a similar manner although, due to its period being further from the resonant period of the Gulf, it is not amplified as much as the semi-diurnal tide, and the phase increase is only about 3° (12 mins) from west to east (Fig. 2(b)).

The amplitudes of the main semi-diurnal tide constituents (M_2 and S_2) in Gulf St Vincent are almost identical (Table 2). This remarkable circumstance (in most regions of the ocean S_2 has about half the amplitude of M_2) means that at neap tides the semi-diurnal tide is virtually absent, and the diurnal tide dominates. Near the equinoxes the diurnal components also vanish, allowing the water level to remain almost constant for a whole day. This condition is known as a dodge or dodging tide (Chapman 1892) (Fig. 3).

Tidal currents are associated with the tidal elevations. Very few continuous current observations have been made in Gulf St Vincent and apparently none in Investigator Strait. Information from ships' logs and divers' observations

TABLE 2

Major tidal constants at ports in the South Australian Sea

Port	M_2		S_2		K_1		O_1	
	g°	A m	g°	A m	g°	A m	g°	A m
Stenhouse Bay	033	.17	091	.16	035	.18	009	.13
Emu Bay	091	.24	160	.21	049	.22	021	.15
American River	092	.23	165	.19	048	.22	020	.15
Hog Bay	080	.19	140	.19	041	.22	013	.15
Antechamber Bay	035	.14	095	.14	042	.20	007	.15
Victor Harbor	343	.14	043	.17	040	.19	009	.13
Cape Jervis	092	.18	160	.19	044	.18	015	.15
Second Valley	102	.33	171	.31	049	.23	021	.15
Edithburg	091	.42	161	.41	045	.26	015	.16
Port Noarlunga	100	.40	171	.39	053	.25	012	.14
Brighton	104	.44	174	.43	049	.24	020	.16
Outer Harbor	105	.50	176	.49	050	.25	024	.16
Inner Harbor	114	.53	186	.54	056	.26	029	.17
Port Vincent	090	.56	161	.49	047	.26	018	.19
Ardrossan	105	.62	174	.63	047	.27	019	.18

A m   Amplitude in metres.
g°   Phase in degrees.
Data from A.N.T.T. (1976).

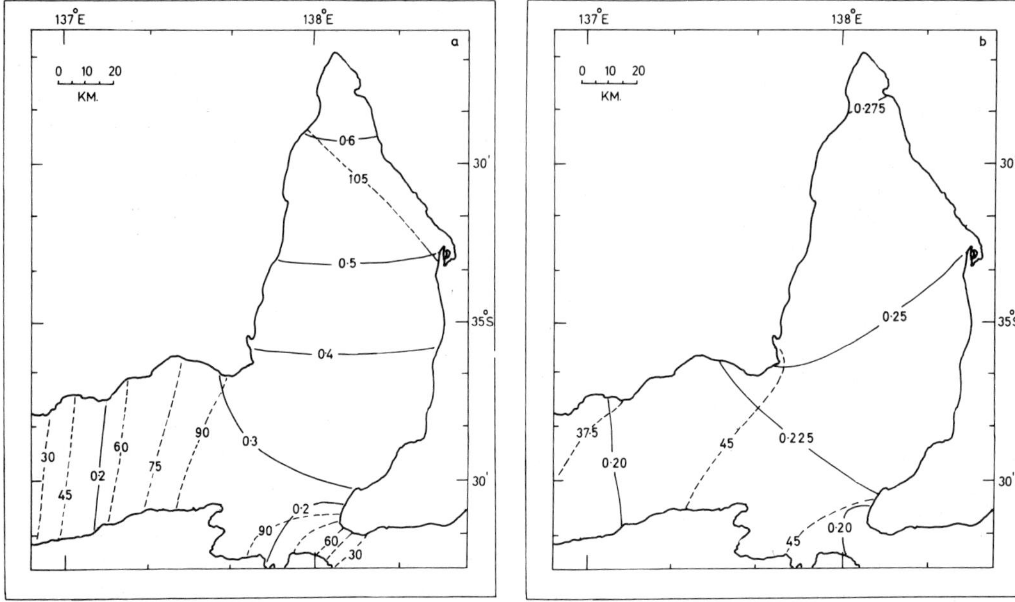

Fig. 2. (a) Co-range and co-tidal chart for the M_2 constituent: ———— amplitude in metres; - - - -
phase in degrees (1 mean solar hour = 28.98°).
(b) Co-range and co-tidal chart for the K_1 constituent: ———— amplitude in metres; - - - -
phase in degrees (1 mean solar hour = 15.04°).

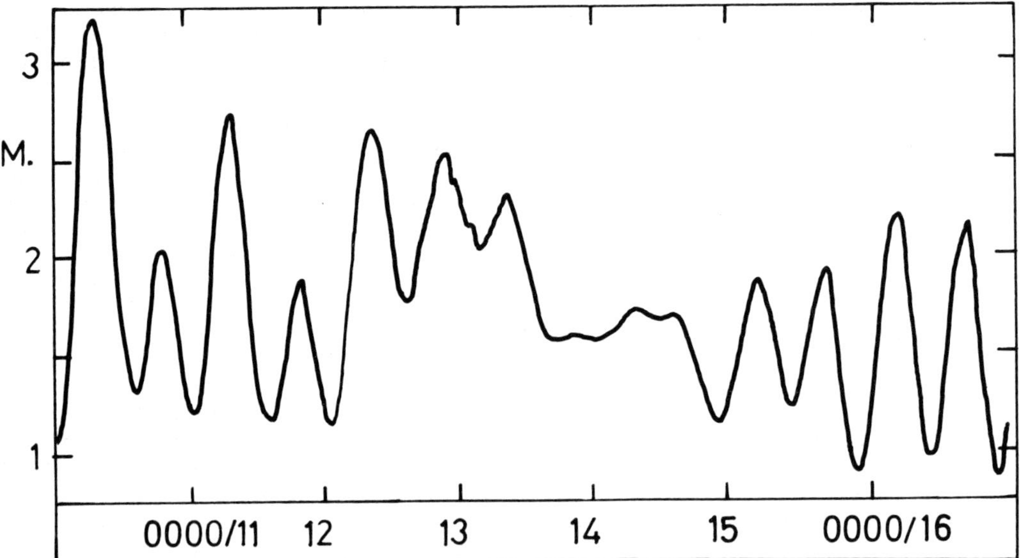

Fig. 3. Tidal record at Outer Harbor for the period 10/11/68-16/11/68 illustrating the phenomenon
of the dodge tide. Elevations in metres.

(Shepherd & Sprigg 1976), however, suggest tidal streams of 1–2 m/s in Backstairs Passage and 1 m/s near Troubridge Shoals.

Data series in the vicinity of Adelaide (Radok 1973, Bye & Suter 1974, Steedman 1975) show that the currents are mainly tidal, with an absence of the diurnal inequality shown in the local tide. Analyses of the records at Port Stanvac and Outer Harbor (Table 3) find that the M_2 and S_2 amplitudes are about 0.07

TABLE 3
Tidal currents in the vicinity of Adelaide

Position	M_2		S_2		K_1		O_1	
	$g°$	$\stackrel{\frown}{A}$ m/s	$g°$	$\stackrel{\frown}{A}$ m/s	$g°$	$\stackrel{\frown}{A}$ m/s	$g°$	$\stackrel{\frown}{A}$ m/s
Pt Stanvac[1] (north-south component)	338	.084	59	.044	284	.014	27	.01
Outer Harbor[2] (north-south component)	344	.065	27	.114	332	.023	37	.005
(east-west component)	80	.010	209	.021	328	.037	100	.003

$\stackrel{\frown}{A}$ m/s amplitude in metres per second.
$g°$ phase in degrees.
[1] Data from the end of Port Stanvac Jetty (Bye & Suter 1974).
[2] Data from wave recorder site about 5 km off Outer Harbor (Steedman 1975).

m/s, and the K_1 and O_1 amplitudes 0.02 m/s and 0.01 m/s respectively. The difference between the phases of the currents and elevations indicates that the maximum flood precedes high water by more than half a tidal period in this region.

In order to obtain a better understanding of the tidal progression, mathematical models have been used to simulate the tidal elevations and currents in Gulf St Vincent and Investigator Strait, using only information at the entrances to Investigator Strait (Tronson 1973[1]). Co-tidal and co-range charts (Fig. 5) predicted by a similar model (cf. Appendix) for the M_2 and K_1 tidal constituents are in approximate agreement with the coastal tidal data. However, the simulated phase variations in the Gulf are greater than observed. The predicted tidal streams show maximum M_2 tidal currents of about 0.7 m/s in Backstairs Passage, with about 0.25 m/s in Investigator Strait and the western part of Gulf St Vincent, whereas the K_1 tidal currents are only about 0.05 m/s. The ratio between the currents of each tidal species is in accord with the data, although the magnitudes in the Adelaide region are about double those observed.

In summary, the following pattern emerges: the semi-diurnal tide enters as a progressive wave (high water and maximum flood occurring almost simultaneously) through both entrances to Investigator Strait, and is subsequently retarded by bottom friction on entering Gulf St Vincent. The retardation of the time of high water relative to the maximum flood, suggests that the tidal energy tends to enter along the central axis, and then propagates towards the coasts. In the Adelaide region the data are consistent with a southward energy flux; however, this is only found in the model further south. Most importantly, there is no appreciable outward propagation of semi-diurnal tidal energy through the entrances to Investigator Strait, so that the interior dissipation process is complete.

On the other hand, the diurnal tide is dissipated to a much lesser extent. Throughout, the currents and elevations are almost 90° out of phase, and the model indicates that the flow of tidal energy is inwards on the south of Investigator Strait, and Backstairs Passage, and outwards in the north of Investigator Strait. The high water is delayed by over half a period following the flood along the Fleurieu Peninsula indicating a southward energy flux here, as is found in the data from Port Stanvac and Outer Harbor.

The rate of dissipation of tidal energy at spring tides (estimated from the models) is about 2×10^6 kW, which corresponds to a rate of inflow of tidal energy into Investigator Strait/unit width of approximately 30 kW/m.

The detailed history of tidal studies in South Australia before 1970 is given in Easton (1970).

Circulation

The general (non-tidal) circulation of Gulf St Vincent and Investigator Strait is caused by three factors. First, the local wind; second the local exchange of heat and water across the sea surface; third, circulation in the deep ocean adjacent to the South Australian Sea.

It is probable that the first effect is the most important, although as indicated in the previous section, there are very few observations of currents; however, the temperature and salinity distributions in Gulf St Vincent have been observed during four consecutive years (1973–1976) (Bullock 1974, Walter 1975,

---

[1] Tronson, K. (1973).—The hydraulics of the South Australian Gulf System. Ph.D. thesis, Flinders University of South Australia (unpubl.).

                JOHN A. T. BYE

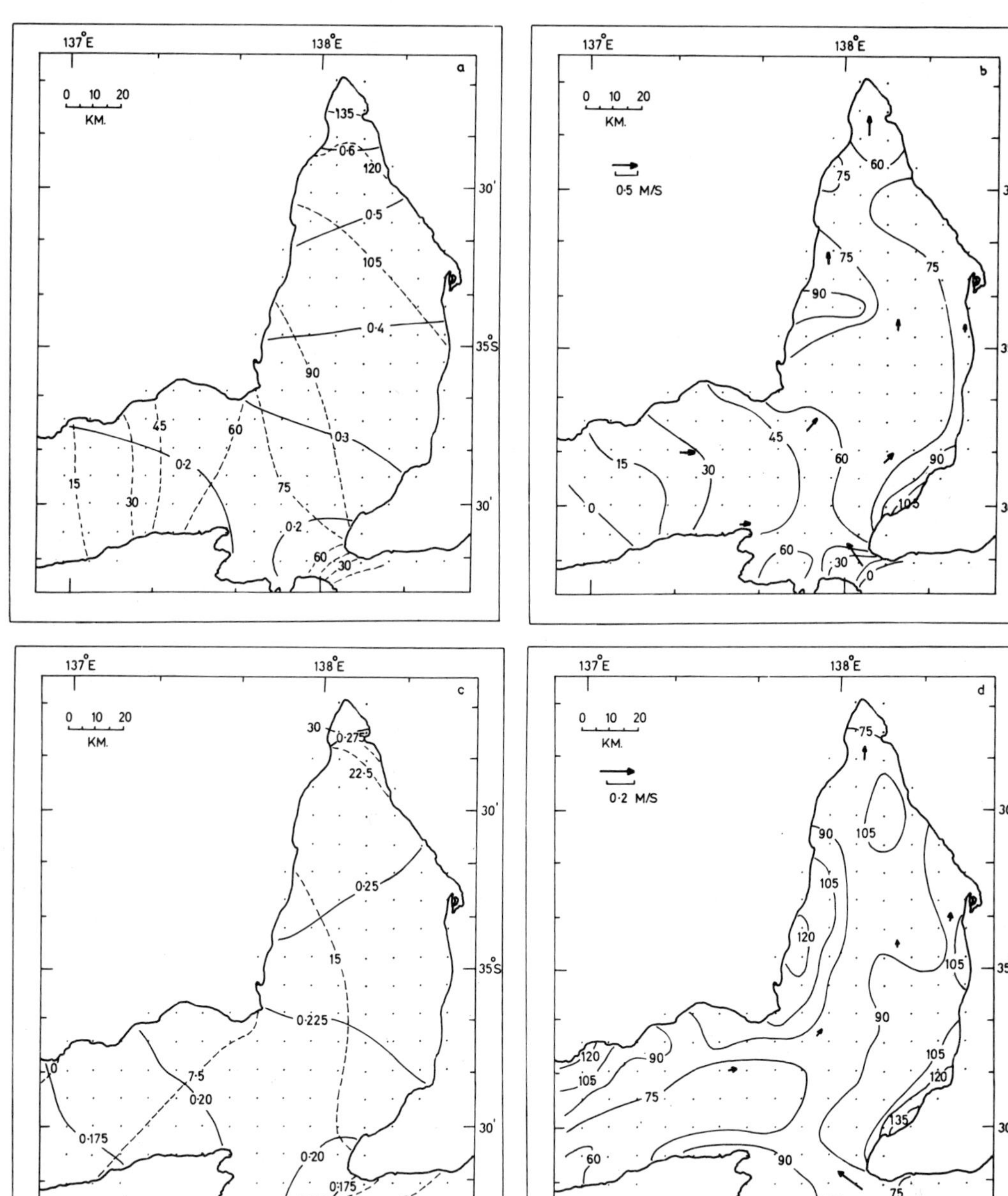

Fig. 4. Simulated tidal circulations.

(a) Co-range and co-tidal chart for the M_2 constituent.
—— amplitude in metres
- - - - phase in degrees

(c) Co-range and co-tidal chart for the K_1 constituent.
—— amplitude in metres
- - - - phase in degrees

(b) Chart of phase lag (in degrees) between maximum flood and high water for the M_2 constituent. Arrows indicate magnitude of maximum flood currents.

(d) Chart of phase lag (in degrees) between maximum flood and high water for the K_1 constituent. Arrows indicate magnitude and direction of maximum flood currents.

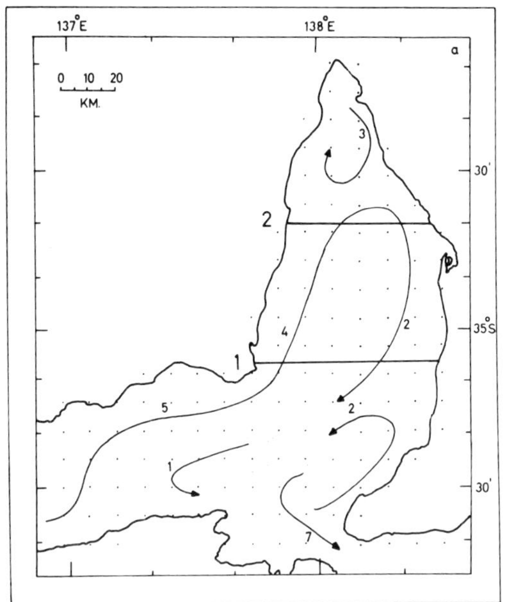
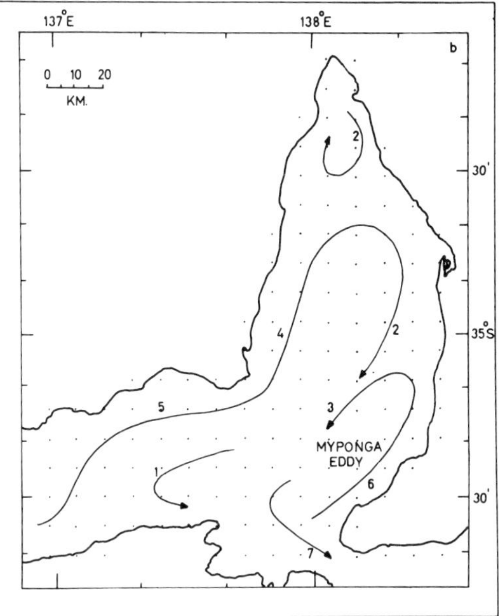

Fig. 5. Simulated mean annual circulation in Gulf St Vincent and Investigator Strait due to (a) wind, and (b), wind and density gradients.
Components of wind stress are .0482 N/m² towards east and .0107 N/m² towards north.
Figures indicate average horizontal velocities in water column in c/s.

TABLE 4

Simulated seasonal circulation parameters for the circulation in Gulf St Vincent and Investigator Strait

Volume transport in m³/s x 10⁻³	Jan.	April	July	Oct.	Annual
Through Investigator Strait (towards east)	—6	8	53	26	17
Through section 1 in Fig. 5a (towards north) by W coast current	11	13	18	18	15
by E coast arm of Myponga Eddy	12	6	0	6	5
Exchange through section 2 in Fig. 5a	2	4	8	6	5

King 1975[2]), and records of tide heights are available from which the meteorological effects on sea level can be studied.

One of the earliest records (Witton 1838) states that "I observed the tides were more influenced by the direction of wind than the moon, and that the time and height of tide could not be accurately discerned".

The mean annual wind stress for the South Australian Sea (Eyre 1973), and a representative density field derived from Figs 11 and 12 have been used to simulate the circulation with the model. It is apparent (Fig. 5) that the circulation pattern depends essentially on the wind field. The density gradients only have a significant effect at the head of Gulf St Vincent.

Throughout the year, the model shows that the Gulf St Vincent circulation (Fig. 6) consists of an inflow along the west coast, and an outflow in the central regions, with an anticyclonic circulation adjacent to the Fleurieu Peninsula, which will be called the Myponga Eddy. The evolution of the Myponga Eddy is remarkable, being very sensitive to the seasonal wind, and most extensive when the westerlies are weakest. It is a very important variable for the metropolitan coastline as will be shown below.

---

[2] King, M. O. (1975).—Water movements in St Vincent Gulf. Hons thesis, Flinders University of South Australia (unpubl.).

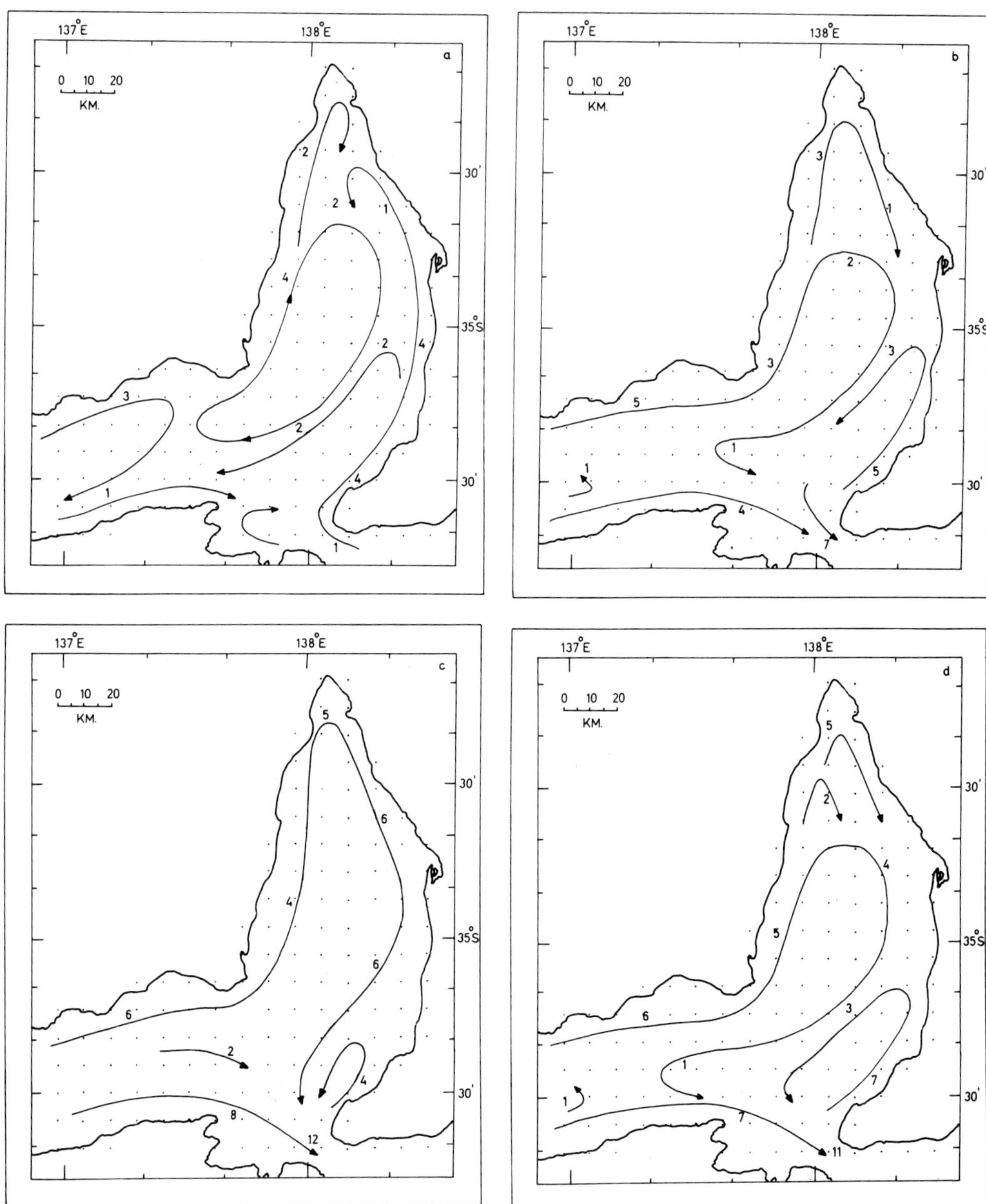

Fig. 6. Simulated seasonal mean circulation due to wind, and density gradients.
(a) January—Wind stress components (.0053, .0167); (b) April (.0350, .0129); (c) July (.0950, −.0255); (d) October (.0694, .0189).
Components of wind stress in N/m² towards east and north respectively are shown in brackets.
Figures indicate average horizontal velocities in water column in c/s.

The inflow of the Gulf is always continuous with an eastward flow on the north coast of Investigator Strait, which is the main link of the Gulf St Vincent waters with the deep ocean. The outflow either turns eastward through Backstairs Passage, or returns as a counter-current in Investigator Strait, especially in summer, so that in January there is a predicted net westward transport through Investigator Strait (Table 4). The annual mean

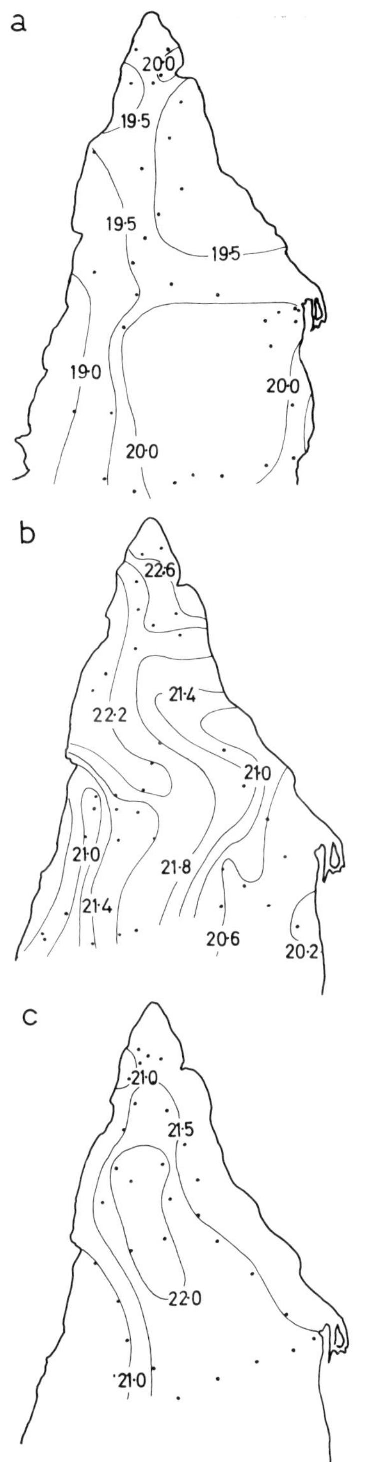

transport, however, is eastward and corresponds approximately to an average current out through Backstairs Passage of 0.1 m/s.

The basic feature of the hydrology is that the salinity and temperature, and hence the density, in the water column are nearly independent of depth (Bullock 1974). This is due to mixing in the vertical by wind and by tidal currents.

The average salinity (Figs 9–10) of the water column reveals features consistent with the water circulation deduced above, and it is clear that the variation of salinity in the horizontal is mainly responsible for the distribution of water density (Figs 11–12). In particular, the inflow on the western coast is evident in a low salinity tongue, and the closed circulation in the north is consistent with the rapid increase in salinity at the head of the Gulf. The distributions result from a balance between the water exchange and the net evaporation, and contain salinities as high as 47‰ at Port Wakefield in summer (Thomas & Edmonds 1956).

The water transports across the two sections of Gulf St Vincent simulated by the model enable the net evaporation to the north of each section to be estimated using the formula

$$E = \frac{1}{AS} \int_{x_1}^{x_2} SV \, dx$$

where S is the average salinity in ‰ in the water column in the section, V is the northward transport, A is the sea surface area, and S is the mean salinity to the north of the section, and the integral is taken across the section. Fig. 13 indicates the estimates so obtained, compared with the net evaporation rates reported for Adelaide (Aust. Bur. Met. 1973, 1974, 1975). The agreement is fair and suggests that the simulated transports are correct within a factor of two. Discrepancies may be due to errors in using the Adelaide meteorological data as representative of the Gulf.

The diffusion of salt, using an estimate of the horizontal coefficient of eddy diffusion of salt of 100 m²/s from the open waters of Spencer Gulf (Holloway 1974) is probably an order of magnitude less than the advection and has been

Fig. 7. Distribution of temperature (°C) in Gulf St Vincent. (a) April 1973; (b) March 1974; (c) March 1975. Values are the average for water column.

[3] Marshallsay, A. (1975).—The temperatures and salinity structure of the Adelaide metropolitan coastline. Hons thesis, School of Earth Sciences, Flinders University of South Australia (unpubl.).

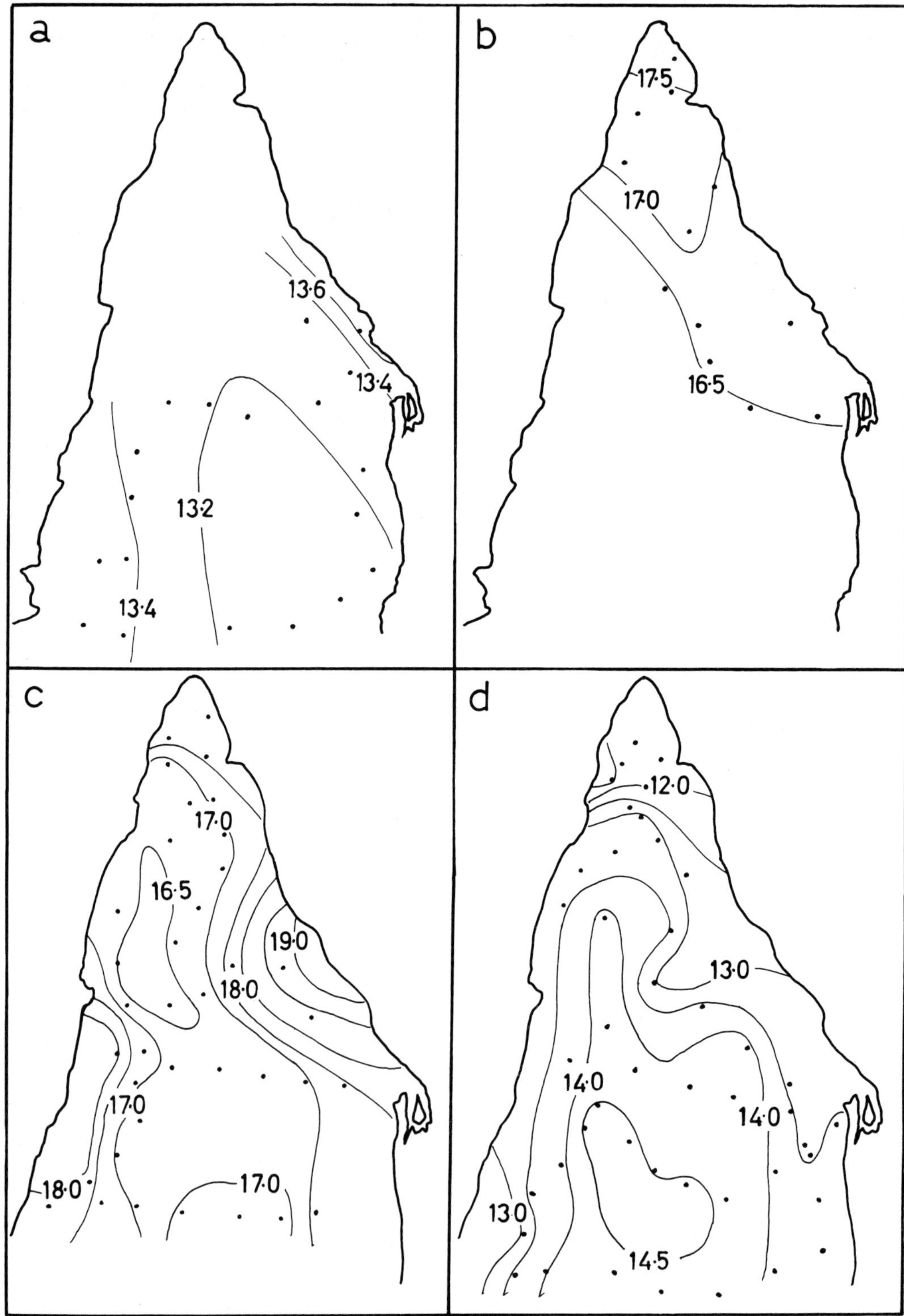

Fig. 8. Distribution of temperature (°C) in Gulf St Vincent. (a) August 1973; (b) October 1973; (c) August 1974; (d) June 1975. Values are average for water column.

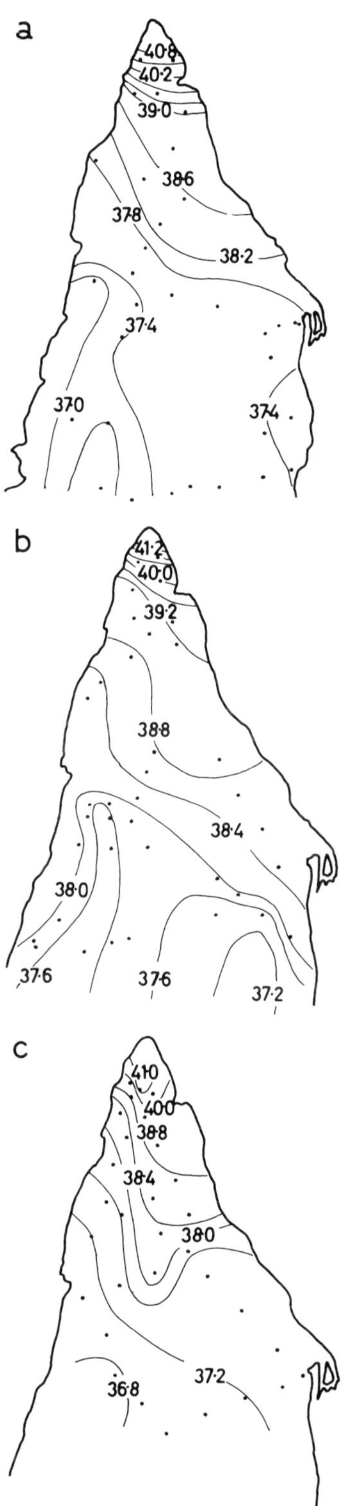

Fig. 9. Distribution of salinity (‰) in Gulf St Vincent. (a) April 1973; (b) March 1974; (c) March 1975. Values are average for water column.

neglected in the oceanographic calculations; and the results were found not to be very sensitive to changes in the wind field, and hence it is considered that not much error has been introduced using monthly wind stress averages.

Along the metropolitan coastline the variation in water properties in 1975 was examined from six sets of temperature and salinity data obtained at a depth of one metre for twenty stations between Hallett Cove beach and Largs North (Marshallsay 1975[3]). It was found that the water temperature increases northwards by about 2°C in summer, and varies over a range of about 1°C in winter. The salinity distribution, however, usually showed a minimum in the vicinity of Brighton or Grange, and the coastal values were in general lower than the offshore values. The minimum and the offshore increase in salinity is interpreted as due to run-off in the Adelaide region. The systematic rise in salinity to the north in summer months probably results from the deflection of the outflow in the Gulf away from the coast by the Myponga Eddy.

Table 5 summarises the observations, and the more extensive 1949–1951 data series obtained by Thomas & Edmonds (1956). A strong annual cycle of salinity, with a summer maximum and a winter minimum is only apparent at Port Wakefield.

A very interesting data series of temperature was obtained from a lighthouse at Wonga Shoal (Fig. 1), between 1869 and 1912 (Hunt 1918). The monthly mean temperature (Fig. 14) has an annual range of about 10°C, and the highest and lowest daily observations were 26°C and 9°C respectively. The mean monthly sea temperature versus the monthly mean air temperature for Adelaide (Fig. 15) indicates that on average in summer the mean sea temperature is about 3°C less than the mean air temperature, whereas in winter it is about 1.5°C higher. These conclusion, however, may be in part due to the time of the sea observations. Higher temperatures in the nearshore zone than at Wonga Shoal in summer (Table 5) are certainly caused by the strong diurnal heating in the shallow water. A statistical analysis of the two time series of data shows a dominant annual cycle, with a maximum in February and a minimum in July, and the

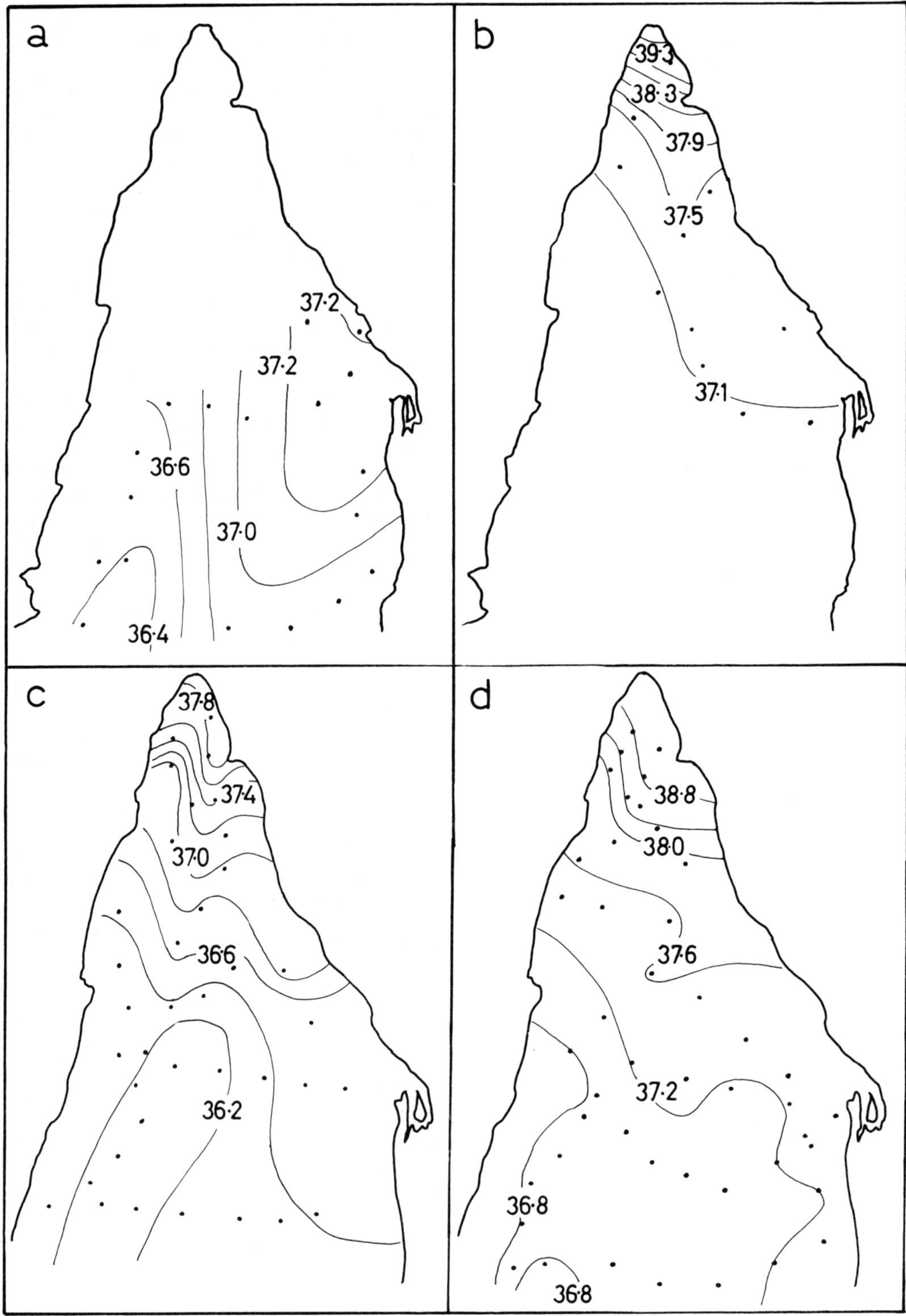

Fig. 10. Distribution of salinity (‰) in Gulf St Vincent. (a) August 1973; (b) October 1973; (c) August 1974; (d) June 1975. Values are average for water column.

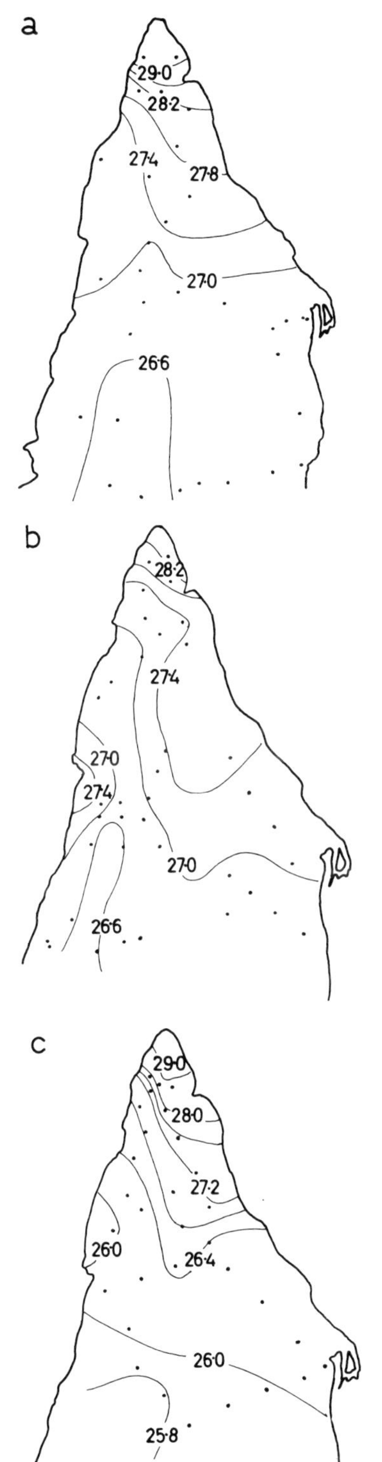

Fig. 11. Distribution of σ_T in Gulf St Vincent ($\sigma_T = \rho - 10^3$ in which ρ is water density in kg/m³). (a) April 1973; (b) March 1974; (c) March 1975. Values are average for water column.

cross-correlation indicates that the sea temperature lags the air temperature by only 5 days.

The approaches to Outer Harbor have been used extensively for oceanographic purposes in recent years, notably in studies of turbulence in a tidal stream (Steedman 1975) and as a site for a wave recorder by the Department of Marine and Harbors, South Australia (Table 6).

As an example of the effect of the passing of a storm, a study has been made of a period of strong winds from the west which occurred between 13 and 17 May 1975. The tidal levels at Outer Harbor (Fig. 16) indicate that a surge ranging to 0.5 m occurred during the period. Predictions based on the geostrophic wind over the South Australian Sea (Vecchio 1975[4]), and on the surface wind at Mt Lofty (Lyons 1973) both show features similar to the recorded surge (Fig. 16). In general, the geostrophic wind simulation lacks structure, and the simulation based on the point surface wind contains too much structure. At Victor Harbor, the surge was similar; however, the oscillations with period of about 10 hours apparent in the Outer Harbor record were absent. These oscillations are probably the resonant response of the Gulf to the wind (cf. Tides, p. 145).

The major cause of the discrepancies between the observed surge and the simulations at Outer Harbor is possibly the truncation of the simulation at the entrances to Investigator Strait, or more probably the lack of data on the wind field over the region, which has a significant mesoscale structure (Schwerdtfeger 1976). The remarkable similarity and impressive magnitude of the surge patterns at long periods (100-300 hrs) at ports in the South Australian Sea is presently being extensively investigated (Krause & Radok in press).

Finally, the possibility of an additional circulation due to the presence of a pressure head across Investigator Strait should be mentioned. Deep sea observations in the period 1967–1971 (Bye 1972) indicate that there was an average level of decrease of about 0.10 m at the edge of the South Australian Sea between the east and west approaches to Investigator Strait. Any

[4] Vecchio, G. (1975).—Some meteorological aspects of ocean waves and storm surges along the South Australian coast. Hons thesis, School of Earth Sciences, Flinders University of South Australia (unpubl.).

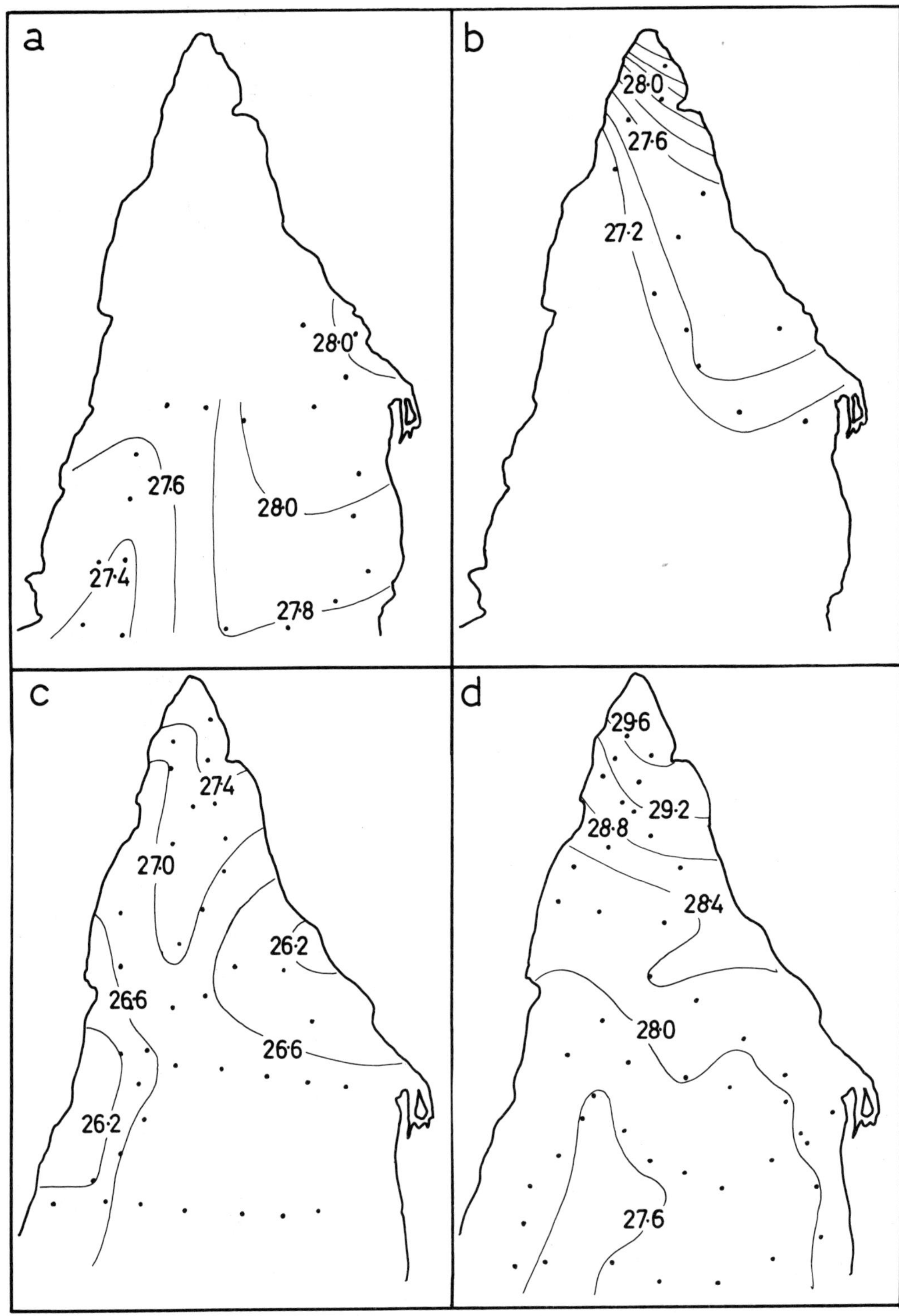

TABLE 5

Maxima and minima of temperature and salinity at coastal stations

	Temperature (°C)		Salinity (‰)	
	Max.	Min.	Max.	Min.
1975				
Marino	23.5 Feb.	13.1 Sept.	37.2 Mar.	36.6 Sept.
Brighton	24.5 Feb.	13.5 Sept.	36.8 Mar.	36.5 Sept.
West Beach	24.0 Feb.	14.2 Sept.	37.0 May	36.5 Mar.
Semaphore	25.9 Jan.	13.2 Sept.	37.6 Mar.	36.7 Apr.
1949–1951				
Brighton			37.5 Dec., Jan., Mar.	36.6 Jun.
Port Wakefield			47.0 Dec.	36.6 Jul.
Kingscote			37.3 Jan., Dec.	35.5 Aug.

1975 data from six sets of observations in Jan.–May and Sept. (Marshallsay 1975); 1949–1951 data from observations at approximately fortnightly intervals between Dec. 1949 and April 1951 (Thomas & Edmonds 1956).

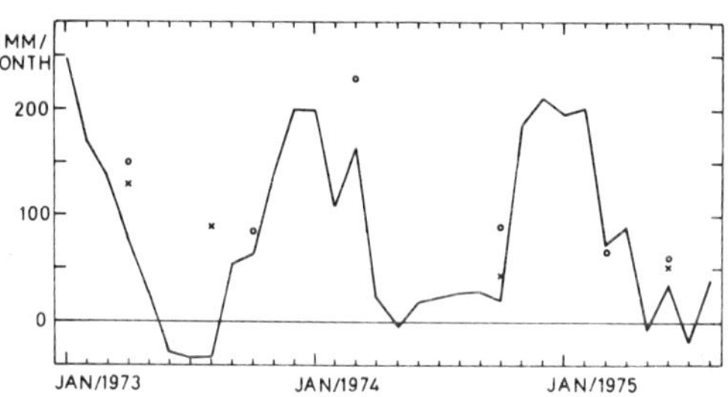

Fig. 13. Net monthly evaporation rate (evaporation — rainfall for Adelaide) for years 1973–1975. X and O indicate estimates derived from oceanographic data for Gulf St Vincent north of Sections 1 and 2 in Fig. 5.

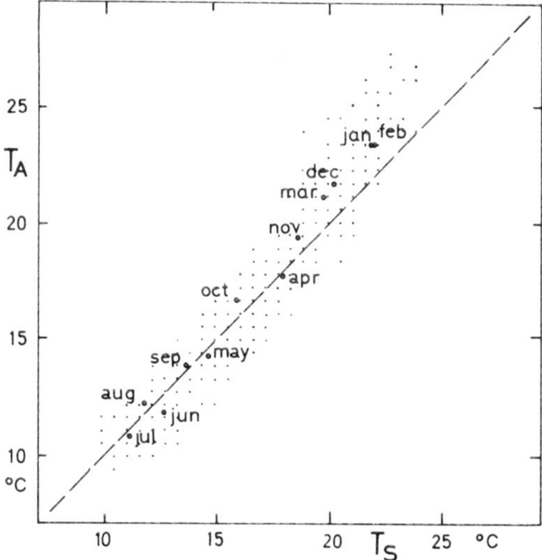

Fig. 15. Regression of mean air temperature at Adelaide (T_A) and mean monthly sea temperature (T_S) at Wonga Shoal. Circles indicate average values for each month for period of observation.

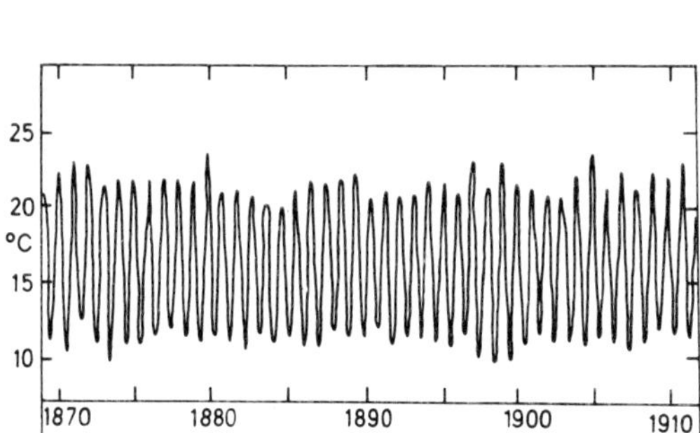

Fig. 14. Mean monthly water temperature at Wonga Shoal for period 1869–1912. Means based on observations taken daily at depth of approximately 0.5 m at 0900.

TABLE 6

Wave records in Gulf St Vincent and Investigator Strait

Port	Period	Recorder
Port Giles	Nov. 1968–Feb. 1970	Wavestaff
Penneshaw	Feb. 1971–Nov. 1973	Wavestaff
Cape Jervis	1971–1973	Wavestaff
	June–Nov. 1974	Wave buoy
Outer Harbor	May 1972–	Wavestaff

Fig. 12. Distribution of δ_T in Gulf St Vincent ($\sigma_T = \rho - 10^3$ in which ρ is water density in kg/m³). (a) August 1973; (b) October 1973; (c) August 1974; (d) June 1975. Values are average for water column.

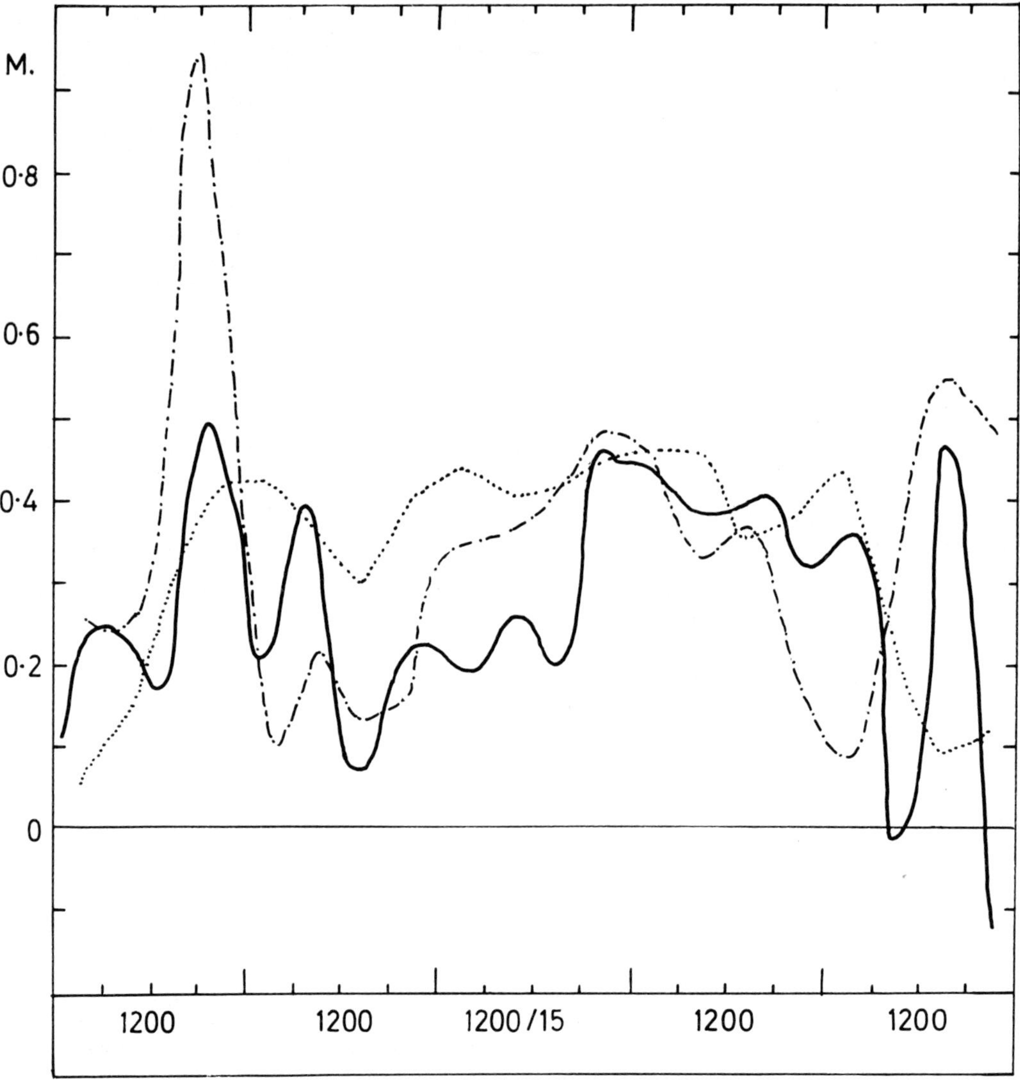

Fig. 16. Observed and simulated storm surges at Outer Harbor for period 13th May–17th May 1975. ————— observed surge; - . - . - . simulated surge for the observed wind at Mt Lofty; . . . . . . simulated surge for the geostrophic wind over the South Australian Sea.

coastal circulation associated with this phenomenon is at present a matter of speculation.

The Spencer Gulf circulation appears to be simpler than in Gulf St Vincent, consisting of one major cyclonic gyre in the southern portion in which the density induced forces are relatively much more important compared with the wind forces than in Gulf St Vincent (Bullock 1975; Bye & Whitehead 1975).

Waves

Several continuous series of wave heights have been obtained in Gulf St Vincent and Investigator Strait (Table 5). Most of the data have been obtained by automatic wave staffs operated by the Department of Marine and Harbors, South Australia, with the analysis made at the University of Adelaide (Noye, Berris & Culver 1973). Other data have been obtained using waverider buoys (Bye, Gunn & Nikpalj 1975).

The two parameters, besides wind strength, which determine the local wavefield are the fetch, i.e. the distance between the point of observation upwind to the nearest coast, and the direction of the wind. These parameters

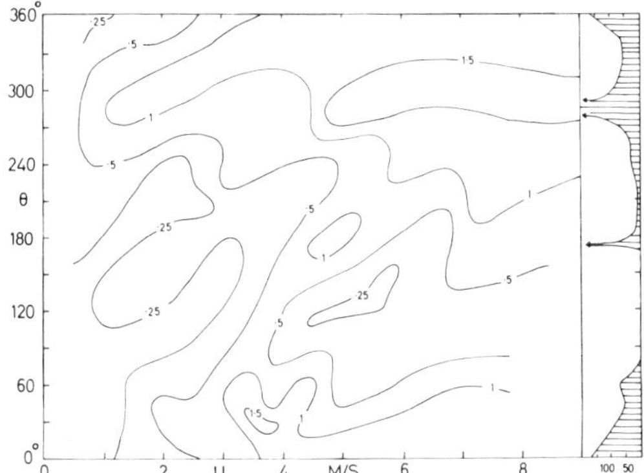

Fig. 17. Dependence of wave height on wind speed (U) and direction (θ) off Cape Jervis. Contours are significant wave height (H_s) in metres.

obviously vary greatly with position and the weather in Gulf St Vincent and Investigator Strait. Wave observations off Cape Jervis correlated with wind observations at North Brighton (Lyons 1973) for the period June–November 1974 (Fig. 17) illustrate this dependence. It is clear that the largest waves occur when the wind is blowing along Investigator Strait ($\sim300°$) or from south of Kangaroo Island ($\sim150°$). Under these conditions, the significant period (T_s) for a wind of speed (U) are

$$H_s = 0.24U \text{ m}; \ 0 \leqslant U \lesssim 10 \text{ m/s}$$
$$T_s \sim 5 \text{ s}.$$

The average transport of wave energy off Cape Jervis is about 1.3 kW/m, which is approximately 1/50th the energy (77 kW/m) in the region west of the Hebrides in the North Atlantic Ocean (Salter 1974), and 1/20th of the spring tidal inflow.

Conclusions

The physical oceanography of Gulf St Vincent and Investigator Strait holds much of interest for the marine scientist, and many questions yet remain to be posed and answered. The continuing need is for good data. Simulation of the fluid dynamics is also a very important technique for understanding the physical processes, and for making predictions on future conditions. The two approaches should be made together.

Present shortcomings seem to be a lack of detailed knowledge of the wind field over the region, and of the bottom friction, so that the energy dissipation processes have had to be oversimplified; this can be improved by incorporation of data on bottom sediment distributions. It is not inconceivable, in the near future, however, that accurate forecasts of sea conditions, including currents, waves, temperatures and salinities could be made as a routine service.

Some other interesting developments should include experiments made jointly with marine biologists to investigate the interaction between physical oceanographic events, e.g. the dodge tide and marine life.

Appendix

A model for flow simulation in a sea

The model predicts the integrated transport components (and hence the average current velocity in the water column), and the surface elevations in the sea produced by various driving mechanisms, including the surface wind stress, horizontal density gradients, deep sea tides (which are applied at the open sea boundaries), and atmospheric pressure gradients. The bottom friction is represented by a quadratic friction law, with a constant drag coefficient of magnitude 0.0025. The resolution of the model is 10 km, and the finite-difference nodes on which the simulations are based are indicated by points on the charts.

The mathematical technique is based on a finite-difference representation of the equations of motion and continuity (Bye 1974).

References

A.N.T.T. (1976).—Australian National Tide Tables 1976. Australian Hydrog. Pub. **11**. (Aust. Govt Publ. Serv.: Canberra.)

AUST. BUR. MET. (1973).—Australian Bureau of Meteorology. Monthly Climate Data—Surface 1973. (Aust. Govt Publ. Hse: Canberra.)

AUST. BUR. MET. (1974).—Australian Bureau of Meteorology. Monthly Climate Data 1975. (Aust. Govt Publ. Hse: Canberra.)

AUST. BUR. MET. (1975).—Australian Bureau of Meteorology. Monthly Climate Data 1975. (Aust. Govt Publ. Hse: Canberra.)

BULLOCK, D. A. (1974).—Cruise Rep. **1**. (Flinders Institute for Atmospheric and Marine Sciences: Flinders University of South Australia.)

BULLOCK, D. A. (1975).—The general water circulation of Spencer Gulf, South Australia, in the period February to May. *Trans. R. Soc. S. Aust.* **99**, 43-53.

BYE, J. A. T. (1972).—Oceanic circulation south of Australia. Antarctic Oceanology II. *Ant. Res. Ser.* **19**, *Am. Geop. Un.*, 95-100.

BYE, J. A. T. (1974).—The flow series of thal-
lasso-models. *Sel. Top. Atmos. Mar. Sci.* **6.**
(School of Earth Sciences, Flinders University
of South Australia.)

BYE, J. A. T. & SUTER, E. F. (1974).—Observations of currents in St Vincent Gulf. *Res.
Report* **11.** (Flinders Institute for Atmospheric
and Marine Sciences: Flinders University of
South Australia.)

BYE, J. A. T. & WHITEHEAD, J. A., JR. (1975).—
A theoretical model of the flow in the mouth
of Spencer Gulf, South Australia. *Estuarine &
Coastal Mar. Sci.* **3,** 477-481.

BYE, J. A. T., GUNN, B. W. & NIKPALJ, C. V.
(1975).—The wave climate off Cape Jervis,
South Australia, between June and November,
1974. Res. Report **17.** (Flinders Institute for
Atmospheric and Marine Sciences: Flinders
University of South Australia.)

CHAPMAN, R. W. (1892).—The tides of the coast
of South Australia. *Aust. Assoc. Adv. Sci.,
Report of Comm.* **2.**

EASTON, A. K. (1970).—The tides of the continent
of Australia. Horace Lamb Centre for
Oceanographical Research Res. Paper **37.**
(Flinders University of South Australia.)

EYRE, W. S. (1973).—The spherical harmonic analysis of global wind stress field and atmospheric angular momentum. *Res. Report* **2.**
Flinders Institute for Atmospheric and Marine
Sciences, Flinders University of South Australia.)

FLINDERS, M. (1814).—"A voyage to Terra Australia." (Nicol: London.)

HUNT, H. A. (1918).—Results of rainfall obser-
vations made in South Australia and the
Northern Territory. Bur. of Met. Comm. of
Aust.

HOLLOWAY, P. (1974).—Determination of eddy
diffusion coefficients for Northern Spencer
Gulf. Cruise Report **4.** (Flinders Institute for
Atmospheric and Marine Sciences, Flinders
University of South Australia.)

INGLETON, G. C. (1944).—Charting a continent.
(Angus & Robertson Ltd: Sydney.)

IRISH, J. D. & SNODGRASS, F. E. (1972).—Australian-Antarctic tides. *Antarct. Oceanol.* II,
Antarct. Res. Ser. **19,** *Am. Geop. Un.* 101-
116.

KRAUSE, G. & RADOK, J. R. M. (in press).—Long
waves on the Southern Ocean. Proc.
I.U.T.A.M. Symp.—Surface gravity waves on
water of varying depth. Aust. Acad. Sci. Canberra 20-24 Jul. 1976.

LYONS, T. J. (1973).—Data held in computer
compatible form—Adelaide Region. *Comput.
Res.* **5,** Met. Data. (Flinders Institute for
Atmospheric and Marine Sciences, Flinders
University of South Australia.)

NOYE, B. J., BERRIS, K. V. W. & CULVER, R.
(1973).—A system for recording and analy-
sing coastal waves. *Proc. 1st Aust. Conf. Sci.
Tech.* 17-21 Nov. 1973, Flinders University of
South Australia.

RADOK, J. R. M. (1973).—Water transport past
Adelaide. *Proc. 2nd S. Aust. Regnl Conf. Phy.
Oceanog.,* Aug. 1973, University of Adelaide.

SALTER, S. H. (1974).—Wave power. *Nature,* **249,**
720-724.

SCHWERDTFEGER, P. (1976).—Climate. *In* C. R.
Twidale, M. J. Tyler & B. P. Webb (Eds):
"Natural History of the Adelaide Region."
(Royal Society of South Australia: Adelaide.)

SHEPHERD, S. A. & SPRIGG, R. C. (1976).—Sub-
strate, sediments and subtidal ecology of Gulf
St Vincent and Investigator Strait. *In* C. R.
Twidale, M. J. Tyler & B. P. Webb (Eds):
"Natural History of the Adelaide Region."
(Royal Society of South Australia: Adelaide.)

STEEDMAN, R. K. (1975).—Transport phenomenon
of the lower boundary of a shallow sea. *Res.
Report* **19** (Flinders Institute for Marine and
Atmospheric Sciences, Flinders University of
South Australia.)

THOMAS, I. M. & EDMONDS, S. J. (1956).—
Chlorinities of coastal waters in South Australia. *Trans. R. Soc. S. Aust.* **79,** 152-166.

TRONSON, K. (1975).—The hydraulics of the
South Australian Gulf System II Resonances.
Aust. J. Mar. Freshwater Res. **26,** 363-74.

WALTER, N. R. (1975).—Temperature and salinity
measurements in St Vincent Gulf. Cruise
Report **5.** (Flinders Institute for Marine and
Atmospheric Sciences, Flinders University of
South Australia.)

WITTON, D. (1838).—*S. Aust. Gaz.* and *Col. Reg.*
Dec. 8, 2d.

CHAPTER 12

SUBSTRATE, SEDIMENTS AND SUBTIDAL ECOLOGY OF GULF ST VINCENT AND INVESTIGATOR STRAIT

by S. A. SHEPHERD* & R. C. SPRIGG†

Introduction

Gulf St Vincent is a shallow marine basin connected to the eastern Indian Ocean by Investigator Strait and Backstairs Passage. The Gulf and its approaches together cover about 13,000 km².

There have been few geological or biological investigations of the region. Sir Joseph Verco pioneered the study of the marine life by his dredging expeditions between 1890 and 1912, providing material for many later taxonomic studies (Verco 1935). The sediments of the region were examined by Cooper (1960)[1], using samples taken by the hydrographic survey vessel HMAS *Warrego* in 1957.

The survey on which this account is based was made by the authors between 1964 and 1969 and involved over 560 diving stations, principally from MV *Saori* occupied at intervals of about 2–3 km, along traverses spaced at intervals of about 8 km in the Gulf and Investigator Strait. These data were supplemented by the results of two later surveys, one of Investigator Strait by Mrs J. E. Watson and the other of a small area off Outer Harbor by Adelaide Cement Co. Ltd. Preliminary results of the principal survey were given by Sprigg (1965) and Cooney (1965), and biological information on some animal groups has been published by Shepherd (1968) for seastars, and Kott (1972a, 1972b, 1975) for ascidians.

Geology and substrate

The bathymetry of Gulf St Vincent and its approaches is shown in Bye (1976). The region is pre-eminently a "carbonate" sedimentary province. This relates primarily to an almost complete lack of significant quantities of terrestrial erosion products entering the basin in rivers. An additional factor is that tidal and wave movements cause littoral drifting of the restricted products of coastal erosion in a direction parallel to the margins of the basin.

Consolidated Cambrian and Precambrian rocks form most of the hard-rock cliffs that line the southerly and south-easterly margins of the Gulf and approaches and also the sector south of Ardrossan on Yorke Peninsula. Most of the remainder of the coast is of soft, sandy and clayey sediments of Permian, Tertiary or Quaternary age. Hard-rock cliffs and outwash provide the principal coastal terrigenous erosional debris.

Tertiary uplift, and former interglacial high sealevels, have left a succession of stranded cliffs, raised shore platforms, and prominent erosional and/or solution nick points, about most of the coastline from the western extreme of Kangaroo Island, via Cape Jervis, to Marino, near Adelaide.

Boulder and cobble littoral deposits occur at or above modern beach level at the bases of many of the cliffs. Submerged, relict boulder beach deposits in Yankalilla Bay are widespread up to 4 to 5 km or more offshore, and northwards to a point opposite Carrickalinga Head. Other submerged shorelines occur off Hallett Cove South, extending 1 km out to sea; for several kilometres off Marino and, several tens of metres offshore, opposite Glenelg. The most southern of the Glenelg relict generally sand-covered, submerged beaches overlie ancient mangrove swamps with heavy roots still in situ. However, the source of the typically Marinoan (see Chap. 1) boulder constituents in this generally low to medium energy situation (mangrove swamp to fine sandy beach) is

---

* Department of Agriculture and Fisheries, 183 Gawler Place, Adelaide, S. Aust. 5000.
† c/- Arkaroola Pty Ltd, Adelaide, S. Aust. 5000.
[1] Cooper, R. M. (1960).—Recent sedimentation in Gulf St Vincent, South Australia. Hons B.Sc. thesis, University of Adelaide (Unpubl.).

difficult to explain. River action can not be invoked here, and there has been no known dumping of such boulders by modern seafarers.

Immediately adjacent to principal block-faults (Ochre Cove and Willunga), Tertiary sediments are exposed underwater for some distances. Off Ochre Cove, Tertiary limestones extend over the succeeding fault block in a pitching monocline. Off Noarlunga, other submarine outcrops of these limestones extend seaward approximately 5 km and form gently warped structures exposed on the sea bottom. Similar limestones off Christies Beach emerge at low tide as "Horseshoe Reef".

At Port Noarlunga, the spectacular linear offshore reef is probably a consolidated beach deposit that has been abandoned by retreating Tertiary marl cliffs. Another such gently arcuate, but completely drowned, beach occurs only a few hundred metres further to seaward. Both are probably of Quaternary age.

The only known submarine basement rock, other than granite pedestals at the western entrance to Investigator Strait, is a submerged reef lying about 5 km offshore from Little Gorge, southwest of Normanville. Steeply dipping basal Proterozoic beds here strike towards Cape Jervis.

Permian shales outcrop in the 71 m deep channel of Backstairs Passage. It is the erosion of these Permian sediments that provides the coarse lag sands present at depth in this area. They tongue out to the north.

Consolidated and calcrete cemented shell beds floor much of Investigator Strait and extend onto the Orontes shelf. Principal molluscs are *Ostrea*, *Bittium*, *Cardium* and *Katelysia*. In Investigator Strait, extensive areas of this broad, low profile platform carry a thin veneer of coarse shelly sand. In places, the calcreted shell beds are undermined, providing ideal shelter for lobsters. Near the western entrance to the Strait the oscillatory motion of the ocean swell during windy weather develops coarse sand ripples with wave lengths of 1–2 m, and amplitudes of 20 cm or more on the sea bottom.

Along the north coast of Kangaroo I., particularly north of Snug Cove, and over much of Eastern Cove in Nepean Bay, aeolianite (consolidated calcareous sand) dunes form prominent sea floor eminences and rough bottom topography. They provide an excellent substrate for a rich variety of algae, sponges, crinoids, reef-molluscs and other benthic organisms, and also a varied fish population, including excellent lobster and snapper grounds.

The 10–20 km wide Orontes Shelf that skirts the eastern coastline of Yorke Peninsula owes its existence partly to block faulting and/or monoclinal flexuring and to the consolidation and calcrete development on Quaternary shell-beds. This calcreted Pleistocene shell-bed occurs extensively along the outer ledge of the platform, usually at depths of 15 m or more. Its best development lies east of Troubridge I., where it forms extensive flat rocky surfaces, or low erosional escarpments, 5 m or more in height. It is an excellent substrate for sponges, ascidians, and reef molluscs, and is also an outstanding mixed fishing ground.

Between Troubridge I., and Sultana Point, at the "heel" of Yorke Peninsula, a string of tidally exposed reefs represent remnants of consolidated Quaternary sand dunes. Channels between them are swept by strong tidal currents which cause severe bottom erosion and transport of sand-sized debris far to the north.

Sand bars, or "spits" as they are known locally, are prominent features of the east coast of Yorke Peninsula, and also occur along the south coast. They usually extend from 2–4 km diagonally and northeast away from prominent cliffs, and continue to grow in the direction of flow of the strong tidal current. They are developed as extensions of stranded Quaternary beach deposits on which stand Stansbury, Port Vincent, Sheoak Flat and Black Point. Formed of fine to medium-grained, shelly, calcareous sands, they protect deeper waters lying on their leeward sides. The associated raised sea beaches at the coastal ends owe their preservation to resistant, underlying, consolidated Quaternary beach deposits.

Sea bottom sediments

Waters (1976)[2] analysed samples of sea bottom sediments and some of the results are given here. Figure 1 shows the distribution of mean grain size and the percentage of terrigenous (quartz and clay) material.

Silica sands are well developed in Gulf St Vincent only in littoral to shallow sub-littoral situations opposite south-central Yorke Peninsula, and opposite the Adelaide metropolitan

---

[2] Waters, C. (1976).—Sedimenation in Gulf St Vincent. Ph.D. thesis, Flinders University of South Australia (Unpubl.).

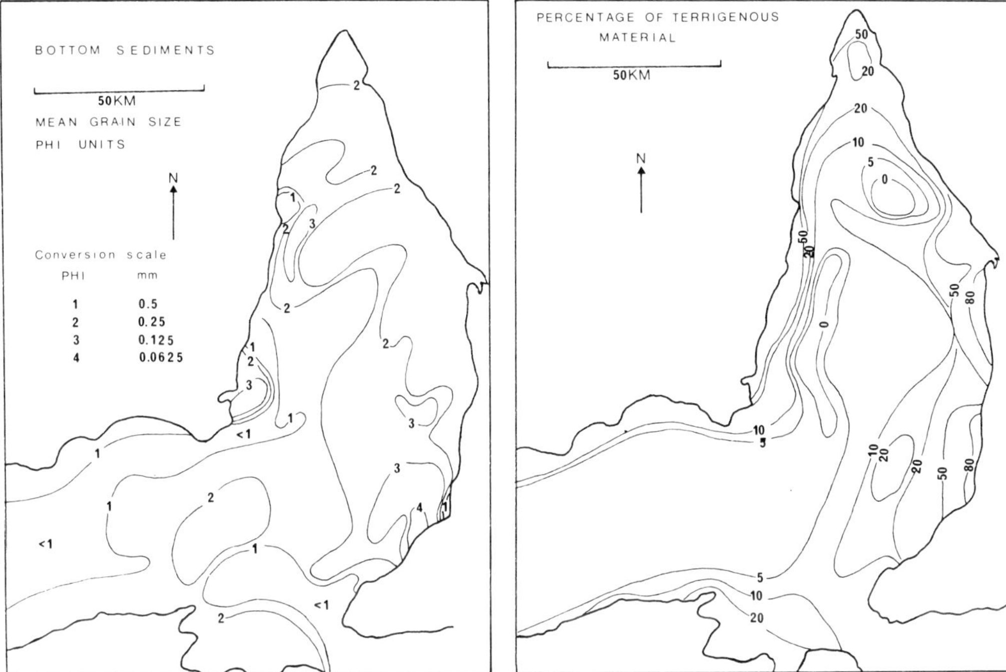

Fig. 1. Left—Distribution of mean grain size in Phi units. Right—Percentage of terrigenous material (quartz and clay) in bottom sediments. The remainder is mainly calcium carbonate. Contours are drawn by hand (after Waters 1976).

beaches. Submarine bars formed off headlands east of Penneshaw (Kangaroo I.) may also be of this nature. In the former cases, erosion of coastal cliffs developed in late Cainozoic sandy clays, and clayey sands are the prime sources of silica supply. At Christies Beach and Aldinga, generally coarser sands have been eroded from local cliffs formed in Tertiary clastic sediments.

The sea floor sands of Investigator Strait are principally coarse, shelly lag sands. The finer products of their erosional comminution are mostly swept around Troubridge I. into Gulf St Vincent, where they settle out into quieter waters as calcareous muds and silts. The shells contributing to this are principally of the *Cardium/Katelysia* association, but include *Dosinia, Mactra* and a range of gastropods, foraminifera, sponge spicules, and crinoid and echinoid plates.

The prime source of the calcium carbonate in both the mud and sand fractions of Gulf St Vincent and Investigator Strait is biologic. They are derived from destruction of molluscs in shallower, wave-agitated waters, but from bryozoa, calcareous algae and foraminifera in deeper reaches. The coarsest sands occur about the sub-littoral approaches into the Gulf via Sultana Channel, Troubridge Shoals and off Cape Jervis.

The texture of deeper water sediments is controlled increasingly by biological factors, involving particularly the nature and size of the original organisms. However, even in these deeper waters, strong tidal currents and ocean-propagated wave action is quite important about the southern reaches of the Gulf. Pelecypods, gastropods, echinoids, brachiopods, sponge spicules, solitary corals, crustacea, foraminifera and faecal pellets are all well represented. Filamentous algae, sponges and molluscs are the principal borers (Cooper 1960)[2] and largely explain the generally high degree of comminution of these deeper water sediments. However, observations made during scuba diving by the authors have demonstrated an even greater role of burrowing organisms (especially holothurians) in the physical destruction of coarser calcareous remains. In the deeper reaches, much of the sea floor is heavily pock-marked by mounds and craters like a desolate, miniature lunar landscape. At night, these areas become alive with crustaceans (notably crabs and prawns) as well as gobies and scallops. The *Pinna*, scallop and basket-bryozoan "meadows" are strongly reworked in this manner.

Aerial observations have confirmed an almost overriding importance of tidal action in the transport of fine carbonate material into the Gulf from the south. Successive clouds of turbid water, daily introduce many tonnes of material to settle out in quieter waters. This is most noticeable along the whole of the outer edge of the Orontes platform, north from about Troubridge I. One result is that quieter waters at the head of the Gulf are almost continuously kept turbid by the fine suspended limey matter.

Lesser turbid plumes extend away from the tidal rip opposite Cape Jervis, and also from a rip operating via a prominent channel east of the North Spit shoal beacon in the northerly reach of the Gulf. Newer sources of supply are from quarries at Rapid Bay, Christies Beach and Kleins Point, the harbour works at Port Giles, and the dredge dumping site off Outer Harbor.

Posidonia australis and *Amphibolis antarctica* are important in stabilising the sediments and promoting sedimentation. The seagrass blades baffle wave action and reduce water movement to such an extent that fine suspended particles settle out, and are trapped by the root mesh system of the seagrass. The sediments of these seagrass banks consist largely of carbonate skeletal debris from epiphytic coralline algae, bryozoans, molluscs and foraminifera, together with aggregates of aragonite and quartz grains. These banks have been studied in detail in upper Gulf St Vincent, where they have been shown to have caused the shoreline to migrate seaward several kilometres since sea level stabilised at its present level approximately 6,000 years ago (Bayly 1973;[3] Brooks 1973;[4] Hansen 1973;[5] Zimmerman 1973[6]).

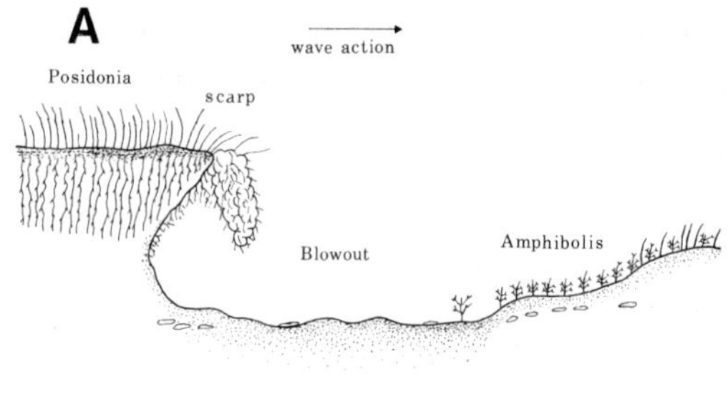

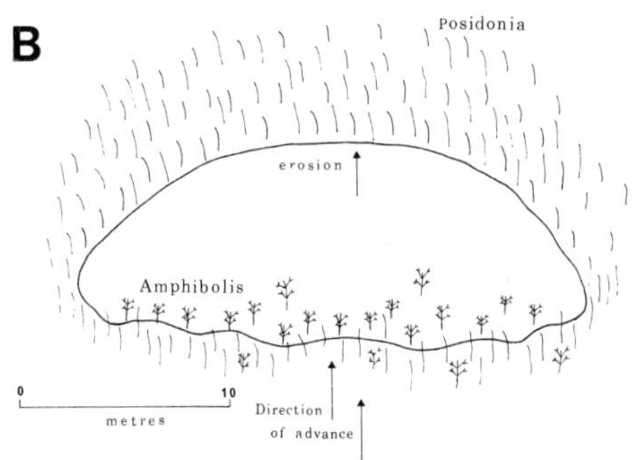

Fig. 2. Features of blowouts and cusps. A. Cross section of a blowout in a *Posidonia* meadow. B. Plan of a cusp and its re-colonisation by *Amphibolis*.

Shepherd (1970)[7] describes the presence of steep undercut, erosional scarps exposing seagrass rhizomes at the edge of crescent-shaped, grass-free cusps within the confines of *Posidonia* meadows (Figs 2, 4). These areas, termed erosion gutters or "blowouts", are believed to be formed by storm waves, which erode the *Posidonia* matte on the seaward side of the cusp, eventually uprooting the seagrass. However, on the leeward side of the cusp *Amphibolis antarctica* readily colonises the bare sand, stabilising it sufficiently for the subsequent invasion of *Posidonia*. The final result is the gradual seaward progression of the cusp. These blowouts have only been observed on

[3] Bayly, M. R. (1973).—Sedimentary environments of northern Gulf St Vincent near Port Arthur, South Australia. Hons thesis, School of Earth Sciences, Flinders University of South Australia (Unpubl.).

[4] Brooks, L. J. (1973).—Stratigraphic mineralogic study of a recent carbonate environment at the head of Gulf St Vincent. Hons thesis, School of Earth Sciences, Flinders University of South Australia (Unpubl.).

[5] Hansen, P. S. (1973).—A recent carbonate environment of the northernmost Gulf St Vincent. Hons thesis, School of Earth Sciences, Flinders University of South Australia (Unpubl.).

[6] Zimmerman, D. (1973).—Carbonate sediment studies in the northern region of Gulf St Vincent on sublittoral seagrass zones and in intertidal mud flats. Hons thesis, School of Earth Sciences, Flinders University of South Australia (Unpubl.).

[7] Shepherd, S. A. (1970).—Preliminary report upon degradation of seagrass beds at North Glenelg. S. Aust. Dept Fisheries Report, 15 pp.

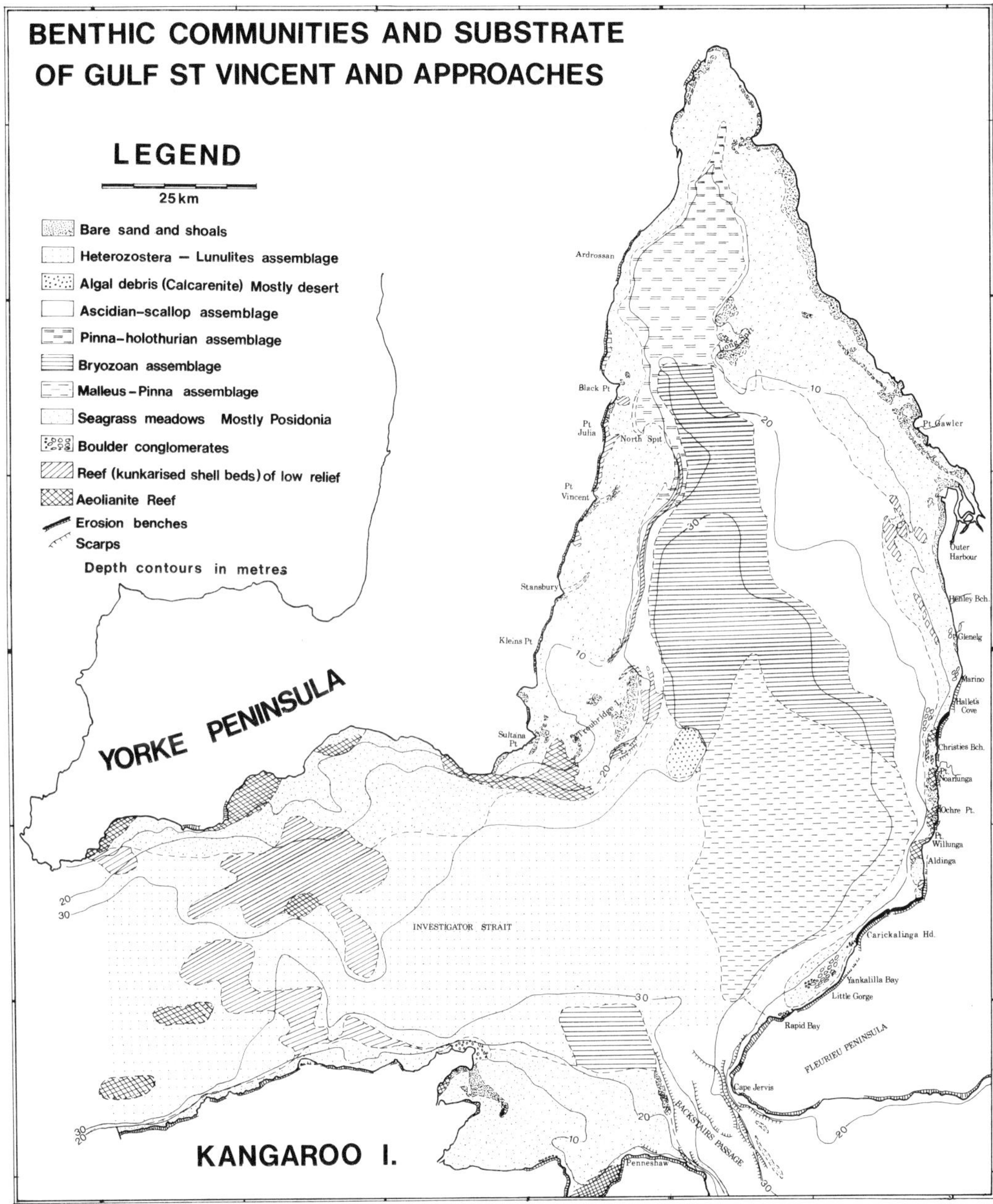

Fig. 3. Substrate and seagrass and faunal zones of Gulf St Vincent and Investigator Strait. Enlargements of this map may be obtained from the Fisheries Branch, Department of Agriculture & Fisheries, Adelaide.

the east coast of the Gulf south of Semaphore; elsewhere wave energy is presumably too low to uproot the matte.

Calcium carbonate in the form of calcite and aragonite is the dominant sedimentary constituent within the Gulf (Fig. 1) with unstable magnesium-rich calcite providing a much smaller, but significant, component particularly where calcareous remains dominate. Details of sedimentary analyses may be found in Cooper 1960)[1], Cooney (1965), Olliver (1964, 1965), Waters (1976)[2] and Zimmerman (1973)[6].

In general, away from the margins of the Gulf, sediments become finer grained and increasingly muddy. One notable exception is the occurrence of coarse and calcareous sand, and contained nodules of calcareous algae (*Lithothamnion*) over the subcircular MacIntosh Bank, approximately 20 km east of Troubridge I. The shoal rises to about 15 m depth from a general depth of 26 m, and appears as a virtual "desert", except for occasional molluscs and seapens. Currents active across the shoals may exceed 1.6 m/sec. (3 knots) so that the finer products of erosion are rapidly winnowed away.

The overall pattern of sediment distribution in terms of grain size diameters is that the finest products, i.e. those less than 0.1 mm, have settled principally into the deeper waters of the central northern Gulf basin, and at the eastern end of Investigator Strait. The coarser sediments (1–2 mm diameter) are concentrated about the southern headlands of the Gulf, across the western sweep of Investigator Strait, and in the sub-littoral zones of the eastern coastal beaches. Tidal currents, wave action and proximity to headlands or erosional rip channels are the controlling features in coarser sand and silt supply. The finer products of erosion typically settle out from turbid tidal currents, or suspension clouds, into deeper quieter waters or are trapped beneath seagrass baffles, around mangrove roots, or on supratidal flats. Coarse bryozoal and other skeletal material in all of these zones is subject to advanced disintegration and comminution by burrowing, boring and crushing activities of a host of predators. In general, the degree of sediment sorting improves significantly in areas of more pronounced wave turbulence or tidal activity, but declines noticeably where mixed sediments are being dumped relatively rapidly into quieter waters. "Salt and pepper" sands of mixed origin are particularly well developed down stream (northwest) from the Investigator tidal rip.

Biotic communities

In this section a descriptive account of the bottom communities is given, based largely on the dominant species, and ignoring many of the smaller and less conspicuous ones. Since the taxonomy of sponges is not well known they have, where present, been lumped together for descriptive purposes. Two approaches to the description of bottom communities are possible. One of these recognises and names communities by those species which are dominant and are constantly present in them. The other emphasizes the individual character of species' distributions and attempts to describe them. The former approach is difficult to apply where there are many species which vie for dominance and the second, although more precise, may become unwieldy and complicate rather than simplify description. The method used here combines both. The distribution of common species of algae is shown in relation to the main environmental factors, depth and water movement, while seagrass and faunal zones are recognised throughout discrete areas and characterised by their dominant species (Fig. 3). The resulting picture is a general one, and since intergradation of zones rather than discontinuity between them is the rule, as a concession to simplicity we have suppressed the detail and sharpened the boundaries beteween zones.

A. SEAGRASS COMMUNITIES

Eight species of seagrass occur subtidally in the region. Four of them, *Amphibolis antarctica*, *Posidonia australis*, *Heterozostera tasmanica* and *Halophila ovalis* occur over extensive areas of the sandy seafloor; the first two species often as dense monospecific meadows, and the latter two as scattered plants or clumps of plants, especially in deeper water. Communities designated by these species are described below. In addition there are two species which are less common. Of these, *Posidonia ostenfeldii* occurs in scattered stands on the north coast of Kangaroo I. and the eastern shore of the Gulf, and *Amphibolis griffithii* usually in stands at depths of 10–15 m where there are moderate currents, and also as isolated plants among stands of *A. antarctica*.

(1) *Posidonia australis* community (Fig. 4)

This is the most extensive community of seagrasses in Gulf waters, covering many square kilometres of sandy bottom. The two forms of this species, broad and narrow leaf morphs, form ecologically distinct communities with the broad leaf form usually restricted to shallow and calmer waters and the narrow leaf form to deeper water. Populations of the species are best developed in the shallower part of its depth range, becoming sparser towards its lower depth limit. In Investigator Strait the lower limit of the seagrass is about 30 m, decreasing to 20 m in the lower Gulf and to about 10 m near the head of the Gulf. The meadows on the western side of the Gulf are usually less dense than those on the eastern side.

Fig. 4. *Posidonia australis* (narrow leaf form) community showing an erosional scarp and blowout in Holdfast Bay, Gulf St Vincent. The seagrass *Amphibolis antarctica* is growing at lower left.

(2) *Amphibolis antarctica* community

This species is better adapted than *Posidonia* to stronger water movements and a substrate of mobile sand, and generally occurs in small stands or clumps rather than as extensive meadows characteristic of *Posidonia*. *A. antarctica* rarely occurs in depths exceeding about 12 m.

Stands of *A. antarctica* are mainly found in the following situations:

(a) on sand spits or bars or as a fringe bordering the shallow or shoreward side of a *Posidonia* meadow.

(b) as an initial community colonising blowouts in *Posidonia* meadows.

(c) on sandy bottom where underlying rock prevents the establishment of a *Posidonia* meadow, or

(d) in water movement conditions intermediate between the preferences of *Posidonia* and *A. antarctica*. The two species may occur as alternating stands forming a mosaic pattern, or less often as intermixed stands.

Generally *A. antarctica* may be considered a primary coloniser of sandy bottom in a successional process, which, if uninterrupted by physical factors such as strong water movement or underlying rock, would lead to a climax community of *Posidonia*.

In addition, long term changes in the proportional abundances of the two species have been observed in Gulf St Vincent over the last decade[8] and these are currently under investigation.

(3) *Heterozostera tasmanica*

This species is widely but sparsely distributed from about low water mark to about 35 m depth. Stands of the species are occasionally seen on submerged sand banks in places of moderate current. The *Heterozostera/Lunulites* assemblage of which this seagrass is a dominant is described in Part C.

---

[8] South Australian Dept Environment & Conservation (1975).—Report of the Port Adelaide sewage sludge outfall working group. No. 5, 42 pp., 7 figs.

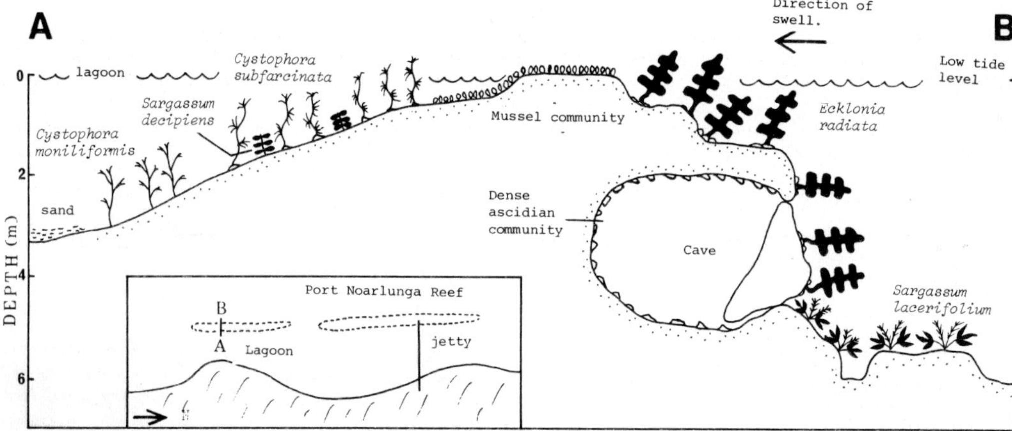

Fig. 5. A vegetation profile of Port Noarlunga Reef.

(4) *Halophila ovalis*

This species is widespread, but nearly always sparse, in the deeper waters (5–35 m depth) of the Gulf in places of slight to moderate water current (Fig. 6). It is a low creeping grass with oval shaped leaves and is often overlooked in communities dominated by more bulky sponges and bryozoa.

Animals in seagrass communities

In general an inverse relation is observed between the density of animals and the density of the seagrass community in which they occur. Thus as seagrass densities decline toward the lower depth limit of the seagrass or in other less favourable conditions for seagrasses there is a gradual transition from a seagrass-dominated to an animal-dominated community. The following species are most common in *Posidonia* meadows: the tube worm, *Chaetopteris variopedatus*, the stalked ascidians, *Pyura spinifera* and *Pyura australis*, razor shells[9] (on which there are commonly epizoic the ascidian *Halocynthia hispida* and the abalone *Haliotis cyclobates*), the seastars *Patiriella brevispina* and, in deeper water, *Anthaster valvulatus*, the grazing sea urchins *Amblypneustes* spp., the blue swimmer crab *Portunus pelagicus*, the sand crab *Ovalipes bipustulatus* and, in calm water conditions, the spider crab *Leptomithrax australiensis*. In places of moderate water currents the solitary ascidian *Polycarpa clavata* and the compound ascidian *Botrylloides leachi* are often epiphytic on the *Posidonia* blades.

Posidonia meadows on the western side of the Gulf are generally less dense than those on the eastern side and here the razor shell *Pinna dolobrata*, the hammer oyster *Malleus meridianus* and the pelecypods *Glycymeris radians* and *Katelysia scalarina* are more common.

B. ALGAL COMMUNITIES

Since the distribution of algae is dependent primarily on suitable substrate for attachment, light intensity reaching the habitat, and on water movement, it is sufficient to describe their relationships with these factors. Algal communities are prominent on rocky bottoms mainly on the subtidal cliffs and rocky shores bordering lower Gulf St Vincent and Investigator Strait from low water mark to a depth of 5–15 m (according to locality) until the rock becomes buried by sand. Algal communities may also be developed offshore on level bottoms, in places where rock protudes above the sand, or on sandy bottoms where dead shell fragments, stones or razor shells provide a suitable substrate for the attachment of algae.

In general character algal vegetation occurs in three strata: an upper stratum or storey of perennial larger brown algae principally of species of the genera *Cystophora* and *Sargassum*, and of *Ecklonia radiata* 30 cm–100 cm in height; a middle stratum of small green, brown or red algae 5–25 cm in height, and a lower stratum of mostly brown or red algae sometimes of prostrate habit up to 5 cm in height. Algal communities of several strata are best developed on rocky coasts in conditions

---

[9] Three species of razor shell occur in the region. *Atrina dumosa*, mostly a deeper water species, *Pinna dolobrata* and *Subitopinna virgata*. They were not distinguished from each other in this survey and are all included in the term razor shell. *Pinna dolobrata* is the most common species.

TABLE 1

Depth water movement chart showing peak occurrences (and depth range in metres) of some common upper stratum species in Gulf St Vincent and Investigator Strait. Each species is placed on the mosaic where it was recorded in greatest abundance. Its range at other depths and water movements is shown in brackets. S = strong; M = moderate; W = weak

Depth (m)	Strong (current >75 cm/sec)	Strong to moderate	Water movement Moderate (current 25–75 cm/sec)	Weak (current <25 cm/sec)
0		*Sargassum linearifolium* *Cystophora moniliformis*		*Cystophora botryocystis* *Cystophora polycystidea*
	Scytothalia dorycarpa		*Cystophora retorta* (0–6) *Cystophora subfarcinata*	
		Seirococcus axillaris		
5	*Sargassum bracteolosum*			*Sargassum decipiens*
			Sargassum sonderi (3–8)	
	Sargassum lacerifolium	*Sargassum varians*		*Scaberia agardhii* (1-10) *Sargassum spinuligerum* (M–W; 3–34)
			Cystophora monilifera (1–10)	
10		*Ecklonia radiata* (S–W; 0–10)		*Sargassum halitrichum*
			Sargassum paradoxum (12–24) *Sporochnus comosus* (S–W; 10–41)	
		Sargassum distichum (10–23)		
15		*Myriodesma quercifolium* (11–43)	*Sargassum biforme* (?) *Caulocystis uvifera* (S–W; 14–20) *Encyothalia cliftoni*	
		Myriodesma integrifolium (23–24)		
20		*Bellotia eriophorum* (26–41)		

* In Investigator Strait *Ecklonia* goes to 40 m.

of moderate to strong water movements, but also occur on the deep rocky outcrops of Investigator Strait. In deeper water on level rocky bottom algal communities are usually comprised of a single stratum of sparse red algae up to 30 cm in height. These communities are generally rich in species and rather than present species' lists for particular localities, the centre and range of distribution of common species is shown along depth and water movement gradients in Tables 1–2. It can be seen that the overlap between species is high, so that on suitable substrate many species are likely to co-occur in one place. A vegetation profile taken at Port Noarlunga Reef in 1968 (Fig. 5) is typical of plant and animal communities to be found on rocky shores of the region. Algal communities of rough-water coasts are described by Shepherd & Womersley (1970).

Several species of the green algal genus *Caulerpa* are also common. These species usually occur as monospecific stands due to their method of vegetative spreading from their rhizomatous attachment to rock or in sand. Stands of *C. brownii*, *C. flexilis*, *C. obscura* and *C. longifolia* are common on rocky shores of the southern approaches, *C. alternans*, *C. cactoides* and *C. trifaria* on sandy bottom, the latter two species being especially common in the *Pinna*-holothurian assemblage in the upper reaches of the Gulf.

Common grazing animals in algal communities on rocky bottom are the sea urchins *Heliocidaris erythrogramma*, *Phyllacanthus parvispinus* and *Goniocidaris tubaria*, the abalone *Haliotis laevigata*, *H. ruber*, *H. scalaris* and the gastropod *Subninella undulata*.

C. ANIMAL ASSEMBLAGES

The distribution of bottom dwelling animals is influenced by the kind of substrate, depth, light and hydrological conditions. Six distinct animal assemblages are described below, their

TABLE 2

Depth water movement chart showing peak occurrence of 61 common middle and lower stratum species (with depth ranges in metres) in Gulf St Vincent and Investigator Strait. Species ranging over strong to weak water movements are listed at the foot of the chart

Depth (m)	Strong	Strong to moderate	Moderate	Moderate to weak	Weak
0					*Jania* sp. 0–5 *Dictyopteris australis*
		Cheilosporum elegans	*Lobospira bicuspidata* 0–6 *Plocamium cartilagineum*		
		Gelidium australe 3–10	*Halopteris ramulosa*	*Zonaria spiralis*	
5			*Halopteris funicularis*	*Distromium flabellatum*	
		Heterosiphonia struthiopenna 4–24		*Protokuetzingia australasica* 5–20	
		Hypnea episcopalis 10–34 *Zonaria crenata* 10–34 *Plocamium mertensii* 5–15	*Medeiothamnion halurum* 10–15 *Wrangelia velutina* 10–18 *Heterosiphonia gunniana* 10–18		
10	*Brongniartella australis*		*Solieria robusta* 8–21	*Botryocladia obovata* 10–21 *Peyssonnelia gunniana* *Doxodasya bulbochaete* 10–33	
	Haloplegma preissii		*Gonatogenia subulata* 13–17		
		Chiracanthia arborea 13–17 *Amansia pinnatifida* 15–23 *Cladurus elatus* 15–23 *Dictymenia tridens* 10–35 *Echinothamnion hystrix* 14–27	*Gloioderma fruticulosum* *Gloioderma speciosa* 14–18 *Dictymenia harveyana* 10–43	*Erythroclonium muelleri* *Laurencia filiformis* 10–20 *Lophothalia* sp. 10–18	*Areschougia congesta* 10–20 *Thamnoclonium dichotomum*
15		*Jeannerettia lobata* 10–23	*Euptilocladia spongiosa* *Heterosiphonia curdieana* *Platysiphonia victoriae* 18–22		
20	*Kallymenia cribrosa* 20-43	*Amansia kuetzingioides* *Apoglossum tasmanicum* 14–35 *Ptilocladia australis* 14–23	*Gulsonia annulata*	*Gracilaria furcellata* *Rhodymenia australis*	*Asperococcus bullosus* *Cryptonemia undulata* *Erythroclonium sonderi*
			Jeannerettia frondosa		*Heterosiphonia australis* *Thaumatella disticha*
25	*Dictymenia sonderi*	*Herposiphonia versicolor* 18–24 *Gloiosaccion brownii* 20–34 *Dictyota furcellata* 23–34 *Phloeocaulon spectabile* 23–31 *Rhodophyllis ramentacea* 30–34			*Coelarthrum muelleri*
30		*Crouania muelleri* 30–34			

Species (with depth range in metres) which range from strong to weak water movements:

Dictyosphaeria sericea (8–23)
Lobophora variegata (1–15)
Dictyopteris muelleri (1–23)
Dasya villosa (15–24)
Claudea elegans (10–31)
Cliftonaea pectinata (15–41)
Mychodea carnosa (10–18)
Osmundaria prolifera (12–33)
Rhabdonia verticillata (10–23)
Warrenia comosa (13–33)

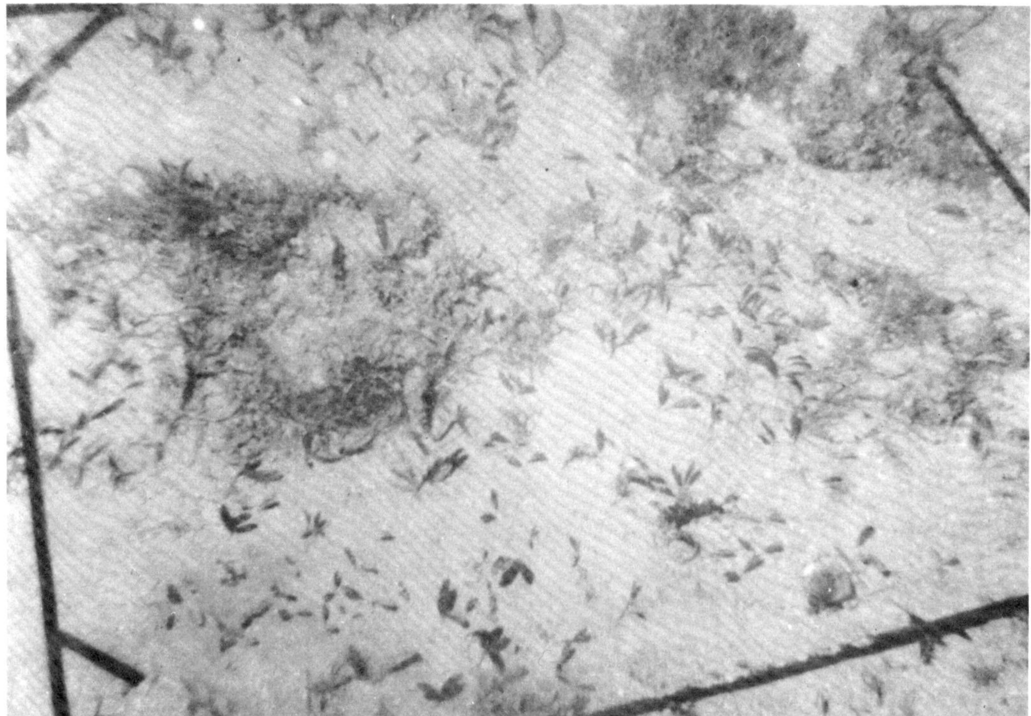

Fig. 6. A patch of the seagrass *Halophila ovalis* in an ascidian-scallop assemblage about 10 km off Outer Harbor. The quadrat is 1 m².

distribution mapped in Fig. 3 and the common animals present in them listed in Table 3.

(1) *Pinna*-holothurian assemblage

The razor shell *Pinna* and burrowing holothurians (sea cucumbers) dominate the muddy bottom of the upper reaches of the Gulf. Razor shells reach densities of 5/m² or more and each shell (whether dead or alive) is a kind of micro-reef supporting a rich epizoic assemblage of small sponges, ascidians and bryozoans. The "milk bottom ascidian" *Phallusia*, usually white in colour, but also brown or yellow, is also conspicuous on the silty bottom.

(2) Ascidian-scallop assemblage

This assemblage (Fig. 6) occurs at intermediate depths between 15–30 m in places of slight to moderate water current. *Pinna* occurs (but less abundantly than in the more northerly reaches of the Gulf) and its epizoic fauna is not so rich. Here ascidians are generally attached to shell fragments buried in sand. Scallops and other bivalves, together with their predators, the seastars *Coscinasterias* and *Luidia,* are usually the most common species.

Fig. 7. The seapen *Sarcophyllum grande* in unstable bottom of sand in Investigator Strait.

TABLE 3

Common species of five faunal assemblages in Gulf St Vincent and Investigator Strait

	Pinna/ Holothurian assemblage	Ascidian/ Scallop assemblage	Bryozoan assemblage	Malleus/ Pinna assemblage	Heterozostera/ Lunulites assemblage
Substrate	very silty calcareous sand	fine calcareous sand	very silty calcareous sand	muddy-silty calcareous sand	coarse shelly sand
Depth range (m)	9–18	15–30	25–35	30–42	27–35
Water current	slight; occ. moderate	slight to moderate	slight	slight	strong
Species					
Sponges	C	OCC	C	FC	OCC
Bryozoa					
Adeona grisea basket bryozoan	—	—	C	OCC	—
Celleporina sp. } *Cellepora* sp. }	C	R	C	—	—
Petralia undata	—	OCC	C	OCC	—
Triphyllozoon spp.	OCC	OCC	C	OCC	—
Parmularia sp.	—	FC	—	OCC	C
Lunulites sp.	—	—	—	—	VC
Coelenterates					
Sarcophyllum grande seapen	—	R	—	—	OCC
Echinoderms—seastars					
Astropecten preissi	—	OCC	R	—	OCC
Coscinasterias calamaria	OCC	FC	R	—	OCC
Luida australiae	—	FC	OCC	—	—
Tosia australis	OCC	R	—	—	OCC
Brittle stars					
Ophiothrix spongicola	C	OCC	—	—	—
Ophionereis semoni	—	FC	—	—	—
Sea urchins					
Goniocidaris tubaria	C	OCC	—	—	—
Sea cucumbers					
Ceto cuvieri	—	—	—	—	FC
**Holothuria hartmeyeri* } **Stichopus mollis* }	C	OCC	OCC	?	—
Stichopus ludwigi	C	OCC	OCC	?	—
Featherstars					
Antedon incommoda	R	OCC	C	—	FC
Molluscs-Gastropods					
Cassis (Hypocassis) cimbriata	C	OCC	OCC	—	—
Haliotis cyclobates	C	R	—	—	—
Pelecypods					
Chlamys bifrons queen scallop	FC ($<$0.5/m^2)	C (0.5–4/m^2)	OCC	—	OCC
Dosinia (Kereia) victoriae	FC	P	—	—	—
Glycymeris (Tucetilla) radians	—	—	OCC	—	—
Katelysia scalarina (?)	—	P	—	—	—
Laevicardium (Fulvia) tenuicostatum	—	OCC	—	—	—
Malleus meridianus hammer oyster	OCC	OCC	C	VC	—
Mimachlamys asperrimus scallop	C	C	OCC	—	OCC
Modiolus inconstans	C	—	OCC	—	—
Ostrea angasi native oyster	OCC	R	FC	—	—
Pecten meridionalis king scallop	OCC ($<$1/m^2)	C (0.5–2/m^2)	OCC ($<$0.2/m^2)	OCC	—
Pinna dolobrata razor shell	C (0.5–9/m^2)	C (0.1–3/m^2)	FC ($<$1/m^2)	OCC ($<$0.2/m^2)	—
Venericardia sp.	FC	—	—	—	—
Brachiopod					
Magellania flavescens	C	—	OCC	—	—

TABLE 3—*continued*

	Pinna/ Holothurian assemblage	Ascidian/ Scallop assemblage	Bryozoan assemblage	Malleus/ Pinna assemblage	Heterozostera/ Lunulites assemblage
Crustaceans					
Metapenaeopsis crassissima— Strawberry prawn *Penaeus latisulcatus—* Western king prawn *Portunus pelagicus—* Blue swimmer crab	These species are nocturnal and seldom seen by divers. They are commonly taken by prawns trawlers throughout the region.				
Leptomithrax australiensis—spider crab	—	OCC	OCC	—	—
Ovalipes bipustulata—sand crab	OCC	OCC	—	—	—
Ascidians					
Ascidia sydneyensis	—	OCC	—	—	OCC
Halocynthia hispida	C	C	—	—	—
Herdmania momus	OCC	OCC	OCC	Not known	R
Phallusia depressiuscula— milk bottle ascidian	C	C	OCC	—	R
Polycarpa papillata	OCC	OCC	—	—	—
Polycarpa pedunculata	C	C	OCC	—	FC
Polycitor giganteum	—	—	—	—	OCC
Pyura australis	OCC	OCC	R	—	FC
Pyura scoresbiensis	—	—	—	—	FC
Pyura spinifera	OCC	OCC	R	—	OCC
Sycozoa cerebriformis	C	—	—	—	—

VC = very common; C = common; FC = fairly common; OCC = occasional; R = rare.
* These species are difficult to distinguish in the field, and the differences were not recognised in this study.

Red algae and the seagrass *Halophila ovalis* are also common.

(3) Bryozoan assemblage

In the deepest reaches of the Gulf, large erect bryozoans (such as the "basket" bryozoan *Adeona grisea* and others, identified to genus only) and massive sponges become dominant. Ophiuroids and crinoids are commonly epizoic on them. Here *Pinna* is marginally, and often patchily, distributed.

(4) *Malleus/Pinna* assemblage

In the lower eastern part of the Gulf about Rapid Bay where the water depths are greatest, water currents are sluggish and the bottom is of a fine muddy silt. The infauna is sparse, but judging from the mounds and craters which pock-mark the bottom, burrowing organisms are especially abundant. Our knowledge of this area is incomplete and much more collecting needs to be done. The hammer oyster *Malleus meridianus* is probably the most common species, followed by *Pinna* and the scallop *Pecten meridionalis*. In addition there are some small sponges and erect bryozoans.

(5) *Heterozostera/Lunulites* assemblage

The bottom of Investigator Strait is swept by strong tidal currents of 50–200 cm/sec (1–4 knots); currents are weakest near the coast of Yorke Peninsula and Kangaroo I. and increase with increasing distance from shore, becoming strongest in the middle of the Strait. Sandy areas are colonised by sparse *Heterozostera tasmanica* at depths of 25–35 m. Animal species are also few and sparse, and limited to species like the "button bryozoan" *Lunulites* sp. and the brachiopod *Magadena cumingi*, both of which lie on the surface of the sand: to those species which can find anchorage in the unstable bottom such as the seapen *Sarcophyllum grande* (Fig. 7), and the stalked "fan bryozoan" *Parmularia* sp. and the stalked ascidians *Pyura australis* and *P. spinifera*. Rocky outcrops above the sand are colonised by sponges, ascidians, reef-molluscs and crinoids, as well as algae. Towards the Kangaroo I. side of the Strait on rocky bottoms, massive, erect, orange sponges are everywhere, together with large populations of its predator, the striking black and white psolid holothurian *Ceto cuvieri*.

(6) Sponge/bryozoan assemblage

The sea floor of Backstairs Passage has been little studied and generally the water depths (35–70 m) are too great, and the current speeds too high (up to 250 cm/sec.) to allow easy study by diving. According to Richardson & Watson (1975) at 40 m depth "the substrate consists of a planed-off surface of bedrock covered with a thin layer of sediment made up of coarse sand, small flat pebbles and biogenic carbonates. These sediments form drifts in the lee of the sparse reef outcrops. . . ." The reef outcrops are exposed Permian shales and our observations at other places in the Passage generally confirm this picture. At greater depths (>50 m) in the tide race, massive sponges (more than 1 m high and across) and large erect bryozoans (such as *Adeona grisea*), dominate the bottom; epizoic crinoids and ophiuroids are especially abundant on them. In the sediment drifts the brachiopod *Magadena cumingi* reaches densities of up to $80/m^2$ (Richardson & Watson 1975).

Acknowledgments

The authors are indebted to Beach Petroleum N.L. for permission to incorporate in this paper the results of extensive marine exploration by MV *Saori* in Gulf St Vincent and Investigator Strait. In particular, appreciation is expressed to Mr D. Von Sanden, Operations Manager, Geosurveys of Australia Pty Ltd, for his joint participation in these diving surveys, and for designing the diving bell that accompanied many of the divers, particularly at night; also to Ben Ranford and Arnold Mittler, skippers of the MV *Saori*.

We are grateful to Messrs R. E. Cooper, Anthony Von Sanden and Tony Cooper, who carried out extensive laboratory examinations of samples; to Mrs J. E. Watson for permission to use her data upon Investigator Strait; to Mr C. Waters for the use of sediment distribution maps; and to Adelaide Cement Co. Ltd for permission to publish its underwater photographs. Prof. H. B. S. Womersley kindly identified the marine algae and Dr P. M. Mather the ascidians.

We are also grateful to Mr A. M. Olsen, Prof. C. C. von der Borch and officers of the Geological Survey of South Australia for improvements to the manuscript.

References

BYE, J. A. T. (1976).—Physical oceanography of Gulf St Vincent and Investigator Strait. *In* C. R. Twidale, M. J. Tyler & B. P. Webb (Eds). "Natural History of the Adelaide Region." (Royal Society of South Australia: Adelaide.)

COONEY, A. M. (1965).—Submarine geological exploration Gulf St Vincent, South Australia. *J. Aust. Petroleum Explor. Assoc.*, 1-5.

KOTT, P. (1972a).—The ascidians of South Australia I: Spencer Gulf, St Vincent Gulf and Encounter Bay. *Trans. R. Soc. S. Aust.* **96**, 1-52.

KOTT, P. (1972b).—The ascidians of South Australia II: Eastern Sector of the Great Australian Bight and Investigator Strait. *Trans. R. Soc. S. Aust.* **96**, 166-196.

KOTT, P. (1975).—The ascidians of South Australia III: Northern sector of the Great Australian Bight and additional records. *Trans. R. Soc. S. Aust.* **99**, 1-20.

OLLIVER, J. G. (1964).—Reconnaissance sampling —floor of Gulf St Vincent. *Mining Review* **116**, 62-65.

OLLIVER, J. G. (1965).—Dredge sampling—floor of Gulf St Vincent. *Mining Review* **117**, 81-86.

RICHARDSON, J. R. & WATSON, J. E. (1975).— Locomotory adaptations of a free lying brachiopod. *Science* **189**, 382.

SHEPHERD, S. A. (1968).—The echinoderm fauna of South Australia I. The asteroids. *Rec. S. Aust. Mus.* **15**, 729-756.

SHEPHERD, S. A. & WOMERSLEY, H. B. S. (1970). —The sublittoral ecology of West Island, South Australia I. Environmental features and the algal ecology. *Trans. R. Soc. S. Aust.* **94**, 105-138.

SPRIGG, R. C. (1965).—Seafloor sedimentation in St Vincent Gulf and its approaches. 38th ANZAAS Congress (Hobart) Aust. 1965 (abstract).

VERCO, J. (1935).—"Combing the southern seas." *In* B. C. Cotton (Ed.) 174 pp. (Mail: Adelaide.)

CHAPTER 13

INTERTIDAL ECOLOGY

by H. B. S. WOMERSLEY & I. M. THOMAS

Introduction

The coasts from Port Wakefield north of Adelaide, south to Cape Jervis and east to Victor Harbor present a great variety of topography and environment; correspondingly the ecology and types of organisms present are varied and especially rich in number of species on the southern coasts. General features of the environment and the intertidal zonation of organisms are well known, and have been covered in the general account of Womersley & Edmonds (1958) of the intertidal ecology of South Australian coasts. Unfortunately, however, there are few accounts of particular coastal areas within the Adelaide region, apart from those of Baldock (1971) of the Aldinga reef and Johnston & Mawson (1946) of the region between Outer Harbour and Port Willunga.

Causes of intertidal zoning

Tidal fluctuations are the prime cause of zonation of organisms in the intertidal region; organisms growing at higher levels must resist desiccation and varied temperatures more so than organisms growing at lower levels. Tidal fluctuations produce the clearest zonation where the tide range is greater and more regular and where sea conditions are calmest. Under such conditions steeply sloping rock surfaces are likely to show horizontal bands of organisms adapted to a particular degree of emergence and thus desiccation. On rough-water or "high energy" coasts (Fig. 1A) however, the variable conditions of surge and wave splash (dependent largely on the wind and state of the sea) usually obscure clear zonation, and detailed studies over a wide area may be needed to produce a general picture of the zonation.

While zonation is basically related to the tides, the vertical extent of intertidal organisms depends on the actual degree of desiccation in the particular zone. Thus on rough-water coasts with strong wave wash and splash, the zonation

may be spread vertically for up to two or three times the tide range (about 1 m), as seen in areas from Granite Island (Fig. 1A, B) at Victor Harbor along the coast to Cape Jervis.

Intertidal zone nomenclature

Although there is considerable variation in nomenclature of the intertidal zones, the following has come into general use in recent years.

Supralittoral (or Littoral Fringe) Zone, corresponding to the splash zone above average high tide level, i.e. not subject to cover by water for more than very short periods. This zone is characterised by littorinid snails on rocky shores in most parts of the world, and above these various lichens commonly occur; thin sheets of blue-green algae may also be present.

Eulittoral Zone, including the main part of the intertidal region from the supralittoral down to the sublittoral zone. This zone has often been referred to as the Littoral Zone or even as the Midlittoral Zone. Organisms in this zone are covered by water to various degrees as the tides fluctuate and on many coasts three subzones, dependent on the organisms, can be distinguished.

Upper eulittoral zone, often characterised on rocky coasts by small barnacles.

Mid eulittoral zone, with either larger barnacles or molluscs such as limpets and often also blue-green algae.

Lower eulittoral zone, characterised by an algal mat or turf, with "sea grapes" (*Hormosira*) commonly present on South Australian coasts.

Sublittoral Zone, from mean low tide down, which does not have other than momentary emergence between waves or during very low tides. It is often marked on rocky coasts by a distinct upper boundary to the occurrence of larger brown algae (0.5–1 m long), and this zone descends fairly uniformly for several metres.

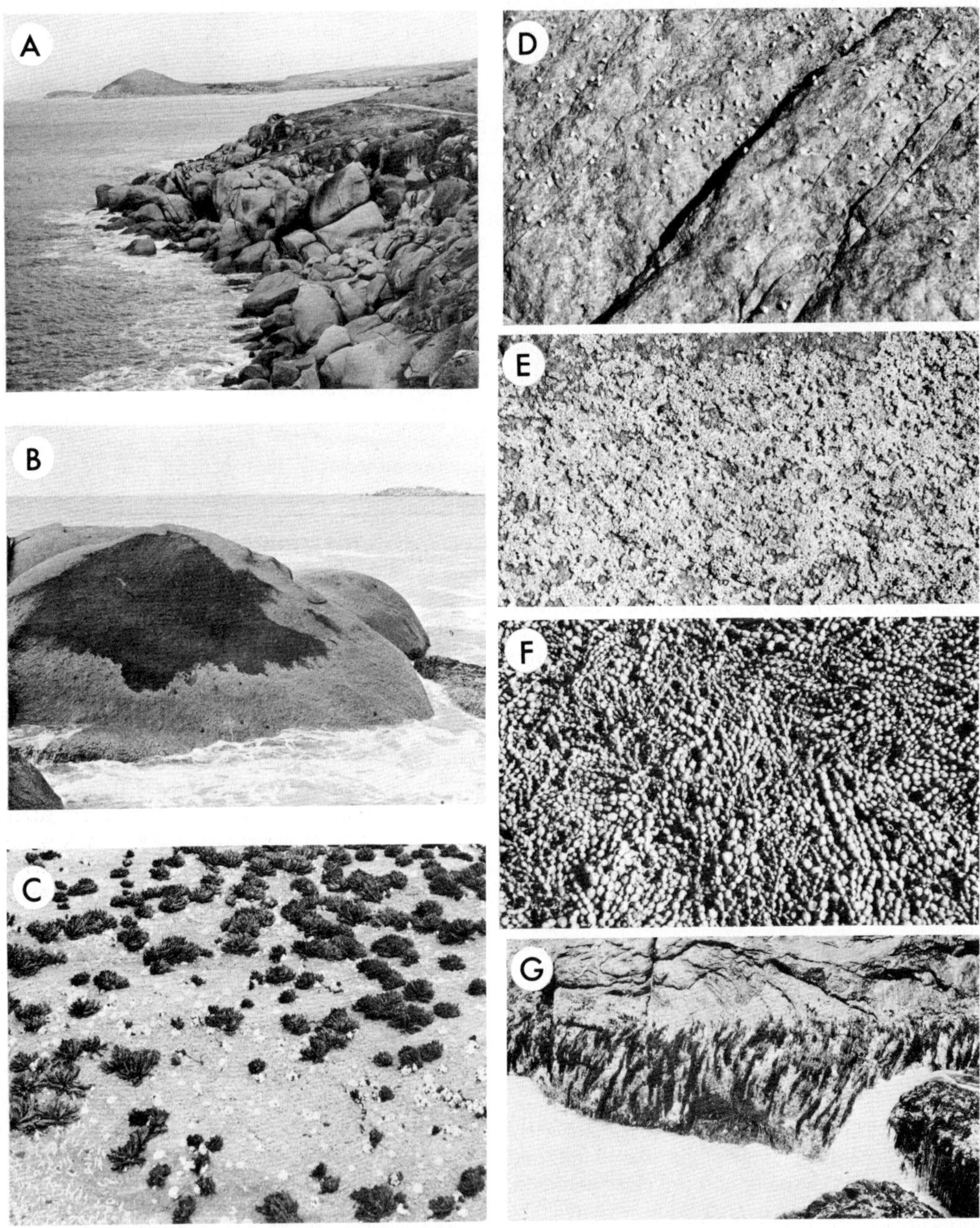

Fig. 1. High energy coasts: A, "high energy" coastline; B, similar coast with a patch of *Calothrix*; C, brown alga, *Splachnidium*, surf barnacles, *Catophragmus polymerus* and polychaete worms *Galeolaria caespitosa* in mid eulittoral zone; D, *Littorina unifasciata* in supralittoral zone; E, honeycomb barnacles, *Chamaesipho columna* in upper eulittoral zone; F, brown alga *Hormosira banksii* in lower eulittoral zone; G, dark brown alga *Cystophora intermedia* forming a sublittoral fringe.

On some coasts, especially those with strong wave action, a distinctive zone may occur at the top of the sublittoral, where it is emergent momentarily during recession of waves at low tide. This zone, rarely more than 0.5–1 m in height, is termed the "sublittoral fringe zone", but is not present on all coasts.

While these zones are dependent on the tidal fluctuations (and other environmental factors as discussed above), they are delimited for convenience by the dominant organisms present; these are the "indicator organisms" referred to above and in more detail in the following account. The best indicator organisms are those fixed in position on the rock or other substrates, such as the barnacles and the algae. Animals which move (such as littorinid snails and other molluscs) have to be used with caution, but most of these return to their normal intertidal position under the usual tidal conditions.

High energy coasts

These are usually coastlines open to the oceans and occur from about Port Elliot westward to Cape Jervis and north to about Second Valley. These coasts are mostly of steeply sloping rock cliffs or boulders, often much dissected (Fig. 1A) and with occasional sandy beaches between the headlands. In some bays, as for instance between Granite Island and the Bluff, conditions are considerably calmer and can be compared with the conditions described under "Moderate Energy Coasts". Along most of the coastline, conditions can be very hazardous; wave action is normally strong, washing over rocks well above the actual tide level, but occasionally much higher waves originating far out in the Southern Ocean can wash 3–4 m higher than normal. The basic zonation is given in Table 1.

The *supralittoral zone* is characterised by the small blue-grey littorinid snail, *Littorina unifasciata*, which occurs sparsely on rock exposed to the sun (Fig. 1D) and often more densely in shaded or moister areas. *Littorina* feeds on the thin cover of algae on the rock, but only occasionally does this algal cover become obvious—such as the dark blue-green patches (Fig. 1B) of *Calothrix fasciculata* on boulders on the southwest side of Granite Island at Victor Harbor. In hot weather in summer, *Littorina* is often found to have migrated to lower levels and occurs amongst the barnacles of the upper eulittoral. A closely allied species, *L. praetermissa* is often found associated with

L. unifasciata, but while the latter is usually found where desiccation is more severe, the former is found in moister surroundings.

Lichens may also be conspicuous in the supralittoral zone. The lowest-growing one is *Lichina confinis,* which forms almost black patches of small branched plants (only 2–3 mm high). Higher, above most of the spray effect when conditions are calm, are other crustose lichens—the grey *Buellia* sp. and the yellow *Caloplaca* sp.

Other animals are not conspicuous in the supralittoral, but in moist, shaded places the isopod *Ligia australiensis* and the orange-yellow crab *Leptograpsus variegatus* may be found; such organisms also occur at upper eulittoral levels.

The eulittoral zone. As on most temperate rocky coasts of the world, the eulittoral zone is dominated by barnacles in its upper part, barnacles and molluscs in the mid part, and an algal mat in the lower part.

Upper eulittoral of rocky shores bears a sparse to dense (with greatest wave action) cover of the honeycomb barnacle, *Chamaesipho columna* (Fig. 1E), with another small barnacle, *Chthamalus antennatus,* often mixed with it.

Mid eulittoral is dominated under strong wave action by the larger surf barnacles, *Catophragmus polymerus* (Fig. 1C), but in slightly calmer (and lower) situations the black mussel *Austromytilus rostratus* forms prominent masses. Both the barnacles and the mussel are filter feeders. Scattered amongst the barnacles, and dominant when *Catophragmus* is absent, are various limpets such as the large "Chinaman's hat", *Cellana ariel,* and the smaller *Patelloida latistrigata,* the pulmonate, *Siphonaria diemenensis* and the predatory *Thais orbita* (formerly *Dicathais textilosa*). White masses of calcareous polychaete worm tubes (*Galeolaria caespitosa*) often coat the rock in the lower part of this zone (Figs 1C, 2C), especially where shaded and slightly sheltered. Blue-green algae such as the 0.5–1 cm hemispheres of *Rivularia firma* and the flat patches of *Isactis plana* may be common (but variable in occurrence), and in summer, the brown, highly mucous *Splachnidium rugosum* (branched, cylindrical, to 10 cm high—Fig. 1C) and the brownish purple worm-like *Nemalion elminthoides* may occur (e.g. on Granite Island and at Port Elliot). Such algae are adapted better than most others to the moderate degree of desiccation in this zone.

Lower eulittoral. Here an algal mat covers the rock. In extreme wave action, this is dominated by branched, calcareous coralline algae (*Corallina cuvieri, C. officinalis*), and the very large barnacle *Balanus nigrescens* may also occur, with old shells overgrown by the *Corallina.* Other algae present include stunted *Laurencia filiformis, Wrangelia plumosa,* and *Dasya clavigera,* and amongst the animals large chitons such as species of *Poneroplax* are most common. A variety of crustaceans and worms live in the algal mat.

Under less severe conditions of wave action, the brown alga *Hormosira banksii* ("sea grapes") occurs (Fig. IF), as a form adapted to strong water flow.

Upper sublittoral is marked by a distinctive fringe zone of the dark brown alga *Cystophora intermedia* (Fig. 1G), which is adapted in its branched form, toughness, and smooth surface to withstand the most turbulent conditions; in slightly less turbulent conditions it does not occur, and its vertical range is usually less than 1 m.

The coralline algal mat of the lower eulittoral does, on very rough-water coasts, descend 1–2 m below low tide level, and occurs as a lower stratum below the *Cystophora.* Other larger algae occur where conditions are less turbulent—e.g. the kelp *Ecklonia radiata,* other species of *Cystophora,* and *Sargassum,* as well as species of the green *Caulerpa* (especially *C. brownii*) as an understorey. Any bare areas of rock are likely to be covered by pink encrusting coralline algae ("lithothamnia").

Animals are usually not conspicuous in this algal dominated upper sublittoral zone, but the abalone (*Haliotis* spp.) and the stalked ascidian *Pyura pachydermatina,* as well as seastars, occur.

While this account of the "intertidal" ecology has for convenience been extended into the uppermost subtidal region, recent studies have greatly extended our knowledge of the deeper subtidal. This region is described in Chapter 12.

Rock pools are common on most rocky coasts, and provide a specialised type of habitat wherever they occur. In general it is preferable to separate discussion of them from the zonation on sloping intertidal rock, and to refer to them, e.g. as "pools at a mid eulittoral level".

Apart from organisms around the pool margin, submerged organisms are mostly of a sublittoral nature, especially those which can withstand greater variation in temperature and salinity. Study of the biota of rock pools is complex and pools are usually very variable in their conditions and organisms. A detailed survey is beyond the scope of this account.

Moderate energy coasts

These extend from about Second Valley northwards to about the level of Outer Harbor. Farther north, wave energy diminishes to the level of a low energy coastline. Of course a moderate energy coast may occasionally be subjected to full, high energy buffetting by southwesterly storms, but these occasions are relatively rare. The substrate here varies from steeply sloping boulders or cliffs (e.g. Second Valley) to more gently sloping, dissected rock (e.g. at Marino), and to almost horizontal rock platforms at a low intertidal level (e.g. Aldinga reef—Fig. 2A). Aldinga reef is one of the largest and most accessible platform areas in South Australia and the offshore ribbon reef at Port Noarlunga is also noteworthy as an example of this type of coast. Both Aldinga and Port Noarlunga reefs have been proclaimed as aquatic reserves to protect their flora and fauna.

The tidal rise varies from just over 1 m at Cape Jervis to about 2 m at Port Adelaide, wave height is normally less than 1 m with calm conditions frequent, and the range of water temperature is greater than on high energy coasts, from as low as 12°C in winter to about 22°C in summer, depending on depth and shelter. The basic zonation (Table 1) is spread over approximately the height of the tidal rise except where splash on a suitable substrate occurs. In many areas, sandy beach backs the rocks at a mid intertidal level (e.g. Aldinga), and where rock does occur at and above high tide level exposure to the sun and severe drying may restrict the number of organisms present.

The supralittoral zone is still dominated by *Littorina,* but much of the rock is bare and *Littorina* are restricted to more sheltered microhabitats. The lichen *Lichina* may occur and extend to the upper eulittoral but is not common, and blue-green algae are rarely conspicuous at this level (except in partial shade on the soft cliffs at Whitton Bluff, between Christies Beach and Port Noarlunga). The higher growing grey and yellow lichens may occur but are sparse.

The eulittoral zone bears many of the same organisms as on rough-water coasts, but with some notable exceptions.

TABLE 1

Summary of intertidal zonation on South Australian coasts

| Zone | High energy | | Moderate energy | | Low energy |
	Rocky shores	Sandy shores	Rocky shores	Sandy shores	Sand or mud flats
Supralittoral (littoral fringe)	*Calothrix* *Lichina* and other lichens *Littorina* spp.		*Lichina* *Littorina* spp. *Ligia*	*Talorchestia* *Orchestia*	Samphires
Upper eulittoral	*Chthamalus* *Chamaesipho*	*Actaecia euchroa* *Cirolana concinna*	*Chthamalus* *Chamaesipho* *Bembicium nanum*		Mangroves *Helograpsus* *Bembicium melanostomum*
Mid eulittoral	*Catophragmus* *Cellana* and other molluscs *Galeolaria* Blue-green algae		*Cellana* *Modiolus pulex* *Galeolaria* Blue-green algae	*Venerupis galactites*	*Enteromorpha* *Philyra laevis* *Modiolus inconstans* *Flavomala biradiata*
Lower eulittoral	Coralline mat *Balanus*	*Plebidonax** *Conuber encei* *Exoediceras* sp.	*Corallina* *Gelidium* *Hormosira* Chitons	*Amphidesma angusta* *Katelysia* spp. *Conuber*	*Hormosira* *Zostera* *Austromytilus†* *Pinna†*
Sublittoral fringe	*Cystophora intermedia*	*Tellinota albinella*	*Cystophora* spp. *Ecklonia* *Sargassum* and other algae	*Mactra* spp. *Ovalipes*	*Hypnea* *Spyridia* and other algae *Portunus pelagicus*
Upper sublittoral				*Posidonia†*	*Heterozostera* *Posidonia†*

* Species extending into upper zone.

† Species extending into lower zone.

The *Upper eulittoral zone* is often sparsely populated, but patches of the barnacles *Chamaesipho* and to a lesser extent *Chthamalus* may occur.

The *Mid eulittoral zone* is essentially a mollusc, *Galeolaria*, blue-green algal zone (Fig. 2B), with the molluscs such as *Cellana ariel* and the tube-dwelling polychaete, *Galeolaria* (Fig. 2C) better developed than on rougher-water coasts. The surf barnacle, *Catophragmus*, is absent from such coasts, but in heavy shade around boulders the barnacle, *Tetraclita rosea*, often occurs.

As well as numerous limpets, three other gastropods are common in this zone: the black periwinkle *Melanerita melanotragus*, the conniwinkle *Bembicium nanum*, and *Austrocochlea concamerata*. Masses of the mussel *Modiolus pulex* are frequent on gently shelving rock. Blue-green algae may or may not be conspicuous; the commonest are probably *Rivularia australis*, *Isactis plana* and *Brachytrichia quoyi*.

Lower eulittoral zone. In contrast to higher zones, this zone is dominated by algae, mostly mat-like forms, 1–3 cm high. With moderate wave action, *Corallina* is dominant (Fig. 2B) and in calmer conditions *Gelidium pusillum* (Fig. 2D) may form an almost pure community (e.g. near the outer margin of Aldinga reef); in many areas these two algae occur intermixed.

One of the commonest algae in the lower eulittoral is *Hormosira banksii* (Fig. 2E), especially on gently shelving platforms such as at Aldinga. The lowest growing plants are often not emergent or in pools at low tide, but *Hormosira* can both tolerate and requires exposure to the air periodically. The form on sheltered coasts differs both from that on rough-water coasts (little branched, oblong bladders) and that on calm sand-mud flats (large spherical bladders), and the plants on Aldinga reef are rather slender and often stunted.

In the upper part of the lower eulittoral and extending into the mid eulittoral, thin brown

 H. B. S. WOMERSLEY & I. M. THOMAS

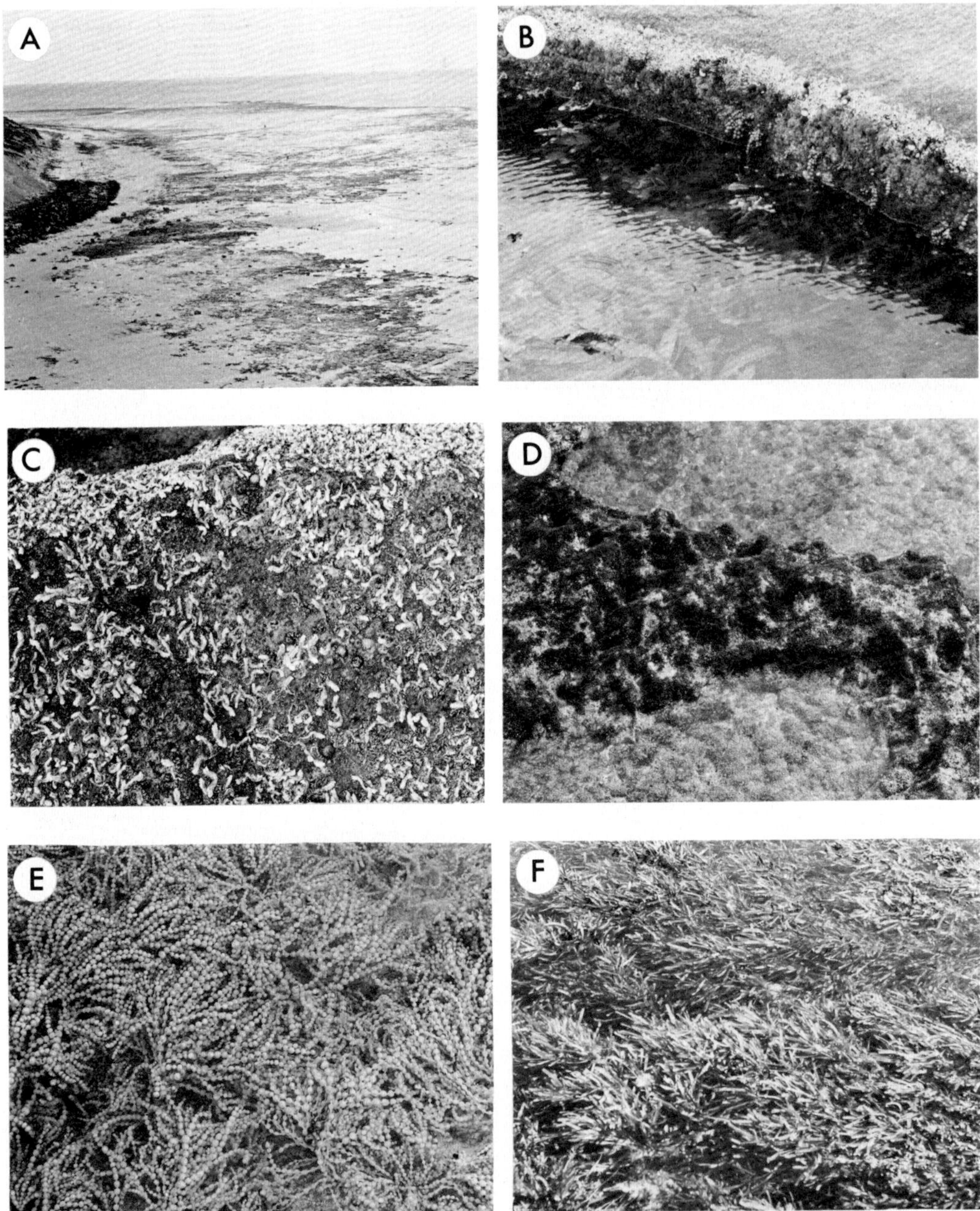

Fig. 2. Moderate energy coasts: A, rock platform at Aldinga reef; B, molluscs, *Galeolaria* and algal mat of mid and lower eulittoral zone; C, tube dwelling polychaete, *Galeolaria* in mid eulittoral zone; D, *Gelidium pusillum* community in lower eulittoral zone; E, *Hormosira banksii*, a common brown alga in lower eulittoral zone; F, sea grass, *Amphibolis antarctica* on sand covered rocks in upper sublittoral zone.

patches often occur on the rock in summer. These have not been investigated fully, but are probably the crustose form of the erect, cylindrical *Scytosiphon lomentaria,* which grows to 10 cm or more high in winter. Large animals are not common in the lower eulittoral, but the algal mat provides a home for large numbers of crustaceans, polychaete worms and other small organisms. Some molluscs of the mid eulittoral descend to this level and chitons may also occur. Pools at the lower eulittoral level are common and frequently contain green algae such as *Ulva* and *Enteromorpha,* small brown algae and slender red algae such as *Spyridia, Ceramium, Polysiphonia* and *Laurencia.*

The *upper sublittoral zone* is characterised by relatively large (to 1 m) brown algae (Fig. 2B), especially by the broad-leaved kelp *Ecklonia radiata* and the much-branched fucoid algae *Cystophora* (several species), *Sargassum* (several species), *Scaberia agardhii* and *Cystophyllum onustum.* These may occur intermixed or on occasion may be almost pure. On sand-covered rocks, the sea grass *Amphibolis antarctica* (Fig. 2F) may form a pure and dense community.

On these coasts of moderate wave action, the rough-water *Cystophora intermedia* is notably absent.

The above brown algae extend fairly uniformly from low water level down for 5–8 m. They often form a canopy, under which smaller green (e.g. *Caulerpa* spp.), brown (e.g. *Dictyota, Zonaria* spp.) and red algae occur, as well as starfish, sea-urchins and many molluscs and fish. This sublittoral zone is discussed further in Chapter 12.

Rocky, low energy coastlines rarely occur in the regions we are considering. The rather flattened rock platforms which may be present, are usually covered with deposited sand or silt. Such low energy coasts are found northwards from about Outer Harbour to the head of Gulf St Vincent. These will be considered below.

Sandy and muddy beaches

These can be classified into high, moderate and low energy coasts as can rocky shorelines. Their fauna and flora have been little studied in South Australia while their meiofaunas and meiofloras (the very small organisms which live in the interstices between the sand grains) have received even less attention.

High energy sandy beaches. These, being exposed to the waves of the open ocean, usually consist of fairly coarse sand which is highly mobile and is able to support only relatively few highly organised species well adapted for burrowing rapidly to avoid the scour of heavy waves. These shores are often backed by dunes which themselves support and are even maintained by a specialised array of plant species.

In the upper eulittoral the isopods *Actaecia euchroa* and *Cirolana concinna* are found, while in the mid eulittoral there is a sparse fauna of polychaete worms which reach larger numbers in the lower eulittoral. Here also is found the commercially important Goolwa cockle *Plebidonax deltoides* and an amphipod *Exoediceras* sp. which can be found in much smaller numbers in higher zones to the upper eulittoral. At low tide, the convoluted trails of the sand-dwelling gastropod *Conuber encei* can be seen on the wet sand. The animal itself might be found buried where the trail ends.

Macroscopic algae are absent in the intertidal region, but on firmer sand, especially in the lower eulittoral zone, brown patches indicate the presence of diatoms that can occur in vast numbers. These diatoms are, however, sporadic in their occurrence.

Moderate energy sand beaches. These can be found between Second Valley and Outer Harbor. They may be backed by sand dunes or rocky cliffs, but the dunes have, in recent years, been largely obliterated by housing development.

In the upper eulittoral zone, the drift of organic debris left by the receding tide, covers numerous small animals such as the sand hoppers *Orchestia* and *Talorchestia* and the isopod *Cirolana concinna.* In the summer and autumn months in addition, large numbers of dipterous kelp flies belonging to the family Ephydridae breed here.

The mid eulittoral zone harbours a variety of polychaete worms and some bivalve molluscs including *Venerupis galactites.* The density of the fauna in this zone varies, being denser in the more protected beaches. Where there is some protection afforded by outcrops of rock, there may be pools with sandy bottoms and mats of eel grass (*Heterozostera tasmanica*) with their associated fauna.

In the lower eulittoral, the fauna is usually denser, the small cockle, *Amphidesma angusta* and species of *Katelysia* being the commonest bivalves. *Conuber encei* is also present.

Sandy areas below low tide level are usually colonised by sea grasses. The distribution of

Fig. 3. Low energy sand and mudflats: A, intertidal flats with mangrove pneumatophores; B, mangrove trees in upper eulittoral zone; C, regenerating mangrove trees; D, molluscs, *Bembicium* in mangrove community; E, *Hormosira banksii* in lower eulittoral zone; F, eel grass, *Heterozostera tasmanica* extending into sublittoral zone from lower eulittoral; G, sea grass *Posidonia australis* in sublittoral zone.

these major components of the Gulf ecosystem is described in Chapter 12, but two in particular reach to low tide level. The commonest, often found in sandy-mud patches on reefs such as at Aldinga, is the eel grass, which forms a dense and pure community that provides a habitat for many small animals. At a deeper level (usually from 0.5 m below low tide down, but again in very shallow water on Aldinga reef) occurs the "tape weed" *Posidonia australis*.

Low energy sand and mud flats

From Outer Harbor northwards to the top of Gulf St Vincent, conditions of water movement are slight and sandy-muddy intertidal flats have developed (Fig. 3A). On the east coast of Gulf St Vincent, the flats are backed by mangrove woodland and samphires, which then grade into true land vegetation.

Apart from the upper zones of mangroves and samphires, eulittoral zonation is not well defined owing to the gradual environmental change over the gently shelving flats. Near low tide level, however, more distinct bands of the sea grasses *Heterozostera* and then *Posidonia* (at least 0.5 m below low tide level) occur.

A striking feature of these areas is the dominance of a few types of plants forming well defined communities—the samphires, the mangrove *Avicennia*, and the sea grasses, *Heterozostera* and *Posidonia*. The absence of hard substrates reduces the variety of macro-algae, but microscopic algae, especially diatoms, are abundant both on the mud of the flats and on other organisms. Animal life is prolific both on and in the mud, though often (especially when in the mud) not conspicuous. The calm water also has a great variety of animals, from fish of various species to small zooplanktonic creatures, and the mangrove woodland is home for a variety of birds.

The supralittoral zone usually bears samphires (species of *Salicornia* and *Arthrocnemum*) growing from where the lower stems may be covered by high tides to well above high tide level, but subject to high salt levels in the substrate. Occasionally the lowest-growing plants may be completely covered at high tide. This samphire community is discussed by Specht (1972). The mud and lower stems of the samphires often appear fairly bare, but always have a film of microscopic algae such as diatoms. In shade, algae such as *Enteromorpha*,

Chaetomorpha, and often *Gelidium* and *Bostrychia* occur, together with molluscs such as *Bembicium melanostomum*.

The eulittoral zone is characterised in its upper part (Fig. 3A, B) by the woodland of mangrove trees (*Avicennia marina* var. *resinifera*), which extend from an upper eulittoral level into the supralittoral. In most areas, the base of the trunk is covered at high tide, as are the finger-like pneumatophores (for gaseous exchange) which project through the mud (Fig. 3A) in lines, showing their origins from the radiating roots. The rest of the tree is normally completely aerial, but is adapted to growing in a high salt substratum and for excretion of the large amounts of salts absorbed by the roots.

While much is known about the one species of mangrove found in South Australia, little is known of the ecology and food chains of such coastal woodlands. Mangroves elsewhere in the world have been shown to be breeding areas for certain coastal fish, and this is also likely to be so in the Gulf region of South Australia. The South Australian occurrences are some of the most southerly for this species, which is in general a hardy species which regenerates readily (Fig. 3C). Good stands occur along considerable distances of the Gulf shores. Unless the water relations of the plants are altered by constructional work, they are resistant to much pollution and where subject to heated water effluence they grow better—as would be expected from their distribution. They also recover well from temporary setbacks, as shown in the Port Pirie region.

While most of the South Australian mangroves are only small trees, there are several noteworthy older stands—such as near the Gawler River and at Fishermans Bay near Port Broughton. These latter trees show considerable regeneration from old branches. Some of them are healthy old trees with deep accumulation of sand around their bases, so much so that pneumatophores are not apparent.

Mangrove woodlands (or forests as they become in the tropics, often with up to 6 genera involved), have an interesting associated flora and fauna. In South Australia, the associated flora is not conspicuous apart from green algae such as *Ulva, Enteromorpha,* and *Cladophora,* but a film of blue-green algae usually occurs and the diatom flora on the mud and in the water is quite rich.[1] In most parts of the

---

[1] Butler, A. J., Depers, A. M., McKillup, S. C. & Thomas, D. P. (1975).—The conservation of mangrove-swamps in South Australia. Report to the Nature Conservation Society of South Australia.

world (including Western Port Bay in Victoria), the red algae *Bostrychia* and *Catenella* very characteristically form a felt on the lower stems and pneumatophores, but these appear to be absent from South Australian mangrove woodlands.

The fauna of mangrove communities in South Australia is also listed in the above-mentioned report. Molluscs, including gastropod genera such as *Salinator, Bembicium* (Fig. 3D) and *Austrocochlea,* and bivalves such as *Modiolus* and *Katelysia,* are common in or on the mud or on the trees, and polychaete worms and crustaceans are also plentiful. The mud crab *Helograpsus haswellianus* is numerous. Nectonic animals and the birds of mangrove woodlands are also listed.

The mid and lower eulittoral zones

The mangrove woodland is confined to the upper (occasionally extending to the mid) eulittoral region, and the mid and lower zones usually comprise sandy-muddy tidal flat with a sparse cover of macroscopic algae (*Ulva, Enteromorpha*) and the microscopic diatoms and blue-green algae, together with a variety of molluscs (the mussel *Modiolus,* and *Bembicium* are common) and worms in the mud. Crabs and small crustaceans are common.

The lower eulittoral becomes a more conspicuous zone in that the "sea-grapes", *Hormosira banksii* (a spherical bladder form—Fig. 3E) become prominent and in some areas form a well defined community. While apparently on the mud, each plant is attached to a solid substrate—such as a pebble or more usually to the large mussel, *Austromytilus erosus.*

Apart from this mussel, if the substrate in the low eulittoral is sandy then cockles (*Katelysia*) occur just buried (often with a plant of the green *Enteromorpha* attached); if the substrate is high in mud, the razor shell, *Pinna dolabrata,* may be prominent,, projecting 10–20 cm above the substrate.

Other animals are common on or in the mud, and reference can be made to Womersley & Edmonds (1958) for a brief account of them.

At low eulittoral levels, the sea grasses occur and often form pure communities covering considerable areas.

The highest occurring sea grass is usually *Zostera,* of which patches of (probably) *Z. mucronata* occur on the coast north of Port Adelaide. This is a short, sparse leaved (to 8 cm) species, and occurs in scattered patches rather than extensive areas. Slightly lower, and extending into the sublittoral, grows *Heterozostera tasmanica* ("eel grass"). This forms much more extensive and denser patches (Fig. 3F), with erect stems bearing leaves to 20 cm long. Both these sea grasses may bear algal and animal epiphytes, and the dense rhizome systems in the mud provide a home for many burrowing or mud-living animals including enteropneusts in some places.

The *sublittoral zone,* in contrast to that on rocky coasts with greater wave action, is marked by the virtual absence of large brown algae (unless there is some hard substrate) and the presence of extensive beds of the sea grass *Posidonia australis* (Fig. 3G), commonly known as "tape weed".

From Port Adelaide northwards, the broad-leaf variety of *Posidonia* occurs along most of the coast, extending from just below low tide level to several metres deep, as either a dense, continuous community or scattered patches. The stiff leaves of plants in shallow water just project above water at low tide, giving a characteristic appearance to the beds.

At about low tide level, several green and red algae are common, often situated between the *Heterozostera* and *Posidonia* zones. These include the green algae *Ulva lactuca, Enteromorpha* spp. and *Cladophora* spp. and the red algae *Hypnea musciformis, Spyridia filamentosa* and *Centroceras clavulatum.* On dead cockle shells, *Acetabularia peniculus* may occur. The algae of this zone in a region on Kangaroo Island similar to that at St Kilda are discussed in Womersley (1956).

Posidonia australis has hard, somewhat rough textured leaves and bears a wealth of epiphytes (see Womersley 1956). It provides a home for many animals, from crustaceans to fish such as whiting (*Sillaginodes punctatus*) and is by far the most important primary producing plant in these areas. However, few animals feed directly on the *Posidonia* leaves, and the food chain is probably dependent on the use of rotted *Posidonia* by protozoa and filter feeding animals.

Acknowledgment

The authors are indebted to Mr M. G. King for information on the fauna of some sandy shores.

References

BALDOCK, R. N. (1971).—Notes on the biology of some South Australian coastal areas and their use in excursions: Part 2. *S. Aust. Sci. Teachers Assoc. J.*, March 1971, 4-25.

JOHNSTON, T. H. & MAWSON, P. M. (1946).—A zoological survey of Adelaide beaches. Handbook of the twenty-fifth meeting of the Australian and New Zealand Association for the Advancement of Science. Adelaide, Aug. 1946, 42-47.

SPECHT, R. L. (1972).—"The Vegetation of South Australia." 2nd Edtn. (Govt Printer: Adelaide.)

WOMERSLEY, H. B. S. (1956).—The marine algae of Kangaroo Island. IV. The algal ecology of American River inlet. *Aust. J. Mar. Freshw. Res.* **7**, 64-87.

WOMERSLEY, H. B. S. (1959).—The marine algae of Australia. *Bot. Rev.* **25**(4), 545-614.

WOMERSLEY, H. B. S. & EDMONDS, S. J. (1958).— A general account of the intertidal ecology of South Australian coasts. *Aust. J. Mar. Freshw. Res.* **9**(2), 217-260.

CHAPTER 14

MARINE FISHES AND MAMMALS

by C. J. M. Glover* & J. K. Ling*

The waters of Gulf St Vincent and Encounter Bay provide habitats for a number of fish species of recreational and commercial importance as well as a few aquatic mammals which visit occasionally.

Recreational fishing in the Gulf waters adjacent to the Adelaide metropolis is undoubtedly of far greater significance, both economically and socially, than the limited commercial operations which persist. It is from this viewpoint that data should be gathered, analysed and used for fisheries management purposes. Adequate protection of the marine environment will require much more study of the Gulf's ecosystems than have hitherto been undertaken. This is especially so if quality recreational fishing is to continue for the growing numbers of amateur fishermen, more and more of whom are now using boats and thereby extending their range of operation beyond the immediate shoreline.

Marine mammals

There are no major colonies of seals or herds of whales close to Adelaide, but individuals of many species are regularly seen in these waters and on local beaches.

Both of the two local species of seals: the sea lion, *Neophoca cinerea*, and the New Zealand fur seal, *Arctocephalus forsteri*, haul out on Adelaide beaches occasionally and frequent sightings are made at various places around the Fleurieu coast. Breeding colonies of these seals occur on Kangaroo Island and other offshore islands farther west. Leopard seals, *Hydrurga leptonyx*, and Weddell seals, *Leptonychotes weddelli*, both true Antarctic species, are seen occasionally at Encounter Bay and elsewhere. The last reported sighting of a leopard seal was on 4 December 1975 at Waitpinga. The first reported sighting in South Australia of the southern elephant seal, *Mirounga*

leonina, was on 21 February 1975 at Encounter Bay.

Aitken (1971) has ably summarized the occurrence of whales in South Australian coastal waters from which 18 species have been recorded.

A thriving whale fishery existed at Encounter Bay in the very early days of South Australian history. In fact it was the colony's first industry. This was an inshore operation concentrating almost exclusively on southern right whales, *Eubalaena glacialis australis*, although sperm whales, *Physeter catodon*, were seen and chased—and invariably lost. Fin whales, *Balaenoptera physalus*, Bryde's whale, *B. edeni*, and humpback whales, *Megaptera novaeangliae*, are seen occasionally in Gulf St Vincent.

The smaller toothed whales are more commonly seen today. Of these the bottlenose dolphin, *Tursiops truncatus*, is the most frequently encountered as it comes close inshore—often to the unwarranted consternation of swimmers. *Tursiops* adjusts well to captivity and is the species used in performances at Marineland of South Australia. The common dolphin, *Delphinus delphis*, is also a familiar sight to boat owners farther from shore, where these playful mammals often give chase to vessels at speed. Pilot whales, *Globicephala melaena*, enter Gulf St Vincent occasionally and have gone aground on local beaches. Another visitor to these waters is the pygmy sperm whale, *Kogia breviceps*, the last stranding being recorded at Hallett Cove, but sightings have also been reported from Encounter Bay, where pygmy right whales, *Caperea marginata*, have been seen in the past. The largest recorded mass stranding of whales occurred in 1944 when almost 300 false killer whales, *Pseudorca crassidens*, went aground at Port Prime in Gulf St Vincent.

Little work has been done on the local biology of the whales. However, increasing

---

* South Australian Museum, North Terrace, Adelaide, S.A. 5000.

attention is being paid to population studies of seals in South Australia from a conservation point of view (see Ling & Walker 1976).

Marine fishes

Due to its proximity to the metropolitan area, Gulf St Vincent has undoubtedly been favoured as a collecting site by naturalists. The earliest ichthyological collecting from the Gulf appears to have been initiated in the 1860's by F. G. Waterhouse, first curator of the South Australian Museum, and other members of the South Australian Institute. This early material was examined by Castelnau (1872, 1873), who listed 27 species and described 5 new genera and 15 new species from it.

Further substantial collections were effected between 1880 and the early 1900's by J. W. Haacke and A. C. H. Zietz of the South Australian Museum, in association with J. C. Verco. Zietz's (1908-09) incompleted synopsis of South Australian fish lists 15 species specifically from the Gulf.

Waite (1921) published the first completed list of South Australian fishes but did not specify distributions within the State. This initial list of the State's fish fauna was subsequently updated by Scott (1962) and Scott *et al.* (1974).

Throughout the 1920's and 1930's the Field Naturalists' Section of the Royal Society of South Australia continued to collect from the Gulf and elsewhere.

During the 1950's and early 1960's T. D. Scott and F. J. Mitchell of the South Australian Museum pioneered speargun and rotenone collecting in coastal waters south of Adelaide.

To date 216 valid species, representing 113 families and 33 orders, have been recorded from the Gulf, including a handful of sporadic oceanic strays such as the short and point-tailed sunfishes, *Mola ramsayi* and *Masturus lanceolatus,* the oarfish, *Regalecus pacificus,* and the basking shark, *Cetorhinus maximus.* These large species occasionally enter shallows and are sometimes washed ashore and stranded on beaches. The Gulf's waters constitute the type locality of 26 recognised species, 7 of which appear to be endemic to the Gulf. Those fishes which frequent the Gulf comprise a diverse variety representing groups typical of the Flindersian geographical region of southern Australia. All are essentially marine except a few species which are known to spend at least part of their life cycles in the more dilute waters of estuaries, and further upstream, in waterways entering the sea. In the Gulf these latter include the short-headed lamprey, *Mordacia mordax,* and the native trout, *Galaxias maculatus.*

However, apart from their taxonomy, little is known of the biology of the Gulf's fishes. Indeed, although a Royal Commission on the South Australian Fishing Industry in 1934–35 recommended the "inauguration of practical research into the habits of fish", there has been a pronounced lack of research in this field (at least of scaled fishes) in the State. What has been achieved may be briefly summarized. Scott (1954)[1] established the life cycle of the spotted whiting, *Sillaginodes punctatus,* and Caton (1966)[2] further examined this species' growth rates, and commenced studies of its movements and catch structures. Ling (1958) investigated the breeding and population dynamics of the sea garfish, *Hemiramphus melanochir,* whilst Weng (1971)[3] and Harbison (1974)[4] examined aspects of the biology of the black bream, *Acanthopagrus butcheri,* and Shepherd[5] examined fisheries in the region of Port Vincent. All these studies were based, at least in part, on Gulf populations.

Although the variety of fish in the Gulf is relatively small compared with warmer Australian waters, there is an abundance of commercial species. Some 14 principal species occur in large numbers and are fished in the Gulf, the most important being the spotted and silver whiting, *Sillaginodes punctatus* and *Sil-*

---

[1] Scott, T. D. (1954).—The life history of *Sillaginodes punctatus* in South Australia. M.Sc. Thesis, University of Adelaide (unpubl.).

[2] Caton, A. E. (1966).—Preliminary synopsis of biological data on the Australian (spotted or King George) whiting *Sillaginodes punctatus* (Cuv. & Val.), 1829. Southern Pelagic Project Committee, Technical Session, C.S.I.R.O., Cronulla, N.S.W., 24-29 August 1966. Agenda Item 5. Mimeographed: 1-43.

[3] Weng, H. T. C. (1971).—The black bream, *Acanthopagrus butcheri* (Munro); its life history and its fishery in South Australia. M.Sc. Thesis, University of Adelaide (unpubl.).

[4] Harbison, P. (1974).—The black bream in the Onkaparinga Estuary. Ad. Dip. Biol. Sci. Thesis, Salisbury College of Advanced Education, S. Aust. (unpubl.).

[5] Shepherd, S. A. (undated).—Report on whiting and other fisheries in the region of Port Vincent. South Australian Fisheries Department. Mimeographed: 1-8.

lago bassensis, and the sea garfish, *Hemiramphus melanochir,* taken by handline and net, and snapper, *Chrysophrys unicolor,* taken by handline in deeper waters. Between 1972 and 1975 the quantity of scaled fish taken in the Gulf comprised 7.6–10.4% of the total commercial catch of scaled fish from all South Australian coastal waters. Although this proportion varies, it appears that the fishing effort in the Gulf has remained fairly constant over a number of years. Regretably, it is only since 1971 that commercial fishermen have been obliged to furnish catch returns for scaled fish, from which statistics can now be compiled and analysed.

Fisheries research efforts are carried out exclusively by Government agencies but have been mainly directed at the non-scaled fisheries (crayfish, prawns, abalone) operating in the Gulf and elsewhere.

The Gulf is noted for its sportfishing, providing as it does excellent angling opportunities for thousands of enthusiastic amateurs operating from jetties and small boats. In addition a large body of spearfishermen operate in summer in localized rocky areas, mainly south of Adelaide and around the tip of Yorke Peninsula. Game fishing on the other hand is only engaged in to a minimal degree within the Gulf, mainly for white pointer and cocktail sharks, *Carcharodon carcharius* and *Carcharhinus greyi greyi.*

Few data are available concerning the impact on fish stocks of this steadily growing amateur pressure, or the possibly even greater long term effects, direct or otherwise, of other impinging human activities such as coastal erosion, land reclamation and pollution in its many forms.

It therefore seems prudent to regard the time well overdue to examine this whole issue, and to undertake appropriate biological investigations, if adequate measures are to be formulated to ensure the future survival of the Gulf's fishes and indeed its whole ecosystem.

Acknowledgments

The assistance of Mr S. A. Shepherd (Fisheries Branch, South Australian Department of Agriculture and Fisheries), and Mr P. F. Aitken (South Australian Museum), in the preparation of this paper is gratefully acknowledged.

References

AITKEN, P. F. (1971).—Whales from the coast of South Australia. *Trans. R. Soc. S. Aust.* **95** (2), 95-103.

CASTELNAU, F. L. (1872).—Contribution to the ichthyology of Australia. No. ii. Note on some South Australian fishes. *Proc. Zool. Acclim. Soc. Vict.* **1**, 243-248.

CASTELNAU, F. L. (1873).—Contribution to the ichthyology of Australia. No. iv. Fishes of South Australia. *Proc. Zool. Acclim. Soc. Vict.* **2**, 37-158.

LING, J. K. (1958).—The sea garfish, *Reporhamphus melanochir* (Cuvier & Valenciennes) (Hemiramphidae), in South Australia: Breeding, age determination, and growth rate. *Aust. J. Mar. Freshw. Res.* **9**(1), 60-110.

LING, J. K. & WALKER, G. E. (1976).—Seal studies in South Australia: progress report for the year 1975. *S. Aust. Nat.* **50**(4), 59-68, 72.

ROYAL COMMISSION ON THE FISHING INDUSTRY (1934-1935).—First Progress Report (1934); Second Progress Report (1935); Final Report (1935); *in* Proceedings of the Parliament of South Australia. (Govt Printer: Adelaide.)

SCOTT, T. D. (1962).—The marine and fresh water fishes of South Australia. (Govt Printer: Adelaide.)

SCOTT, T. D., GLOVER, C. J. M. & SOUTHCOTT, R. V. (1974).—The marine and freshwater fishes of South Australia (2nd Edition). (Govt Printer: Adelaide.)

WAITE, E. R. (1921).—Illustrated catalogue of the fishes of South Australia. *Rec. S. Aust. Mus.* **2**(1), 1-208.

ZIETZ, A. H. C. (1908).—A synopsis of the fishes of South Australia: Part I. *Trans. R. Soc. S. Aust.* **32**, 288-293.

ZIETZ, A. H. C. (1908).—A synopsis of the fishes of South Australia: Part II. *Trans. R. Soc. S. Aust.* **32**, 294-299.

ZIETZ, A. H. C. (1909).—A synopsis of the fishes of South Australia: Part III. *Trans. R. Soc. S. Aust.* **33**, 263-269.